Library of Congress Cataloging-in-Publication Data

Kohl, Harold W., 1960-
    Foundations of physical activity and public health / Harold W. Kohl III and
Tinker D. Murray.
        p. ; cm.
    Includes bibliographical references and index.
    ISBN 978-0-7360-8710-0 (hard cover) -- ISBN 0-7360-8710-9 (hard cover)
    I. Murray, Tinker Dan, 1951- II. Title.
    [DNLM: 1. Exercise. 2. Health Promotion--methods. 3. Public Health. QT 255]

613.7'1--dc23

2011045270

ISBN-10: 0-7360-8710-9
ISBN-13: 978-0-7360-8710-0

The web addresses cited in this text were current as of October 2011, unless otherwise noted.

Acquisitions Editor: Myles Schrag
Developmental Editor: Judy Park
Assistant Editor: Brendan Shea, PhD
Copyeditor: Patsy Fortney
Indexer: Bobbi Swanson
Permissions Manager: Dalene Reeder
Graphic Designer: Joe Buck
Graphic Artist: Tara Welsch
Cover Designer: Keith Blomberg
Photographs (interior): © Human Kinetics, unless otherwise noted
Photograph (cover): © Joshua Huber/Aurora Photos
Photo Asset Manager: Laura Fitch
Visual Production Assistant: Joyce Brumfield
Photo Production Manager: Jason Allen
Art Manager: Kelly Hendren
Associate Art Manager: Alan L. Wilborn
Printer: Sheridan Books

Printed in the United States of America   10 9 8 7 6 5 4 3 2

The paper in this book is certified under a sustainable forestry program.

Human Kinetics
Website: www.HumanKinetics.com

United States: Human Kinetics
P.O. Box 5076
Champaign, IL 61825-5076
800-747-4457
e-mail: humank@hkusa.com

Canada: Human Kinetics
475 Devonshire Road Unit 100
Windsor, ON N8Y 2L5
800-465-7301 (in Canada only)
e-mail: info@hkcanada.com

Europe: Human Kinetics
107 Bradford Road
Stanningley
Leeds LS28 6AT, United Kingdom
+44 (0) 113 255 5665
e-mail: hk@hkeurope.com

Australia: Human Kinetics
57A Price Avenue
Lower Mitcham, South Australia 5062
08 8372 0999
e-mail: info@hkaustralia.com

New Zealand: Human Kinetics
P.O. Box 80
Torrens Park, South Australia 5062
0800 222 062
e-mail: info@hknewzealand.com

E4957

To our colleagues and their students, who will improve public health by zealously promoting science and the practice of physical activity in individuals and in populations.

CONTENTS

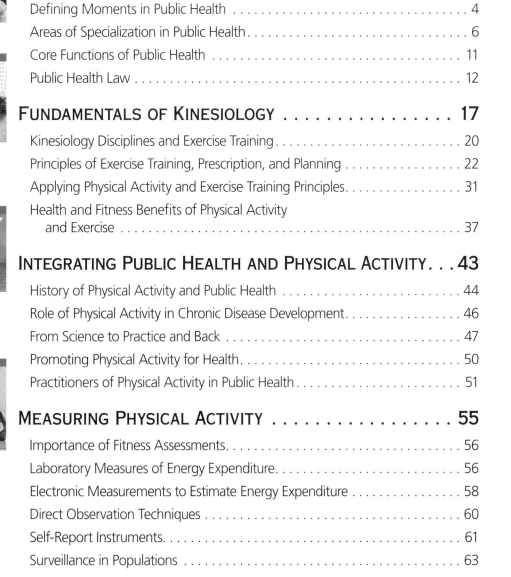

| PART II | HEALTH EFFECTS OF EXERCISE AND PHYSICAL ACTIVITY . . . . 71 |

PREFACE

Welcome to *Foundations of Physical Activity and Public Health*. This text is a collection of the concepts that define the emerging field of physical activity and public health. Much like the more established fields (i.e., the effects of nutrition and smoking on public health), physical activity and public health has its roots in the grafting of two other fields. In this case, public health scientists and exercise scientists have come together to create a window to improve health through research and promotion of physical activity. Methods and evidence from the public health sciences (epidemiology, health promotion, behavioral science, and environmental health) and kinesiology (exercise physiology, the movement sciences, and sport and exercise psychology), combined with a necessary eye on health policy, constitute our field. Although nothing can be substituted for experience, this text offers the background and introduction to tools needed for the planning, implementation, and evaluation of physical activity promotion programs. This is the first textbook of its kind designed for a semester-long course in the field.

Few singular health behaviors can have as broad an impact on the health of individuals and populations as does physical activity. The scientific base is growing and solidifying regarding the effects of physical activity on all-cause morbidity and mortality due to multiple noncommunicable diseases such as heart disease, some cancers, diabetes, and osteoporosis.

It is a very exciting time of growth in physical activity and public health. The seminal scientific works of Dr. Jeremy N. Morris and Dr. Ralph S. Paffenbarger Jr. helped set the stage for what is now a worldwide focus on advancing the science as well as reducing physical inactivity and promoting the benefits of regular physical activity for the prevention and treatment of chronic diseases and other health challenges. As the science and practice advance, physical activity is receiving increased attention from policy and organizational decision makers worldwide, including governmental ministers of health. Professional societies have been created to focus on advancing the research and practice of the field, the scientific literature has expanded dra-matically on multiple fronts, and physical activity and inactivity are becoming parts of health policy decisions at all levels. An outstanding example is the Toronto Charter for Physical Activity: A Global Call to Action (www.globalpa.org.uk/charter). This advocacy tool drives policies worldwide that are supportive of the role of physical activity in promoting health.

HOW THIS BOOK IS ORGANIZED

Foundations of Physical Activity and Public Health is organized into three parts and 16 chapters.

PART I: INTRODUCTION TO PHYSICAL ACTIVITY AND PUBLIC HEALTH

Part I introduces concepts of public health, kinesiology, and measurement. The chapters in this part highlight fundamentals of each and how they have come together.

Chapter 1 introduces the fundamentals of public health and provides information about the various subdisciplines of public health and how public health differs from medicine. Finally, there is a discussion about how public health policy is often linked to the legal and regulatory system as well as discussion of an emerging specialization in public health.

The fundamentals of kinesiology are discussed in chapter 2. In the past, exercise was studied and often promoted as a means of enhancing maximal performance rather than promoting basic health benefits for all. The components of exercise training are presented as well as the methods for applying them to target populations. The general health, fitness, and performance effects of physical activity and exercise are discussed. A final section presents ways to integrate traditional exercise prescription into physical activity and exercise programs.

Chapter 3 focuses on the emergence of the subdiscipline of physical activity and public health. Examples of the interdisciplinary interest in the field of physical activity and public health are reviewed and the knowledge, skills, and aptitudes for careers in physical activity and public health are provided.

In chapter 4, the importance of measuring physical activity is introduced and the strengths and weaknesses of various laboratory and field methods are discussed. Overviews of the following techniques are included: indirect calorimetry, doubly labeled water, accelerometers, pedometers, direct observation, and self-report instruments. Observational techniques such as physical activity surveillance and sources of data-based comparison are also discussed.

PART II: HEALTH EFFECTS OF EXERCISE AND PHYSICAL ACTIVITY

The scientific base of the health effects of physical activity and inactivity is remarkable in its size and complexity. It continues to grow each year, and the overwhelming evidence for the health benefits and risks of physical activity provides much of the rationale for action.

Cardiovascular and metabolic diseases and their relations to physical activity are presented in chapter 5. The chapter starts with a discussion of the prevalence and economic costs of cardiovascular and metabolic diseases. Specific physiological, biomechanical, and behavioral adaptations to physical activity and exercise are also identified. Common testing methodologies for predicting and diagnosing metabolic disease are provided. The evidence for the effect of physical activity on cardiorespiratory and metabolic disease is discussed.

Chapter 6 contains common definitions for overweight and obesity and a discussion about the prevalence (U.S. and worldwide) and the economic costs of these conditions. A discussion of caloric balance is included, and the contributions that physical activity and exercise have on balance are highlighted. The various risk factors associated with overweight and obesity are discussed and specific physiological, biomechanical, and behavioral adaptations to physical activity and exercise are identified. Methods for assessing body composition are provided. The effects of physical activity on weight loss, weight maintenance, and weight regain are discussed along with the physical activity guidelines for achieving caloric balance and a healthy weight.

Chapter 7 focuses on musculoskeletal disorders and functional health. The risk factors, prevalence, and economic costs of musculoskeletal disorders and disability are discussed. Specific physiological, biomechanical, and behavioral adaptations to physical activity and exercise are provided. Common testing methodologies for muscle function and functional health are included. The evidence for the effect of physical activity on musculoskeletal disorders and disability in functional health is discussed.

In chapter 8, cancers related to physical inactivity are discussed and the prevalence of each is highlighted. The mechanism by which physical activity might reduce the risk of some cancers is included along with a discussion of common risks for cancer. Specific physiological, biomechanical, and behavioral adaptations to physical activity and exercise are identified. Included is a discussion of scientific evidence supporting the benefits of physical activity for cancer survivors as well as evidence for the role of physical activity in the prevention of cancer.

Chapter 9 examines the effects of physical activity on mental health. The prevalence, economic costs, and risk factors of mental health disorders are discussed. A framework for studying mental health problems and their response to physical activity interventions is provided along with a discussion about the effects of physical activity on brain function: reaction time, learning tasks, cognitive function, and academic achievement. The recommendations for physical activity complete the chapter.

In chapter 10, adverse events associated with physical activity are discussed. Participation in regular physical activity and exercise may increase the risk of musculoskeletal injuries and sudden cardiac death in some cases. The chapter contains a discussion about defining adverse events, the prevalence of problems, the risks associated with injury, and the adaptive processes that may help prevent injury.

PART III: STRATEGIES FOR EFFECTIVE PHYSICAL ACTIVITY PROMOTION

The chapters in part III introduce evidence-based strategies for increasing physical activity in individuals and populations. Public health is characterized by translating science into action for advancing the health of the population. The strategies presented in part III have been scientifically demonstrated to increase physical activity and can be used for action in a variety of settings.

Methods for promoting physical activity are discussed in chapter 11, which opens with a discussion about the importance of using the Guide to Community Preventive Services as a resource for identifying physical activity intervention programs that work. A discussion about the impact of community-wide campaigns on increasing physical activity is included along with an overview of mass-media campaigns.

In chapter 12, the rationale for school-based physical activity interventions is presented. The scientific benefits of physical activity in youth are reviewed, and commonly used physical fitness tests for school settings are discussed. A section that highlights current U.S. strategies and policies for promoting physical activity via school-based programs is included. The remainder of the chapter focuses on examples of evidence-based school physical activity programs.

In chapter 13, the focus is on evidence-based strategies for behavioral and social approaches to physical activity promotion. The chapter includes a discussion of current behavioral theories and theoretical models that are used to explain physical activity behavior in individuals. Social support strategies for physical activity promotion in communities are defined and highlighted, and examples of both types of approaches are provided.

In chapter 14, environmental and policy influences on physical activity are reviewed, as are strategies for change. The ways in which aspects of the physical and built environment can encourage or inhibit physical activity are reviewed. The role of urban design for physically active populations and evidence-based strategies for change are discussed.

In chapter 15, evaluation of physical activity programs is introduced. The chapter begins with a discussion of the six-step Physical Activity Evaluation Framework developed by the Centers for Disease Control and Prevention (CDC). The concepts of formative evaluation, process evaluation, outcome evaluation, and cost-effectiveness evaluation are covered. Logic models are presented. The chapter also contains discussions about evaluation designs, data collection and analysis, and publishing and communicating results.

Chapter 16 is the final chapter in the text, which focuses on building effective partnerships for physical activity programs. Examples of effective partnering include a state plan (Active Texas 2020), the U.S. National Physical Activity Plan,

and the international Toronto Charter for Physical Activity. Strategies for physical activity advocacy are included, and models for advocacy and effective leadership conclude the chapter.

SPECIAL FEATURES

The content and chapter organization of *Foundations of Physical Activity and Public Health* is based on contemporary teaching principles to maximize learning opportunities for students. Following are the features in each chapter:

- *Objectives* are summaries of take-away messages you should learn by reading and studying the material.
- *Opening questions* help you think about how you can use information in the text.
- *Highlight boxes* are examples of topics covered in the text, which will help you translate theory into practice.
- *Case studies* are real-life examples of selected concepts covered in the chapter, and are found in part III.
- *Key leader profiles* are mini-biographies of world leaders in physical activity and public health. Each leader addresses four key questions about his or her work in the field.
- *What you need to know* is a bulleted review of the chapter to help you study the information provided.
- *Terms to know* are the key terms covered in the text.
- *Study questions* are general questions that represent all the material covered in the text.
- *E-media* are web-based resources that pertain to the material covered in the chapter.
- *Bibliographies* are additional published resources for further study.

NOTE TO STUDENTS

As the field of physical activity and public health expands, an increasing number of job opportunities will be available for those who achieve the core competencies as endorsed by the National Physical Activity Society (www.PhysicalActivitySociety.org), established in 2006 as the National Society of Physical Activity Practitioners in Public Health. Coursework that covers concepts of physical activity and public health will help future graduates in diverse employment settings such as public health and health care, business and industry, the nonprofit sector, education, mass media, urban planning and architecture, and parks and recreation. University students in majors and minors such as kinesiology, athletic training, physical therapy, medicine, nursing, and nutrition, as well as trainers in public services (fire, police, and military), rehabilitation specialists, and wellness instructors will find a natural connection between their professional duties and the need for promotion of physical activity and public health to colleagues and communities.

NOTE TO INSTRUCTORS

This text is targeted to students in exercise science or public health programs who are enrolled in elective courses that expand their understanding beyond what is taught in traditional core courses. The *2008 Physical Activity Guidelines for Americans* (www.health.gov/PAGuidelines), the accompanying Physical Activity Guidelines Advisory Committee Report (www.health.gov/PAGuidelines/committeereport.aspx), and the CDC's *Guide to Community Preventive Services* (www.thecommunityguide.org) are valuable resources that provide much of the framework for the development of this text.

The following free ancillaries are also available to instructors who adopt this textbook:

- The **instructor guide** includes syllabus suggestions, teaching tips, and sample class assignments.
- The **test package** includes over 300 questions, including multiple choice, true-false, and fill-in-the-blank questions. The test package can be downloaded in multiple formats depending on your teaching needs, and can also be modified to include test questions that you create.
- The **image bank** includes all of the figures and tables from the text. You can use these items to create your own Power Point presentations, handouts, or other class materials.

These resources can be accessed at www.HumanKinetics.com/FoundationsOfPhysicalActivityAndPublicHealth. The authors, who have taught courses in physical activity and public health, have helped develop all of the ancillary materials.

We trust that *Foundations of Physical Activity and Public Health* will allow you to develop courses that inspire students to pursue careers in physical activity and public health.

ACKNOWLEDGMENTS

Because life is a journey with many encounters that continually make us who we are, it is nearly impossible to acknowledge all those who have influenced and taught me over the years. Several people do stand out, however. Thom McCurdy and Louis E. Burnett Jr. sparked and fed my early interest in science. Caroline A. Macera introduced me to epidemiology and public health. Milton Z. Nichaman made me an epidemiologist. Steven N. Blair helped me tremendously by showing me how it all fit together and being the role model that we all should have and be. Thanks to each of these mentors who have helped to shape my thinking.

My wife, Ann, has been with me throughout the process and has seen my challenges as no one else can see them. My parents, Harold W. Kohl Jr. and Rose Ann Kohl, gave me every possible advantage and pushed me to challenge myself every day. Virginia Michelli assisted me throughout the process. This project would not have happened without the influence each has had on me.

—HWK

I thank my parents, Bob and Louise Murray, for being role models for active living and for supporting my academic pursuits. I want to also thank Karen Mitchell for encouraging my writing efforts and Bill and Ann Kohl for their friendship and wit, which made the whole process even more worthwhile.

—TDM

We acknowledge Geoffrey P. Whitfield, MS, RCEP, and two anonymous reviewers for their time and comments, which made this text more focused. Their contributions are sincerely appreciated. Mariya Grygorenko provided much needed editorial assistance. At Human Kinetics, several people contributed the right blend of patience, prodding, and talent to help bring this project to completion. Myles Schrag, Judy Park, and Brendan Shea in particular were most helpful and a pleasure to work with.

© Photodisc

PART I

INTRODUCTION TO PHYSICAL ACTIVITY AND PUBLIC HEALTH

FUNDAMENTALS OF PUBLIC HEALTH

OBJECTIVES

After completing this chapter, you should be able to discuss the following:

» The definition and history of public health

» How public health has become specialized and the five main pillars of public health

» The five main principles that guide health promotion and health education efforts in public health

» The 10 essential functions that support the core services of public health

» Why public health policy is often linked to legal and regulatory systems

» The emerging physical activity specialization in public health

OPENING QUESTIONS

What comes to mind when you read the words *public health*?
Screening children for nutritional deficiencies?

» Quarantine practices to isolate a person with tuberculosis to prevent an outbreak of the disease?

» Disaster responses to prevent disease transmission during and after a hurricane or earthquake?

» Prenatal education for expectant mothers?

» Promotion of physical activity to lower the burden of chronic, noncommunicable diseases such as heart disease and diabetes mellitus?

If you answered *yes* to any of these questions, you are correct. Public health is all this and more.

Public health is a field that encompasses many disciplines in an effort to promote and protect health and prevent disease and disability in defined populations and communities. Although medicine and medical training are integral to public health, particularly in understanding the mechanisms of disease transmission, medicine is more interested in the treatment of and cures for diseases and disabilities in individuals. The key difference between public health and medicine is that public health traditionally has focused less on individuals and treatment and more on populations and prevention.

Clearly, then, public health should be focused on problems that affect, or could affect, a substantial portion of the population. For this reason, rare diseases and disabilities and seemingly random health events are often less of a concern to public health than problems that may affect many people in a population. This is not to say that such situations are not important, particularly to the people afflicted, but rather, that the focus of public health is on the health of the population as a whole. Overall, the health of a population is rarely improved by focusing only on rare diseases and health problems that affect the few.

This first chapter offers an overview of the principles and key areas of public health and describes the fundamental services of public health. Happily, public health has grown far beyond its origins and has allowed populations to thrive in the face of new and emerging health problems.

DEFINING MOMENTS IN PUBLIC HEALTH

Although a complete treatment of the history of public health is beyond the scope of this chapter (it could, and does, fill whole books), an understanding of some defining moments in public health is instructive. This understanding helps place the emergence of physical activity and public health as a separate discipline within public health in context. Winkelstein (2011) offers a more complete treatment of the history and evolution of public health.

Although disease and **epidemics** have occurred for thousands of years, the earliest roots of organized public health emerged in the mid-14th century. At the time, the Black Death (bubonic plague) ravaged Europe, killing an estimated 25% of the population. As we know now, the disease was tied to the black rat, the rat flea (*Xenopsylla cheopis*) that lived on the blood of the black rat, and the bacterium *Pasteurella pestis* that helped the flea to seek out additional food by biting warm-blooded humans. At the time, however, an understanding of the germ theory of infection and disease (i.e., that microorganisms are responsible for sickness and not simply "bad air" or other nonbiological reasons) was still 400 to 500 years in the future. Advances in transportation (shipping) around Europe and the Middle East spread the disease to other geographic areas. Although no one knew when or how the disease would strike, public health was advanced

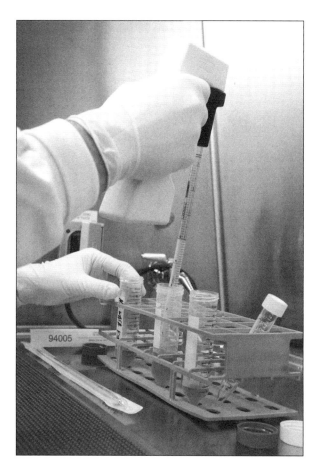

Although the cause of the bubonic plague was unknown in its time, counting the dead was one of the first examples of creating public health statistics. Can you think of examples of tracking modern diseases to the source?

by the creation of health boards and systems for counting and collecting the dead. Unknowingly, this was the first attempt at a vital statistics system, which is now routine in public health organizations throughout the world.

Another advance in public health grew out of concern about the health of workers (particularly children) and the deplorable working conditions that were rampant early in the industrial revolution of the 19th century. Little regulation was in place at the time, and workplaces were polluted, unsafe, and very conducive to disease and injury. Young children were working to support families, and the poor and less advantaged were particularly susceptible. The association between poverty and health was recognized early in the 19th century, and policies and programs to address such disparities began to emerge. Edwin Chadwick in Great Britain

was an early leader in the cause of improving sanitation, housing conditions, worker safety, and garbage disposal practices in poor communities. This is one of the first examples of using policy and legislation to improve health.

The lens of history points to the early 20th century as another critical period in the history of public health. With the legitimization of the germ theory of disease transmission by Robert Koch and Louis Pasteur (working independently) in the late 1800s, new methods for treating (and preventing) disease emerged. Sanitation, quarantine, and other methods for controlling infectious diseases became standard practices in cities. Boards of health were developed to deal with health threats to the community. Vaccines and antibiotics were discovered and quickly resulted in monumental improvements in disease control.

The 20th century represents a bridge between a focus on infectious (communicable) diseases and a focus on chronic (noncommunicable) diseases. Once **infectious diseases** were becoming less influential, **nutritional diseases** (due largely to micronutrient deficiencies) became a priority. Maternal and child health also was a critical piece of the public health puzzle in the 20th century. The infant mortality rate, as well as the maternal mortality rate, was abominable. Advances such as mandating training and licensure of midwives were public health interventions that helped to control this burden.

Finally, following the decline of infectious diseases and nutritional deficiency diseases, the mid-to-late portion of the 20th century was witness to the emergence of **chronic diseases** (noncommunicable) as those that had the largest population reach and thus were a substantial public health concern. Heart disease, diabetes mellitus, cancers, mental health disorders, and musculoskeletal disorders firmly replaced infectious and nutritional diseases as key causes of death and illness in the world. To be specific, a 2011 report by the World Health Organization (WHO; 2011a) detailed that more than 60% of all deaths worldwide in 2008 were due to chronic diseases (see table 1.1).

Only 150 years in the past, infectious diseases were the leading concern and the primary cause of sickness and death. Today, diseases influenced by lifestyle and genetics are the greatest public health

Table 1.1 Ten Leading Causes of Death Worldwide (2008)

Cause	Deaths in millions	% of deaths
Ischaemic heart disease	7.25	12.8%
Stroke and other cerebrovascular disease	6.15	10.8%
Lower respiratory infections	3.46	6.1%
Chronic obstructive pulmonary disease	3.28	5.8%
Diarrhoeal diseases	2.46	4.3%
HIV/AIDS	1.78	3.1%
Trachea, bronchus, and lung cancers	1.39	2.4%
Tuberculosis	1.34	2.4%
Diabetes mellitus	1.26	2.2%
Road traffic accidents	1.21	2.1%

Reprinted, by permission, from World Health Organization, 2011, World Health Organization fact sheet.

concern. This remarkable transition in public health coincides with the beginning of the physical activity story.

AREAS OF SPECIALIZATION IN PUBLIC HEALTH

An important part of the evolution and history of public health has been the emergence of training programs and techniques to address public health challenges. The establishment of the London School of Tropical Medicine and Hygiene in the United Kingdom, and of the Johns Hopkins School of Public Health in the United States, in the early 20th century were key steps to creating a workforce with the skills necessary for handling public health problems. Very rapidly following these early efforts, additional training and certification of academic programs took hold in the United States. In 2011, the United States and Mexico had 50 accredited schools of public health providing leadership and training opportunities for master's and doctoral students. These training programs have evolved over the years, resulting in widely accepted standards for areas of training and specialization in public health.

Figure 1.1 illustrates the five broad areas of specialization, or pillars, of public health, each of which contributes uniquely to the field.

EPIDEMIOLOGY AND DISEASE CONTROL

Epidemiology is the basic science of public health. The word *epidemiology* comes from Greek origins: *epidemia* ("on people") and *-ology* ("to study"). Although several definitions exist, a modern-day definition of epidemiology is "the study of distributions and determinants of disease and disability in populations" (Mausner and Bahn 1974). Notable in this definition, and following from the preceding discussion, is the word *populations*. Epidemiologists are focused on a defined population and how a disease or disability affects that population. What causes the spread of the disease or disability? How can it be prevented? How many people are affected? What types of people or other organisms are pos-

KNOWLEDGE INTO ACTION

Public health science is characterized not just by the accumulation of new knowledge, but also by the application of that knowledge to improve health. Public health research must be able to be translated to action for disease prevention, health promotion, or both.

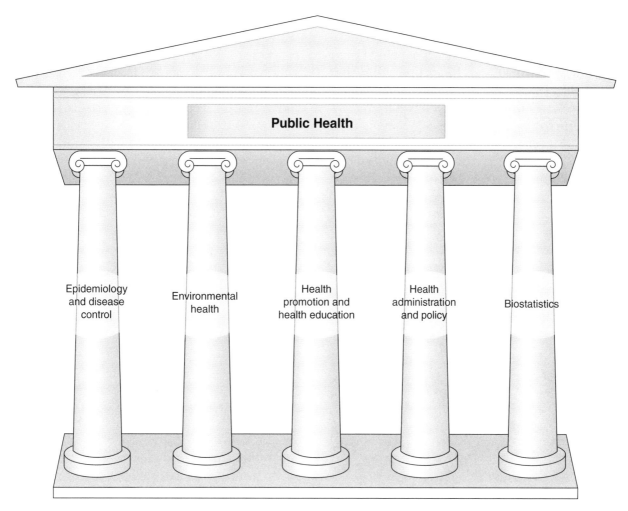

Figure 1.1 Five pillars of public health.

sibly affected more than others? Who is at risk? How many could be affected in the future? These are all questions that epidemiologists are trained to answer.

Epidemiology is a quantitative scientific discipline that relies heavily on statistics and study design. With the transition in the 20th century to disease burden being attributed primarily to noncommunicable diseases, epidemiological methods have evolved to apply not only to infectious disease outbreak investigations, but also to studies of longer-term chronic disease investigations. For example, much of what we know about risk factors for heart disease (e.g., poor lipid and lipoprotein profile, high blood pressure, cigarette smoking, physical inactivity) came from early and ongoing epidemiological studies of (mostly) men with and without these characteristics. Researchers used epidemiological methods to compare and contrast study participants with and without the conditions and then calculated the risk associated with the occurrence of a disease. These techniques have evolved as the need to address more complicated analytical questions has increased.

ENVIRONMENTAL HEALTH

The *environment* can be defined as "all that is external to the host organism" (WHO 2011b)—including physical, biological, and cultural influences. Our physical environment (i.e., where we live, work, and play) has a powerful influence on our health. The air we breathe, the water we drink, the food we eat, the safety of our work environment, our exposure to radiation, and the ways we control those environmental influences can promote or hinder public health. Thus, a large part of public health addresses **environmental health**.

Major advances have been made in public health as a result of environmental health studies. Prohibition of lead-based paint to reduce the risk of learning disabilities in children, fluoridation of water supplies to reduce dental problems in communities, air quality regulations for automobile manufacturers and industrial polluters to promote cleaner air and water, and food safety standards to reduce the risk of food-borne diseases are all examples of public health initiatives that came about as a result of environmental health studies. Can you think of others?

Clearly, environmental influences on health have been known for centuries. Systematic approaches to studying the environmental influences on health, quantifying these influences, and prioritizing resources and approaches to eliminate the health

The construction of bike and pedestrian-friendly routes increases access to places to be active in a community. In addition to physical activity, what are other benefits of such efforts?

hazards have been advanced only relatively recently. We will learn in chapter 14 that our understanding of the role of the environment in promoting or inhibiting physical activity has advanced rapidly since the mid-1990s. We can now identify barriers and correlates in the physical and social and cultural environment that influence physical activity participation. This has been, and will continue to be, a major growth area in the field of physical activity and public health.

HEALTH PROMOTION AND HEALTH EDUCATION

Why do some people exercise consistently, avoid tobacco, eat well, use alcohol responsibly, avoid illicit drugs, see their doctors regularly, and do other things necessary for health maintenance, whereas others do not? How can we best teach basic health concepts for lasting effectiveness? What is the most effective education strategy to improve birth outcomes for teenage mothers? How can population-level health behaviors be changed to maximize life expectancy and quality of life? These are examples of key questions those in the **health promotion** and **health education** pillar of public health routinely address. Although the environment and our genetic makeup contribute substantially to our health status, how we deal with health threats through our behavior has become a major focus of public health.

Much of the basis for health promotion and health education in public health comes from the concept of social justice, which is central to many ideas of public health. Social justice in public health refers to the assumption (some call it an imperative) that the health burdens and benefits in a population should be distributed equitably. We have known for centuries that poverty is a predictor of disease, disability, and poor quality of life. Those who promote health promotion and health education strategies in public health aim to be part of a solution to reduce such disparities.

The WHO has advanced five principles to guide health promotion and health education efforts in public health (WHO 2011b). See the following highlight box for a description of these principles.

WHO PRINCIPLES OF HEALTH PROMOTION

Empowerment and Inclusion
Health promotion should empower all individuals and communities to take some responsibility for the influences on their personal health.

Intersectoral Collaboration
Health promotion programs should be directed at all the relevant determinants or causes of health and should therefore include relevant collaborations among agencies and sectors with influence beyond the health or medical care sector.

Multidimensional
Health promotion initiatives should use all tools possible to minimize health hazards and promote positive health. These dimensions include, but are not limited to, communication, legislation/policy, education, community/organizational change, and finance.

Participatory
Health promotion programs should strive to be inclusive at all steps and seek to maximize participation individually and collectively.

Advocacy
All people with an interest in health, including medical care systems, should take responsibility for health promotion and health education.

Based on WHO 2011b.

HEALTH ADMINISTRATION AND POLICY

Health administration and policy is the fourth pillar of public health and focuses on the delivery of public health services. This area of expertise addresses important skills such as budgeting, policy development and analysis, planning and prioritization, communication, and the like. In keeping with the theme that public health is action oriented, health administration and policy skills support the appropriate implementation of programs that, theoretically, are derived from research in the area. A good example lies in the HIV/AIDS epidemic. Research has shown that needle exchange programs likely reduce hypodermic needle sharing among intravenous drug users and thus the risk of transmission of HIV/AIDS (Palmateer et al. 2010). Health administration and policy experts can use such data and the results of studies to plan community-based projects that promote such programs.

In the field of physical activity and public health, the effects of policies and program administration are an emerging knowledge base. As addressed in part III of this textbook, policies for providing places to be physically active are showing promising results for the promotion of physical activity. Expertise in health administration and policy is critical for the implementation and evaluation of such efforts.

BIOSTATISTICS

Public health relies on both qualitative and quantitative methods to move from knowledge to action. **Biostatistics** provides the basis for the quantitative branch. Is the difference between two interventions to promote breast cancer screening in a community due to the effect of the intervention, or simply to chance? What is the predicted number of cases of influenza in the upcoming year? Based in mathematical theory, biostatistics allows for the practical and rational analysis of data, the interpretation of study results, and the translation of those results into action. Biostatisticians and epidemiologists work closely to advance the science of public health.

KEY LEADER PROFILE

Ross C. Brownson, PhD

Courtesy of Ross C. Brownson.

Why and how did you get into this field?

I grew up in western Colorado and spent much of my youth backpacking, fishing, and cycling. Therefore, physical activity has been an important part of my life for many years. My graduate training is in environmental health and epidemiology, and after finishing college, my first job in public health was with the Missouri Department of Health as a cancer epidemiologist. There were many datasets that had not been analyzed, including cancer registry information with smoking, alcohol, and occupational histories. These data allowed us to conduct studies of cancer risk in relation to occupational physical activity. While the early part of my career focused mainly on chronic disease etiology (Does smoking cause leukaemia? Does physical inactivity cause colon cancer?), I realized that a bigger and more immediate need was in determining intervention effectiveness and learning how to apply effective programs and policies in public health and community settings (so-called *translational research*).

Did any one person drive your research?

Numerous people have influenced my research and career in many positive ways. Among these is my former professor, John Bagby. John was involved in the early efforts to eradicate smallpox. Through this work, I learned about the victories public health can achieve. I have also had the good fortune to work with a number of Centers for Disease Control and Prevention (CDC) scientists in understanding and promoting physical activity, including Mike Pratt, Eduardo Simoes, and Tom Schmid.

What are your current research interests?

I am the codirector (along with Beth Baker from the Saint Louis University School of Public Health) of the Prevention Research Center in St. Louis. Our center is supported by the CDC and is one of 37 centers around the country. We work closely with community members, public health agencies, academic partners, and others to develop, test, and disseminate interventions to reduce the risk of chronic diseases such as cardiovascular disease, cancer, and diabetes. Among my research interests is the exploration of the effects of the built environment and policies on the risk of physical inactivity and obesity. We are also conducting innovative work in understanding and promoting physical activity in Brazil and Latin America. These efforts have shown the value of cross-national partnerships.

What drives you as a researcher and activist?

My drive as a researcher is to make our society a more supportive place for healthy and active living. Through our work with local trails in rural Missouri, we have observed the positive effects of changing the built environment. Unfortunately, the benefits of a healthy lifestyle are not reaching all sectors of society.

What are one or two key issues to be addressed by 2022?

There are many important research issues before us. Among these are (1) How can we take what we know about promoting physical activity (evidence-based interventions) and apply this science to change local policy? and (2) How do we frame our messages to develop more political will for addressing physical inactivity?

CORE FUNCTIONS OF PUBLIC HEALTH

Even with the specializations in public health listed earlier, the core functions of public health professionals and agencies must be defined. What should these professionals and agencies do? How does a public health agency interact with the medical care system to promote and protect the health of communities? Do standards for public health practice exist? In 1999, a joint project led by the U.S. Office of Disease Prevention and Health Promotion (ODPHP; 2011) was created to describe and define the fundamental services of public health. This Public Health Functions Project used this opportunity to strengthen the public health infrastructure in the United States by developing and describing a common set of primary services that define public health. By addressing these functions, in combination with skill sets learned in the topic areas of the five pillars of public health, professionals can effectively promote public health. The 10 essential functions of public health are listed in figure 1.2.

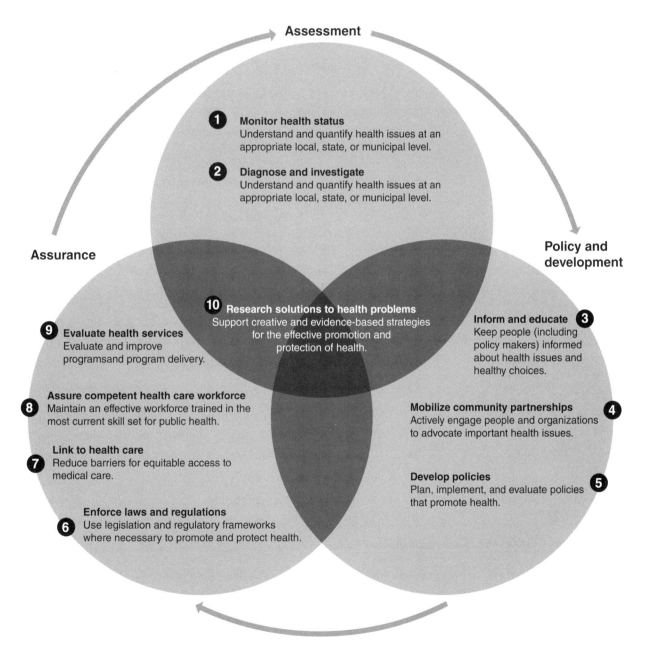

Figure 1.2 Cycle of the 10 essential functions of public health.

The 10 essential functions of public health help define and guide the core services of public health. Competencies in the functions are developed through training in and across the five pillars. Actually, these functions go beyond a simple listing—they have been adopted as the key organizing framework for the CDC's National Public Health Performance Standards Program (NPHPSP; 2011). The purpose of the NPHPSP is to improve the quality of public health practice and the performance of public health systems by providing performance standards, engaging and leveraging partnerships, promoting continuous quality improvement, and strengthening the science base for public health practice improvement.

These 10 essential functions are meant to interact in a cycle (see figure 1.2) to maximize public health. The assessment phase begins the cycle with the monitoring of health and the diagnosis and investigation of important or emerging health problems. Policy development ensues, including informing, educating, and empowering people regarding health threats, mobilizing partnerships to deal with those health threats, and developing policies to minimize threats and promote health. During the assurance phase, laws and regulations are created and enforced, linkages to care are created, a competent workforce is developed, and effectiveness is evaluated. The cycle can then begin anew with ongoing monitoring. Importantly, this cycle revolves around research advances and administration (management) skills. Can you think of a specific public health problem and apply steps to each of the 10 essential functions?

As addressed in this text, physical inactivity is a public health problem that can be viewed as a content area addressed in terms of these essential functions. Each function can and should support physical activity promotion. Public health professionals with interest and expertise in physical activity can make this happen.

PUBLIC HEALTH LAW

Although efforts have been made to promote health and behavior change among individuals in a population, many times the best strategy for making meaningful changes in public health is through the legal system, regulatory system, or both. Indeed, many of our most important public health successes have come from such changes: immunization laws for school children to prevent the transmission of infectious diseases at school, legislation and regulations to prohibit cigarette use in public places to reduce exposure to secondhand smoke, motorcycle helmet laws to reduce traumatic brain injury and death among riders involved in accidents, minimal workplace safety standards and inspections to reduce the risk of occupational injuries, and food safety and transportation standards to reduce the risk of food-borne illnesses. These advances in public health did not rely on individual or community-generated changes, but instead on legal or regulatory action.

Such legal avenues have given rise to an entire area of specialization called **public health law**. Specialists in this field seek to leverage legal and governmental authority to promote and protect the health of populations. Experts in public health law, particularly as it relates to issues of preparedness with the emergence of bioterrorism threats (e.g., anthrax), mass disasters such as earthquakes and hurricanes, and possible disease pandemics such as avian influenza, have developed strategies and approaches as well as research agendas to move the field forward toward optimal health protection. They constantly deal with the issue of balancing individual rights with the moral obligation to minimize risks in a population.

FUTURE OF PUBLIC HEALTH

Public health is a vast discipline encompassing many specialties and functions. At the core, public health seeks to promote and protect the health of populations. With health threats continuing to emerge as developed countries face new challenges and developing countries face the challenges of modernization, public health should be used to predict these threats and respond accordingly. Physical activity and public health is a new subdiscipline in this area that bridges the traditional areas of public health and kinesiology. Large, unexplored areas need to be addressed in physical activity and public health, many of which will be introduced throughout this textbook.

KEY LEADER PROFILE

Kenneth E. Powell, MD, MPH

Courtesy of Kenneth E. Powell.

Why and how did you get into this field?

I recommend that you have a plan. Aside from one year during my early 20s when I had no idea what I wanted to do or where I wanted to go, I have always had a plan.

During my third year in medical school I forsook my plan to become a clinician for a plan to become a public health worker. During a term abroad at a rural hospital in India, I realized that the fine clinical services we provided in our hospital provided only brief respite from the environmental assaults suffered by ordinary Indians in their daily lives. The value of widely available potable water, effective sewage systems, immunizations, and other public health accomplishments rose in my esteem. A second example of a change in plans occurred in 1983 during my seventh year of service at the CDC. Taking note of the rising importance of chronic diseases as major public health problems in developed countries, I shifted my focus from infectious diseases to health-related behaviors and accepted a position as chief of a unit designed to focus on behavioral epidemiology. A focus on the epidemiology of the behaviors (e.g., smoking, overeating, alcoholism) rather than the outcomes of those behaviors (e.g., cancer, heart disease, diabetes) was a relatively new concept at that time, but I was prepared to bring epidemiologic principles to bear. The week before assuming my new duties, my supervisor said to me, "I want you and your staff (there were four of us) to begin with exercise." Huh? It didn't take long for me to decide that I liked the new focus, but it hadn't been part of my plan.

Did any one person have an overriding influence on you?

During the previous 28 years I have learned much from many people. During my first year in the field there were three who helped me find my way: Ralph Paffenbarger, Marshall Kreuter, and Carl Caspersen. I consulted with and studied the papers of Ralph Paffenbarger. He set the epidemiologic table for the rest of us. Marshall Kreuter, an expert on health promotion, was my supervisor. After telling me to focus on exercise, he gave me free rein to plan and conduct our early epidemiologic work. His enthusiasm and commitment to health promotion reminded me that our purpose is to bring about improvements in the mental and physical health of the general population. Carl Caspersen was the first person I hired in my new assignment. He taught me about exercise science, guided us to the definitions of physical activity and exercise to help us understand what we were focusing on, and was a fountain of good ideas for our nascent activities in behavioral epidemiology.

What are your current research interests?

I bring my interest in the topic to bear in the areas I am involved in (i.e., violence prevention and state-level epidemiology of chronic disease) and tools we were using (e.g., public health surveys, surveillance systems, and intervention evaluations). Physical activity influences so many facets of public health that it was easy to include.

What are one or two key issues to be addressed by 2022?

First, the evidence that all levels of intensity and every increment in volume of activity bring better health makes it even more important for us to understand the dose-response relationships that Bill Haskell has been urging us to better understand for years. Second, we need to recognize that epidemiologic studies as we have been doing them show that more physical activity makes us healthier but does not tell us how much activity to do to be healthy. That requires a different approach. Third, we need to place greater emphasis on population-wide interventions. We need interventions that facilitate activity for everyone.

CHAPTER WRAP-UP

What You Need to Know

- Public health is a field that encompasses many disciplines in an effort to promote and protect health and prevent disease and disability in defined populations and communities. The key difference between public health and medicine is that public health traditionally has focused less on individuals and treatment and more on populations and prevention.
- Chronic (noncommunicable) diseases have taken over from infectious diseases as leading causes of death in the world.
- Epidemiology is the basic science of public health.
- The five subject pillars of public health are epidemiology, environmental health, health promotion and health education, health administration and policy, and biostatistics.
- There are ten essential functions of public health that help define and guide the core services of the field. Competencies in the functions are developed through training in and across the five pillars.

Key Terms

public health	chronic diseases	health education
epidemics	epidemiology	health administration and policy
infectious diseases	environmental health	biostatistics
nutritional diseases	health promotion	public health law

Study Questions

1. What is public health?
2. How did the discipline of public health begin and why?
3. Why are the prevention and control of chronic diseases important today in public health?
4. How has the area of public health become specialized?
5. What are the five pillars of public health?
6. What is epidemiology, and how does it advance public health?
7. What is the definition of the term *environmental health*?
8. What are the 10 essential functions of public health, and how does physical activity as a content area fit with them?
9. How can policy development support public health?
10. Why do you think physical activity is a new and emerging area of specialization in public health?

E-Media

Explore issues related to physical activity, exercise, and public health at the following websites:

U.S. Association of Schools of Public Health: What Is Public Health?	www.whatispublichealth.org/what/index.html#Administration
U.S. Centers for Disease Control and Prevention: Physical Activity	www.cdc.gov/physicalactivity/
U.S. Centers for Disease Control and Prevention: National Public Health Performance Standards Program	www.cdc.gov/nphpsp/index.html
U.S. Centers for Disease Control and Prevention: Public Health Law Program	www2.cdc.gov/phlp/
World Health Organization	www.who.int

Bibliography

Mausner J and Bahn AK. 1974. *Epidemiology: An Introductory Text*. Philadelphia: Saunders.

Palmateer N, Kimber J, Hickman M, Hutchinson S, Rhodes T, Goldberg D. 2010. Evidence for the effectiveness of sterile injecting equipment provision in preventing hepatitis C and human immunodeficiency virus transmission among injecting drug users: A review of reviews. *Addiction* 105: 844-859.

U.S. Centers for Disease Control and Prevention. 2011. National Public Health Performance Standards Program. www.cdc.gov/nphpsp/index.html. Accessed 17 June 2011.

U.S. Department of Health and Human Services, Office of Disease Prevention and Health Promotion. 2011. Public Health Functions Project. www.health.gov/phfunctions/Default.htm. Accessed 16 June 2011.

Winkelstein W. 2011. History of Public Health. www.enotes.com/public-health-encyclopedia/history-public-health. Accessed 16 June 2011.

World Health Organization. 2011a. Global Status Report on NCDs. www.who.int/chp/ncd_global_status_report/en/index.html. Accessed 16 June 2011.

World Health Organization. 2011b. Milestones in Health Promotion: Statements From Global Conferences. www.who.int/healthpromotion/milestones/en/index.html. Accessed 16 June 2011.

PHYSICAL ACTIVITY IN PUBLIC HEALTH SPECIALIST

This chapter covers these competency areas as set forth by the National Society of Physical Activity Practitioners in Public Health:

1.4.1, 1.4.2, 1.4.3, 2.1.1, 2.2.1, 2.4.1, 2.4.6, 3.3.1, 3.3.2, 3.4.2, 3.4.3, 4.3.3, 5.1.1

CHAPTER 2

FUNDAMENTALS OF KINESIOLOGY

OBJECTIVES

After completing this chapter, you should be able to discuss the following:

» The field of kinesiology and how its subdisciplines have contributed to our understanding of exercise, fitness, maximizing performance, and the shifting paradigm of the promotion of public health benefits through physical activity

» The concepts and principles of exercise training

» The general health, fitness, and performance effects of physical activity and exercise

» How to integrate the principles of traditional exercise prescription programming for individuals into physical activity and exercise plans for populations

OPENING QUESTIONS

» Do you know what areas of kinesiology positively affect individual health and physical fitness through the traditional exercise training model?

» How do you think knowing about the fundamentals of kinesiology and exercise training can help you promote health and physical fitness for populations in a public health model?

Chapter 2 introduces the field of kinesiology and the basic concepts of exercise training that have been used since the 1960s and 1970s to improve individual human performance and physical fitness (otherwise known as the *traditional training model*). You will learn how the traditional model of exercise training, based on the science of kinesiology, has evolved to also include the promotion of public health. The emerging concept of promoting public health by increasing physical activity through the use of population physical activity plans is introduced at the end of the chapter and described in more detail in chapter 3.

Before we begin our discussion of kinesiology, it would be helpful to define some of the terms we will be using. **Physical activity** is any bodily movement that results in energy expenditure (i.e., burning of calories) (Caspersen et al. 1985). Moving your arm up and down at your desk, skateboarding, lifting sacks of groceries out of the trunk of your car, and running a marathon are all types of physical activity. **Exercise** is a specific type of physical activity that is planned, repetitive, and done for a specific purpose (Caspersen et al. 1985). Walking the dog for 2 miles (3.2 km) every night, training for a soccer

tryout, swimming laps, and working out at a gym are all examples of exercise. All athletes who train to improve components of their physical fitness exercise. Anyone who gets out of bed in the morning is doing some kind of physical activity.

In contrast to physical activity and exercise, **physical fitness** is a set of measurable physiological parameters (Caspersen et al. 1985). Most people have their own views of what physical fitness looks like. What's yours? An Olympic-caliber weightlifter? A marathon runner? The fastest kid on the basketball court? How about your grandparents who are happy and able to physically do everything they want to do without limitations?

Although there are many ideas of what physical fitness is, a very useful list of definitions is shown in the highlight box Taxonomy of Physical Fitness. Physical fitness can be either health related or skill related. Health-related fitness encompasses attributes that are thought to be related to improved health (e.g., reductions in chronic diseases, injuries, and rates of disability). These are usually the attributes we are most interested in changing or improving when promoting physical activity and exercise in public health.

TAXONOMY OF PHYSICAL FITNESS

Health-Related Fitness	Skill-Related Fitness
Aerobic capacity or aerobic stamina	Balance
Muscular strength	Agility
Muscular endurance	Coordination
Flexibility	Power
Body composition	Speed

See U.S. Department of Health and Human Services (USDHHS), Physical Activity Guidelines Advisory Committee (PAGAC) for more in-depth definitions of these terms (2008).

Skill-related fitness attributes are those that are the most important for successful movement and sport participation. Skill-related fitness parameters are also more likely to be determined by genes than by physical activity or exercise, although some can be improved with specific training. However, they may or may not be related to health outcomes. For example, speed is an incredibly important fitness parameter in competition sports (e.g., track and field). People can be healthy, however, without being fast. Good balance, on the other hand, is important in gymnastics, but it is also critical for preventing falls among older people.

Fitness can be measured or estimated in the very tightly controlled setting of a laboratory or in field settings in which groups of people can be tested simultaneously. Many useful techniques have been developed over the years to measure physical fitness very precisely. Although a full treatment of approaches to measure physical fitness is beyond the scope of this text, it is important to know that physical fitness, particularly aerobic fitness and muscular endurance, is frequently used to validate (or compare) measures of physical activity and exercise (see chapter 4 for more information).

The exercise sciences provide the basis for the academic discipline of **kinesiology**, which addresses the interrelationship of physiological processes and the anatomy of the body with respect to movement. The study of kinesiology includes the major exercise sciences of exercise physiology, the movement sciences, and sport and exercise psychology, which have been used for the past 30 years to promote the traditional exercise training model. Some of the other important exercise science areas of kinesiology you may have read about are physical education, health education, anatomy, pedagogy, motor learning, motor control, biomechanics, nutrition, sociology of sport, sport management, athletic training, physical therapy, and special populations.

Figure 2.1 shows how the traditional exercise training model has been applied using the integration of major exercise sciences to maximize performance. When concepts from these same exercise sciences are integrated with physical activity, they can also be used to address public health policy goals.

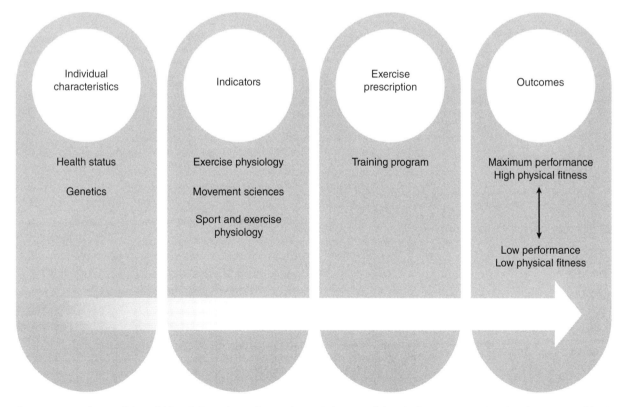

Figure 2.1   The traditional kinesiology-based exercise training model, which promotes the development of high levels of physical fitness, maximal performance, or both. Exercise physiology, the movement sciences (e.g., motor learning, motor control, and biomechanics), and sport and exercise psychology have been the primary kinesiology integrators (connectors) used by professionals to develop training programs that yield positive outcomes.

How many calories would you expend playing golf for 18 holes and carrying your clubs the whole round?

The material presented in this chapter, when combined with the public health concepts discussed in chapter 1, represents a paradigm shift from primarily promoting exercise training and maximal performance (the 1970s and 1980s model) toward also promoting physical activity for positive health outcomes (present-day model). The integration of kinesiology concepts with public health outcomes is derived primarily from the *2008 Physical Activity Guidelines for Americans* (USDHHS 2008) and the *Physical Activity Guidelines Advisory Committee Report, 2008* (USDHHS, PAGAC 2008) and is covered in more detail in part II of the text.

KINESIOLOGY DISCIPLINES AND EXERCISE TRAINING

A complete review of the subdisciplines related to the field of kinesiology is beyond the scope of this text. What follows is an overview that focuses on the areas of exercise physiology, the movement sciences, and sport and exercise psychology. These primary

exercise sciences were chosen because they have been the common integrators for both traditional exercise training models and emerging public health physical activity models (see chapter 3 and part II of the text for more).

Other important subdisciplines of kinesiology, such as nutrition and the sociology of sport (particularly as related to behavioral science), should also contribute to the development of physical activity and exercise public health plans, but they are not a primary focus of this text. Nutrition public health goals and interventions have been covered extensively in other resources, such as the 2010 Dietary Guidelines for Americans (www.cnpp.usda.gov/dietaryguidelines.htm) and Nutrition.gov (www.nutrition.gov), and you should become familiar with these sites if you are not already. Behavioral and social methods for promoting physical activity and exercise are covered in more detail in chapter 13.

EXERCISE PHYSIOLOGY

According to Kenney, Wilmore, and Costill (2012), **exercise physiology** is the study of how the body structures and functions are altered by acute bouts of exercise or physical activity, and how the body adapts to the chronic stress of physical training. Exercise physiology also addresses the integration and coordination of body systems needed to maintain homeostasis, including the musculoskeletal, nervous, circulatory, respiratory, immune, endocrine (hormone-producing), digestive, urinary, integumentary (skin), and reproductive systems. The findings from exercise physiology training studies conducted since the 1960s were initially used for the prevention and treatment of chronic diseases such as cardiovascular disease. Exercise physiology has now become a key discipline that helps explain the role of physical activity and exercise in disease prevention and rehabilitation (see figure 2.2).

The study of exercise physiology evolved from a basic concept (i.e., exercise is medicine) promoted by the ancient Greeks and Romans such as Hippocrates and Galen and other physicians and scientists through time. Eventually, it developed into a complex discipline. Exercise physiology now addresses a spectrum of issues ranging from the molecular mechanisms associated with positive physiological adaptations via physical activity and exercise all the way to developing and promoting public health poli-

Physical inactivity and sedentary lifestyle

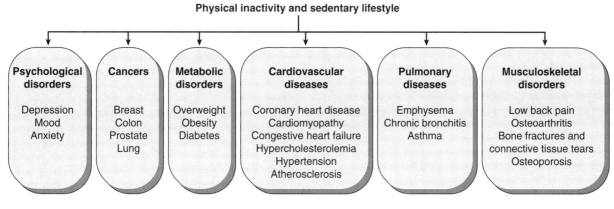

Psychological disorders	Cancers	Metabolic disorders	Cardiovascular diseases	Pulmonary diseases	Musculoskeletal disorders
Depression Mood Anxiety	Breast Colon Prostate Lung	Overweight Obesity Diabetes	Coronary heart disease Cardiomyopathy Congestive heart failure Hypercholesterolemia Hypertension Atherosclerosis	Emphysema Chronic bronchitis Asthma	Low back pain Osteoarthritis Bone fractures and connective tissue tears Osteoporosis

Figure 2.2 Role of physical activity and exercise in chronic disease prevention and rehabilitation.
Reprinted, by permission, V. Heyward, 2010, *Advanced fitness assessment and exercise prescription,* 6th ed. (Champaign, IL: Human Kinetics), 2.

cies that drive the concepts presented in this text. The study of exercise physiology can be applied to a variety of physical activity and exercise settings and research questions, including the following:

- What type of exercise training program can increase maximal oxygen uptake (fitness) ($\dot{V}O_2$max) the most?

- How much exercise is needed for optimizing the benefits of training while minimizing the risk of injury?

- How much physical activity per week is required for good health?

- How much physical activity or exercise is enough to maintain a healthy weight?

MOVEMENT SCIENCES

The **movement sciences** include the study areas of motor learning, motor control, and biomechanics. **Motor learning** is the study of how we learn and perform motor skills such as cycling and dancing. It also addresses the concepts that influence motor skills negatively or positively. **Motor control** is the study of human information processing and the integration of motor movements that involve motor planning and execution. **Biomechanics** is the study of physics applied to the understanding of movement in living organisms.

Movement scientists often study movement efficiency by observing movement patterns (through high-speed filming), measuring forces involved with movement, and developing equipment to maximize performance or protect participants from injury due to excessive movement forces. They also study

the impact of instructional information and feedback on skill development. Information from the movement sciences can be applied to a variety of physical activity and exercise settings and research questions, including the following:

- How can we improve the efficiency or economy (i.e., reduce the energy cost) of walking in individuals?

- How do fundamental motor skills affect physical activity behavior and the development of physical fitness?

- What movement challenges are related to falls, and can physical activity and exercise prevent falls?

- How does aging affect the biomechanics of walking and energy expenditure?

The basic concepts of the movement sciences can help us understand factors such as the economy of physical movements, balance, and overall physical mobility. This understanding can help in the planning of physical activity and exercise interventions for populations.

SPORT AND EXERCISE PSYCHOLOGY

Sport and exercise psychology is the study of behaviors and outcomes related to participation in sports or programs of exercise training. In sports, coaches are interested in sport psychology for a variety of reasons, including a desire to motivate their athletes to optimize performance. Behavioral cues can help athletes focus on key movement components to optimize their performance, and they can learn how to use specific cues to mentally rehearse how to react positively.

Currently, researchers in physical activity and public health often apply behavioral models such as the transtheoretical model of behavioral change and the social cognitive model (see chapter 13), to understand and promote positive health behaviors in individuals. The study of sport or behavioral psychology can be applied to a variety of physical activity and exercise settings and research questions, including the following:

- What factors contribute to competition anxiety?

- What motivates people to become physically active?

- What motivates people to remain physically active for a lifetime?

- How does aging affect the motivation to participate in physical activity or exercise?

- How can we understand a population's attitude toward and motivation to participate in physical activity?

Understanding several sport and exercise psychology theories or strategies is important for promoting behavioral change that can result in the achievement of public health, physical activity, and exercise goals. In fact, the traditional foundations of sport and exercise psychology have become part of what we will refer to as the *behavioral sciences* for the remainder of the book.

Although the integration of the principles of kinesiology (particularly exercise physiology, the movement sciences, and the behavioral sciences) have been used extensively in exercise training programs for maximal performance, they are also essential for effective population physical activity and exercise promotion and planning. As you will learn more in chapter 3 and part II of the text, kinesiology principles can be used to promote physical activity and exercise in the following lifestyle domains for populations: daily living, active transportation, recreation and leisure, and occupation.

PRINCIPLES OF EXERCISE TRAINING, PRESCRIPTION, AND PLANNING

Numerous principles of exercise training theory established in the kinesiology literature are associated with the outcomes of optimizing performance and improving physical fitness. Exercise training theory has provided the basis for individual exercise prescription since the late 1970s. However, an understanding of exercise training theory is needed for developing physical activity and personalized exercise plans. Participation in regular physical activity and exercise programs is essential for attaining basic **functional health** both individually and in populations (this is covered in detail in chapter 7). A *functionally healthy person* has been defined as

HEALTH AND FITNESS PROFESSIONS

In addition to the scholarly fields of kinesiology, other professional subdisciplines have started to embrace the promotion of physical activity as a primary mission related to their public health educational and rehabilitative initiatives. Fields such as health education, physical education, athletic training, physical therapy, and sport management all contain information that can, and should, be applied to physical activity and exercise public health interventions. In fact, the integration of the concepts from the subdisciplines of kinesiology have led to employment opportunities via the National Physical Activity Society (NPAS) and its certification opportunities (see chapter 3 for more). The NPAS was originally known as the National Society of Physical Activity Practitioners in Public Health (NSPAPPH) when it was established in 2006. The ability to translate research into effective clinical applications has become a new requirement for many new job opportunities related to the promotion of physical activity and public health.

"a reasonably (not perfectly) healthy person [with] a lot of health but some disability" (USDHHS 2008, p. G6-1) and includes the maintenance of functional ability and role ability. The following sections highlighting general physical training principles will help you integrate exercise science into long-term physical activity and exercise plans.

TRAINING THEORY AND PRINCIPLES

Exercise training theory includes understanding how to use information from the exercise sciences (data-based outcomes from the research literature) and how to apply it (the clinical practice of using training principles based on knowledge and experiences) effectively to improve performance.

The science of exercise prescription is often based on the FITT concept; FITT stands for the *frequency, intensity, time* (duration), and *type* (mode) of exercise. The FITT concept can be further described as follows:

- *Frequency:* How often an exercise or physical activity is performed. Frequency can be expressed in sessions, episodes, or bouts per week.
- *Intensity:* How hard one works, or the physical effort required to perform a physical activity or exercise. Intensity can be provided in absolute or relative terms such as low, moderate, or vigorous (see the discussion of intensity for more).
- *Time:* The amount of time in which a physical activity or exercise is performed. Duration is usually expressed in minutes.
- *Type:* The specific mode of physical activity or exercise.

A variety of factors or principles of training should be considered in order to realize adaptations. Traditional concepts of training theory were originally applied to individuals but are also pertinent to physical activity or exercise interventions for populations. Following are descriptions of the principles of training (labels and descriptions may vary by reference sources you use).

- *Practical goal setting:* Determining the needs and goals of an individual or population.
- *Genetics and individual variation:* The genetics and potential for change of an individual or population with training.

- *Motivation:* Evaluation of the drive or personal motivation of an individual or population and the need to provide appropriate feedback and reinforcement for success with physical activity programming.
- *Teaching model:* The need to teach individuals or populations how to participate effectively in physical activity or exercise to improve performance while reducing the risk of injuries.
- *Fitness evaluation:* Evaluating physical abilities to participate effectively in physical activity and exercise based on issues such as age, health status, and experience.
- *Progressive overload:* Changing the FITT variables of a plan to improve physiological, movement, and psychological adaptations.
- *Specificity:* The specific physiological, movement, and psychological adaptations that occur as a result of the specific demands applied.
- *Modification:* Adjusting the physical activity or exercise regimen based on factors such as disease, injury, and change in medications.
- *Periodization:* An overall physical training plan that includes cycles of varying training volumes based upon seasonal variations that provides for appropriate rest and recovery periods.
- *Overtraining:* Participating in too much physical activity or exercise, which can result in negative physical performance; if continued, overtraining can produce negative psychological effects and increase the risk of musculoskeletal injury.
- *Detraining:* Discontinuing physical activity or exercise, and the rate and magnitude at which training benefits are lost.
- *Recovery:* How a person recovers from participating in a program of physical activity or exercise and the strategies that might help improve recovery time.
- *Compliance:* Why or how an individual or population continues to participate in or drops out of a physical activity or exercise program.

VOLUME AND DOSE RESPONSE

The concept of **total caloric expenditure** is directly related to the **volume,** or dose, of training (physical activity or exercise) that accumulates over time.

INTEGRATING THE TRADITIONAL EXERCISE SCIENCE PRESCRIPTION INTO PERSONALIZED AND POPULATION EXERCISE PLANS

Professional organizations and associations such as the American College of Sports Medicine (ACSM) have been promoting traditional exercise science methods for developing individualized exercise prescriptions since the 1970s. Although these traditional exercise prescription methodologies can be useful for promoting physical activity and exercise to the public, many people may find them too complex and difficult to comply with. This may be especially true for people and groups who are interested only in improving or maintaining their health and fitness, and not in performance outcomes. The *2008 Physical Activity Guidelines for Americans* (USDHHS 2008) were developed with real-life examples of children, adolescents, adults, and older adults who have become and remained physically active.

Given the large prevalence of inactive people worldwide, it is important to become familiar with the cases presented in the *2008 Physical Activity Guidelines for Americans* (e.g., Harold, Maria, Douglas, Anita, Mary, and Manuel). Understanding the individual and population strategies for the cases described in that report and learning how to integrate the physical training concepts presented in this chapter will help optimize success in physical activity promotion.

Daily voluntary caloric expenditure can account for 15 to 30% of a person's total caloric expenditure. Many of the benefits of physical activity and exercise training are related to the amount of caloric expenditure achieved, which is why it should be considered when developing exercise prescriptions and public health, physical activity, or exercise plans. (Energy, or caloric, balance is described in detail in chapter 6.)

Essentially, when you are physically active, you expend more kilocalories (or kcals) than when you are sedentary. For example, an average person (approximately 70 kg, or 154 lb) expends about 1.2 kcals per minute while sitting at rest. Another common metabolic measure is the metabolic equivalent, or **MET**; 1 MET ($3.5 \text{ ml} \cdot \text{kg}^{-1} \cdot \text{min}^{-1}$) is equal to the resting energy expenditure for the same average person. It is also important to understand the difference between gross and net energy expenditure, because both are used by practitioners to measure or estimate total caloric expenditure. **Gross energy expenditure** (e.g., in METs) combines physical activity or exercise energy requirements with resting energy expenditure, whereas **net energy expenditure** reflects just the physical activity or exercise energy requirement.

The volume, or dose, is the amount of physical activity or exercise performed and is based on frequency, duration, and intensity. **Accumulation** refers to acquiring a specific dose of physical activity or exercise, or achieving a physical activity or exercise goal, by performing several shorter bouts and adding them up (e.g., three bouts lasting 10 minutes each to achieve 30 minutes of daily physical activity or exercise).

Dose-response refers to the amount of physical activity or exercise needed for achieving health, fitness, or performance goals. Dose-responses can be measured in terms of the frequency, duration, and intensity of physical activity or exercise or the total volume of work. Dose-responses related to physical activity or exercise are similar to those related to medications, in that responses vary with the dose of medication. Figure 2.3 shows some of the dose-response curves for physical activity or exercise in relationship to various health outcomes.

TYPE

When considering the FITT variables for exercise prescriptions or physical activity and exercise plans, most exercise professionals recommend selecting the type of physical activity or exercise first, based on a practical goal-setting process for either individuals or populations.

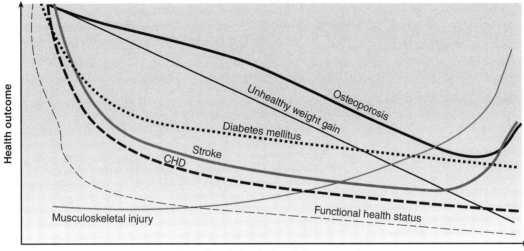

Health outcome

Unhealthy weight gain

Osteoporosis

Diabetes mellitus

Stroke

CHD

Musculoskeletal injury

Functional health status

Quantity of physical activity or exercise

Figure 2.3 Physical activity, exercise, and health outcomes.

The type (or mode) of physical activity or exercise can be categorized as anaerobic, aerobic, or combined, and as either static or dynamic. According to the *2008 Physical Activity Guidelines for Americans* (USDHHS 2008, p. C-2), **anaerobic activities** require the use of nonoxidative energy systems (energy-producing reactions in the body that do not require oxygen) and engaging in these activities can improve the capacity of these systems and increase the tolerance of acid–base imbalance during high-intensity exercise, which allows one to work more effectively at high intensities for short durations. Examples of anaerobic activities are 100-meter sprints, gymnastics, and **resistance training** (e.g., weight training to improve strength, power, and endurance and to increase mass). **Aerobic activities** require the use of large oxidative systems (i.e., heart and lungs); these activities improve cardiorespiratory endurance. Examples of aerobic activities are walking, cycling for 30 minutes, jogging, and rowing.

Combined physical movements take place in sports such as soccer, tennis, and racquetball that require significant contributions of energy from both anaerobic and aerobic energy sources. Some examples of combined physical movements are the bouts of physical activity or exercise that involve **interval training**, which requires working at a higher intensity for a few seconds or minutes followed by working at lower (recovery) intensities for the same amount of time. Interval training can be used for either improving health or maximizing performance.

Static physical activity or exercise (or isometric physical activity or exercise) is anaerobic and requires an increase in force production with limited range of motion (ROM); a test of handgrip strength is an example. **Dynamic physical activity or exercise** requires muscle-shortening (concentric) and muscle-lengthening movements (eccentric). Dynamic physical activities involve greater ROM than static activities do and are usually rhythmic and more continuous.

Physical activity and exercise can be further categorized as creating specific physiological demands for energy based on intensity and duration. **Anaerobic power** (or peak power) activities involve short-burst, high-intensity movements that last less than 15 seconds and stress the anaerobic energy pathways of the body. Anaerobic power abilities are highly related to genetics and the number of fast-twitch muscle fibers the person can recruit. Fast-twitch fibers are recruited primarily with higher work intensities and quickly fatigue, while slow-twitch fibers are primarily recruited more at low to moderate work intensities and are fatigue resistant. **Anaerobic capacity** (also known as mean anaerobic power or peak anaerobic power) activities also involve short-burst, high-intensity movements, but they last from 15 seconds up to 3 minutes. Anaerobic capacity activities produce high levels of metabolic products such as lactic acid and challenge the person's tolerance of an acid–base imbalance.

Aerobic power activities require high levels of oxygen delivery to the working muscles and last from 3 to 15 minutes. Aerobic power is associated with the measurement or estimation of maximal oxygen uptake ($\dot{V}O_2$max), which is the maximal ability of the body to use oxygen to produce energy for performing work. Figure 2.4 illustrates the concept of exercise intensity (running) and $\dot{V}O_2$max for a trained and untrained man. **Aerobic capacity** activities stress the ability to maintain high percentages of $\dot{V}O_2$max for extended periods of time (e.g., 20 minutes or longer). An average untrained person should be able to work for this amount of time at about 50% of $\dot{V}O_2$max aerobic capacity prior to a program of physical activity or exercise, but can increase this to 70% (or higher) after several weeks. Improvements in aerobic capacity can improve cardiorespiratory **economy** (i.e., oxygen cost at a given workload or speed) by recruiting less muscle mass to perform the same amount of work.

INTENSITY

Intensity can be determined for a physical activity or exercise in a variety of ways based on whether the activity is more aerobic or more anaerobic. For example, a person can obviously work at a higher intensity for a shorter period of time (e.g., a few

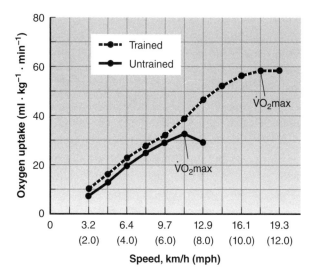

Figure 2.4 Relationship between exercise intensity (running speed) and $\dot{V}O_2$max for a trained and an untrained man.
Reprinted, by permission, from L. Kenney, J. Wilmore, and D. Costill, 2012, *Physiology of sport and exercise*, 5th ed. (Champaign, IL: Human Kinetics), 122.

seconds) than for several minutes. Intensity is the primary FITT variable that affects the total caloric expenditure of physical activity or exercise when the duration is held constant. Intensity level also affects the risk of injury if it is set too high for the current level of conditioning of an individual or a population.

Intensities can be classified as moderate or vigorous and as absolute or relative. The *Physical Activity Guidelines Advisory Committee Report, 2008* (USDHHS, PAGAC 2008) generally classifies physical activity or exercise as **moderate intensity** when the person is working between 3 and 5.9 METs and **vigorous intensity** when the person is working above 6 METs. **Absolute intensity** can be expressed in kcals/min, in METS, or as walking at 4 miles per hour (6.4 km/h) or jogging at 5 miles per hour (8 km/h). For resistance exercise, absolute intensity can be expressed as the amount of weight lifted or force exerted (e.g., in pounds or kilograms). Absolute intensity may also be classified into categories such as low, moderate, vigorous, and maximal. Table 2.1 shows the classification of physical activity levels from the *2008 Physical Activity Guidelines for Americans* (USDHHS 2008).

Relative intensity is usually expressed as a percentage of aerobic power ($\dot{V}O_2$max) or as a percentage of measured heart rate or heart rate reserve (maximal heart rate – resting heart rate). Perceived exertion ratings (i.e., how hard people feel they are working, from light to very hard) can also be used as a relative intensity measure along with a percentage of 1-repetition maximum (i.e., the maximum a person can lift in one trial) for weightlifting activities.

The *Physical Activity Guidelines Advisory Committee Report, 2008* (USDHHS, PAGAC 2008, pp. D-3, D-4, Table D-1) contains classifications of physical activity and exercise intensities that show how the intensity of physical activity and exercise is significantly affected by the differences of $\dot{V}O_2$max of individuals or populations. As shown in figure 2.5, an adult with a higher-than-average $\dot{V}O_2$max of 14 METs who is walking at 3 miles per hour (4.8 km/h) is working at a relative intensity of 24% of her max (low). However, an adult with a low $\dot{V}O_2$max of 4 METs is working at a relative intensity of 83% of his max (hard). An adult with a low aerobic capacity of 4 METs cannot walk at 4 miles per hour

Table 2.1 Classification of Total Weekly Amounts of Aerobic Physical Activity Into Four Categories

Levels of physical activity	Range of moderate-intensity minutes a week	Summary of overall health benefits	Comment
Inactive	No activity beyond baseline	None	Being inactive is unhealthy.
Low	Activity beyond baseline but fewer than 150 minutes a week	Some	Low levels of activity are clearly preferable to an inactive lifestyle.
Medium	150 minutes to 300 minutes a week	Substantial	Activity at the high end of this range has additional and more extensive health benefits than activity at the low end.
High	More than 300 minutes a week	Additional	Current science does not allow researchers to identify an upper limit of activity above which there are no additional health benefits.

Reprinted from USDHHS 2008.

(6.4 km/h) for very long because the intensity exceeds $\dot{V}O_2$max and the activity becomes mostly anaerobic; this workload requires a 5 MET absolute intensity.

A full discussion of the many ways to determine physical activity and exercise intensities is beyond the scope of this text. However, the common aerobic and anaerobic methods used for determining intensity for exercise prescriptions, which also apply

to physical activity and exercise interventions for populations, are discussed next.

PERCENTAGE OF MAXIMAL HEART RATE

The percentage of maximal heart rate (MHR) method is an aerobic method based on the simple exercise physiology assumption that predicted MHR

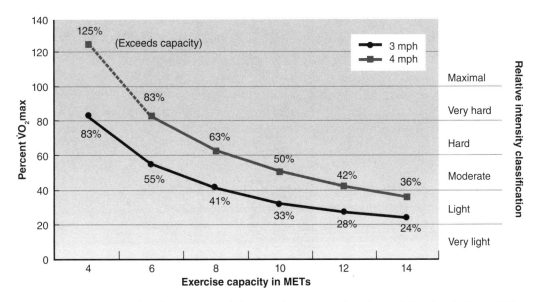

Figure 2.5 Relative intensity of walking at 3 mph (4.8 km/h, 3.3 METs) and 4 mph (6.4 km/h, 5.0 METs) expressed as a percentage of $\dot{V}O_2$max for adults with an exercise capacity ranging from 4 to 14 METs.
Reprinted from USDHHS 2008.

is equal to 220 minus age (generally the error is plus or minus 10 beats per minute, or bpm). You then multiply MHR by the percentage of the intensity desired. For example, if a goal is to have a 30-year-old man work at 60 to 80% of his MHR, the calculation would be as follows: 220 − 30 = 190 × 0.6 and 190 × 0.8, or an exercise heart rate between 114 bpm and 152 bpm. This method is convenient to use with beginners who may be at higher health risk, because it is conservative (when using lower percentages) and yields lower intensities than some other methods do.

PERCENTAGE OF MAXIMAL HEART RATE RESERVE, OR TARGET HEART RATE

The percentage of MHR reserve, or target heart rate (THR), method is an aerobic method also based on simple MHR prediction; it involves subtracting resting heart rate (RHR) according to the following equation:

(1) MHR reserve = MHR − RHR

(2) THR = (MHR − RHR) × (desired percentage of intensity, e.g. 60%) + RHR

A 40-year-old with an RHR of 80 who works at 60 to 70% of MHR reserve would yield the following equation: THR = (180 − 80) × 0.6 + 80 = 140 bpm and THR = (180 − 70) × 0.7 = 157 bpm for a THR range of 140 to 157 bpm. This formula is called the Karvonen formula in honor of the exercise physiologist who first described it in the late 1950s. This method is often used for those who are familiar with regular exercise and at low health risk, because it provides higher intensities than a simple percentage of MHR.

METABOLIC EQUIVALENTS

The metabolic equivalent (MET) aerobic method is based on the concept described earlier in this chapter that, for healthy adults, 1 MET = 3.5 ml · kg^{-1} · min^{-1}. As you have learned, physical activity and exercise can be classified as moderate or vigorous intensity based on the MET level required to perform specific movements. The *Physical Activity Guidelines Advisory Committee Report, 2008* has a listing of activities and exercise that require at least 6 METs' worth of intensity (USDHHS, PAGAC 2008, pp.D-5, D-6).

Scientific studies show us that it is never too late to begin a physical activity program. What are some reasons that physical activity is important throughout life?

Once the $\dot{V}O_2$max of an individual or a population is measured (usually on a maximal exercise test) or estimated (see chapter 5 for more on measuring or estimating $\dot{V}O_2$max), it can be expressed in METs. A percentage of the maximal METs achieved can then be used to calculate an initial or follow-up exercise training intensity. A person with a $\dot{V}O_2$max of 12 METs might train at 50% of his maximum, or at 6 METs. Physical activity practitioners commonly use this method to help people train at a certain intensity.

PERCENTAGE OF MAXIMAL OXYGEN UPTAKE

As with the MET determination just discussed, if you know the individual's or population's maximal oxygen uptake ($\dot{V}O_2$max), you can calculate an intensity for physical activity or exercise based on a percentage of that max. For example, if the goal is to train a group at 70% of $\dot{V}O_2$max, and the group average $\dot{V}O_2$max is 35 ml · kg^{-1} · min^{-1} (or 10 METs), you would use the calculation 35 ml · kg^{-1} · min^{-1} × 0.7 = 24.5 ml · kg^{-1} · min^{-1} (or 7 METs). Figure 2.6 illustrates how to use the linear relationship between HR and $\dot{V}O_2$max to determine a representa-

tive physical activity or exercise HR response (75% of $\dot{V}O_2$max) based on the percentage of $\dot{V}O_2$max prescribed. This method is still one of the most popular methods used by exercise physiologists to determine exercise intensities.

KILOCALORIES PER MINUTE OR PER HOUR

The kilocalories per minute or kilocalories per hour aerobic method has become one of the most popular methods to determine physical activity and exercise intensity, particularly in weight loss and weight maintenance programs. The conversion factor if 1 liter of oxygen consumed being equal to ~5 kcals/min is used to determine the total kcals expended for a bout of physical activity or exercise. For example, if someone weighs 80 kilograms (176 lb) and is working at 35 ml · kg^{-1} · min^{-1} (or 10 METs), you could calculate caloric expenditure as follows: 35 ml · kg^{-1} · min^{-1} × weight [in kg] × 1,000 = oxygen uptake in liters/min, or 35 × 80 / 1,000 = 2.8 liters/min. Then, converting to kcals/min, you would have 2.8 liters/min × ~5 kcals/min, or 14 kcals/min, or 840 kcals/h (14 × 60).

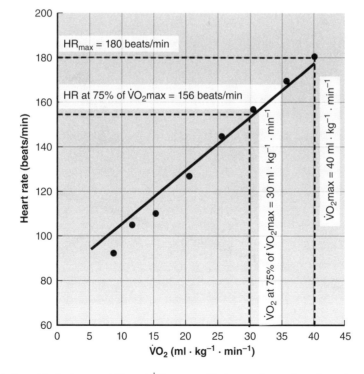

Figure 2.6   Linear relationship between HR and $\dot{V}O_2$max with increasing rates of work and the HR equivalent to a set percentage (75%) of $\dot{V}O_2$max.

Reprinted, by permission, from L. Kenney, J. Wilmore, and D. Costill, 2012, *Physiology of sport and exercise*, 5th ed. (Champaign, IL: Human Kinetics), 511.

PERCEIVED EXERTION SCALES AND THE TALK TEST

The perceived exertion (PE) scale and talk test methods require that people evaluate how hard they are working, or determine whether they can carry on a conversation at a given intensity of work. Figure 2.7 shows a PE scale called the OMNI-Walk/Run Scale for adults; it is a simple way for practitioners to teach people how to rate their physical activity and exercise intensity. For example, an OMNI scale rating of 7 for an untrained adult would indicate that the person is working "somewhat hard" to "hard." Scales such as the OMNI scale can be used to help people focus on achieving moderate to vigorous physical activity or exercise intensities by achieving PE levels ≥4 (PE of 4 to 6 = moderate intensity physical activity or exercise, and PE > 6 = vigorous intensity physical activity or exercise). The talk test basically tells practitioners whether someone is working at a comfortable level (i.e., can carry on a conversation easily) or struggling (i.e., cannot carry on a conversation comfortably).

PE scales and the talk test are very practical because people can report on their feelings about their bodily responses to physical activity or exer-

cise (e.g., breathing, muscle pain, sweating). However, they are complicated by various physiological factors that influence the ratings such as lactate and ventilatory thresholds (see ACSM 2010 for more on the lactate and ventilatory thresholds). Exercise practitioners should make sure that the people using these methods have experience with them. Those who don't (especially youth) often underestimate how hard they are working.

REPETITION MAXIMUM

The most commonly used anaerobic method related to weightlifting as a resistance training mode is the 1-repetition maximum (1RM). For example, a person who can lift 100 pounds (45.5 kg) over his head one time and no more has a 1RM of 100 pounds. This would be called his **absolute strength** for that lift. **Relative strength** would reflect his 1RM divided by his body weight as per table D-1 of the *Physical Activity Guidelines Advisory Committee Report, 2008* (USDHHS, PAGAC 2008) as discussed earlier.

Because performing a 1RM can be challenging for inexperienced weightlifters or older adults, many practitioners have the person perform a 5RM (maximum amount of weight lifted five times) or 10RM

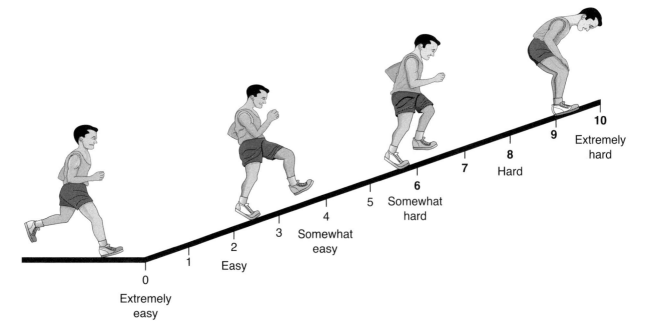

Figure 2.7 OMNI-Walk/Run Scale of perceived exertion for adults.

Reprinted, by permission, from R.J. Robertson, 2004, *Perceived exertion for practitioners: Rating effort with the OMNI picture system* (Champaign, IL: Human Kinetics), 142.

(maximum amount of weight lifted 10 times), which is safer and more comfortable. Normative standards are available (see ACSM 2010) for 1RM by age and sex, which can be used for resistance performance evaluations of individuals or populations. In addition, values from 5RM and 10RM can be used to predict 1RM abilities, which can then be used to develop exercise prescriptions or physical activity or exercise plans.

WORKLOAD

The workload anaerobic method considers factors that affect resistance exercise such as speed, muscular strength, muscular endurance, power, and the specific muscle groups involved. Several resources (e.g., ACSM 2010) detail the specifics of resistance training programs for various individuals and populations. However, all resistance training regimens should consider the following: emphasis (e.g., strength or muscular endurance); participant's level (e.g., beginner versus advanced); percentage of 1RM, 5RM, or 10RM; number of sets and number of repetitions per set (e.g., 10 repetitions per set, for three sets); velocity of movement (e.g., slow, moderate, or fast); rest between sets (see the following discussion of time); and the frequency of participation per week (see the following discussion of frequency).

TIME

Table 2.2 provides some general recommendations based on traditional exercise prescription programming for determining the time (duration) of types of physical activity and exercise activities. The table includes recommendations for achieving basic functional health (a minimal long-term goal) as well as high performance.

FREQUENCY

Table 2.3 provides some general recommendations for the frequency of types of physical activity and exercise activities. It includes recommendations for achieving basic functional health (a minimal long-term goal) as well as high performance.

APPLYING PHYSICAL ACTIVITY AND EXERCISE TRAINING PRINCIPLES

Once you are comfortable using the FITT concept, it would be helpful to learn how to apply exercise science principles and practice them to improve the exercise training adaptations of individuals and populations. The exercise principles previously defined are explained in more detail in this section.

Table 2.2   Recommendations for Time for Achieving Functional Health or High Performance

Type of activity	Functional health	High performance
Aerobic	20-60 min	20-120 min
Anaerobic	20-30 min	20-120 min
Aerobic and anaerobic	Based on practical goal setting	Based on practical goal setting

Recommendations can be achieved continuously or accumulated in multiple bouts per day.

Table 2.3   Recommendations for Frequency for Achieving Functional Health or High Performance

Type of activity	Functional health	High performance
Aerobic	3-5 days/week	5-7 days/week
Anaerobic	2-3 days/week	3-4 days/week
Aerobic and anaerobic	Based on practical goal setting	Based on practical goal setting

Recommendations can be achieved continuously or accumulated in multiple bouts per day.

TOTAL PHYSICAL ACTIVITY PER WEEK

The amount of physical activity and exercise performed for a given period of time (see USDHHS, PAGAC 2008, section D, for more) can be determined by using the absolute intensity, the time or duration, and the frequency. As shown in table 2.4, the dose of physical activity and exercise can be expressed in minutes or hours per week (of moderate-intensity, vigorous-intensity, or moderate-plus vigorous-intensity activity), or in terms of distance walked or jogged or ran per day or week. The amount of physical activity and exercise can also be expressed in kilocalories per day or week, kilocalories per kilogram of body weight per day or week, or MET-minutes or MET-hours per day or week. By using the concept of relative intensity, time or duration, and frequency (e.g., 30 minutes at 70% of predicted MHR five times per week for 24 weeks), one can compare and generalize across various FITT amounts to determine recommendations for physical activity and exercise in population interventions.

Table 2.4 shows that if a person participated in physical activity (walking at 3 mph, or 4.8 km/h, at moderate intensity) for 2.5 hours per week (a minimum recommendation), she would average 495 MET-minutes per week, 8.25 MET-hours per week, or the equivalent of 7.5 miles (12 km) per week. If that person weighed 165 pounds (75 kg), her energy expenditure would be 620 kcals per week. If the same person were active at the same intensity for 5 hours per week, she would average 990 MET-minutes per week, 16.5 MET-hours per week, or the equivalent of 15 miles (24 km) per week, and her energy expenditure would be 1,240 kcals per week.

If the person in the preceding example were able to participate in physical activity or exercise at a vigorous intensity such as jogging at 7 miles per hour (11 km/h) for 2.5 hours per week, she would expend considerably more calories per week (620 kcals versus 2,155 kcals based on gross energy expenditure values), which reinforces the importance of intensity when determining FITT variables and setting goals for individuals and populations.

Table 2.4 Quantitative Measures of Weekly Physical Activity for Walking, Jogging, and Running

Speed (mph)	METs	FOR 150 MIN (2.5 HR) OF PA/WEEK				FOR 300 MIN (5.0 HR) OF PA/WEEK			
		MET-min	MET-hr	Miles	kcals†	MET-min	MET-hr	Miles	kcals†
Rest	1.0	150	2.5	0.0	190	300	5.0	0.0	380
2.5	3.0	450	7.5	6.25	565	900	15.0	12.5	1,130
3.0	3.3	495	8.25	7.5	620	990	16.5	15.0	1,240
4.0	5.0	750	12.5	10.0	940	1,500	25.0	20.0	1,880
4.3	6.0	900	15.0	10.75	1,125	1,800	30.0	21.5	2,250
5.0	8.0	1,200	20.0	12.5	1,500	2,400	40.0	25.0	3,000
6.0	10.0	1,500	25.0	15.0	1,875	3,000	50.0	30.0	3,750
7.0	11.5	1,725	28.25	17.5	2,155	3,450	56.5	35.0	4,310
8.0	13.5	2,025	33.75	20.0	2,530	4,050	67.5	40.0	5,060
10.0	16.0	2,400	40.0	25.0	3,000	4,800	80.0	50.0	6,000

2.5-4.3 mph (4-7 km/h) = walk; 5-10 mph (8-16 km/h)= jog or run

† Kilocalories for a 75 kg (165 lb) adult when exercising at the given intensity for either 150 or 300 minutes.

Note: These are gross energy expenditure values during exercise; thus, they include the energy expenditure at rest and not just the additional energy expenditure as a result of the activity. Kilocalories were calculated using 1 MET = 1 kcals^{-1} · kg^{-1} · h^{-1} and rounded to the nearest 5 kcals. MET values from Ainsworth et al. 2000, p. 9.

Reprinted from USDHHS 2008.

PRACTICAL GOAL SETTING

When determining the needs and goals of an individual or group, a needs assessment should be conducted that includes practical and achievable goals for physical activity and exercise. Use the concepts and strategies of behavioral science theoretical models described in chapter 13 to fine-tune your goal setting skills. A good goal for a group would be to achieve the minimal recommendations of the *2008 Physical Activity Guidelines for Americans* (USDHHS 2008).

GENETICS AND INDIVIDUAL VARIATION

The genetics of an individual or a population can affect the ability to benefit from physical activity and exercise. Genetics also accounts for adaptation rates and the maximal potential for adaptations. Obviously, there are variations for levels of change within individuals and populations for important health fitness variables similar to the variance in the curves such as that for physical activity, exercise, and health outcomes previously shown in figure 2.3.

Figure 2.8 shows five genetic and training variations that you might see in relationship to the benefits of physical activity or exercise training over time (e.g., six months). Curve 3 represents a profile of adaptation to physical activity and exercise that most practitioners expect to see; however, many people do not respond in this fashion. You will need to evaluate your physical activity and exercise plans regularly to adjust them to facilitate further gains in health outcomes.

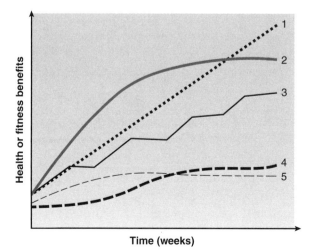

Figure 2.8 Training adaptation curves for five individuals.

MOTIVATION

Motivation refers to the importance of the behavioral reinforcement of intrinsic motivation goals that promote regular participation in physical activity or exercise. Chapter 13 describes behavioral models practitioners commonly use to encourage and maintain motivation levels.

TEACHING MODEL

The *teaching model* refers to the need for practitioners to be effective teachers who use educational strategies or logic models (see chapter 15 for more about logic models) to motivate target populations and reinforce the benefits of physical activity and exercise. Figure 2.9 provides an example of a simple teaching, or interaction, model a practitioner can use to work with an individual or population to minimize barriers that might negatively affect the adoption of regular physical activity. They also can work to maximize those factors that positively promote the adoption of regular physical activity and exercise.

FITNESS EVALUATION

Evaluating fitness levels prior to initiating physical activity and exercise programs is valuable for determining the effectiveness of the programs. The level of evaluation required will vary greatly depending on the physical abilities of the target population and its goals and needs. Performing fitness evaluations in conjunction with determining goals will help ensure that the program meets the needs of the target population.

PROGRESSIVE OVERLOAD

The *progressive overload principle* refers to the fact that improvements in health and physical fitness are directly related to increases in FITT variables by gradually increasing the physical stress on the body. Progressive overload should be considered in conjunction with genetics and individual variation to ensure that people have adapted to current training levels before changing any FITT variables. Too little or too much change in progressive overload will not produce the desired results. In most cases, changing all the FITT variables at once is unwise, because people do not have enough time to adjust to new workloads, which can result in injury or burnout. Applying the progressive overload principle

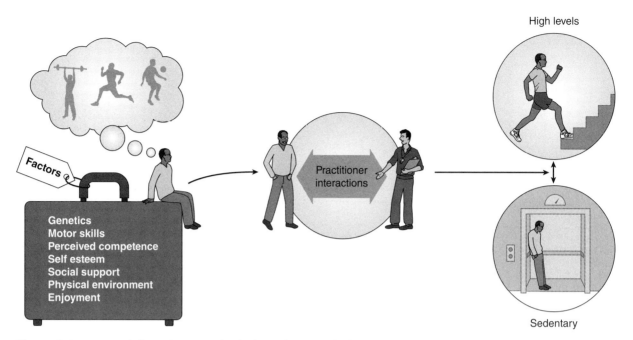

Figure 2.9 Factors influencing an individual to adopt regular low to high levels of physical activity and exercise.

correctly requires a lot of experience because of the great variability of physical adaptations to physical activity and exercise in individuals and populations.

SPECIFICITY

The *specificity principle* refers to the fact that changes in health and fitness depend on adjusting FITT variables for the desired adaptations. Whether working with individuals or populations, it is important to consider the specific physical demands of the physical activity or exercise plan. For example, a plan designed to help sedentary, high-health-risk middle-aged men reach minimal levels of functional fitness would be significantly different from one designed to help young, experienced adult women runners train to complete a marathon. The two goals are obviously very different and will require very different manipulation of FITT variables to achieve success.

MODIFICATIONS

Exercise prescriptions and physical activity and exercise plans often need to be modified based on individual, population, and environmental changes. Variables that may require a modification of a physical activity or exercise plan include injury, illness, medications, lack of recovery, and symptoms of overtraining. A modification might involve changing FITT variables to allow a person or population to continue with the plan, but at

adjusted levels; this can minimize the risk of complete relapse (stopping and not starting again). Learning how to modify physical activity and exercise plans will encourage compliance (see the discussion of compliance for more).

PERIODIZATION

Periodization refers to cycling (or varying) the FITT variables, volume of physical activity and exercise, and recovery time. Figure 2.10 illustrates a simple periodization plan that shows a progression of health and fitness outcomes over 30 weeks of participation. The progression of improvement depends on the starting health or fitness levels of the person or population (usually, the lower the starting health or fitness level, the greater the initial gains). Normal periodization plans include an initial stage, a plateau stage, and an improvement stage.

Training plateaus are periods when little, if any, improvement in fitness occurs. The rate of improvement can be influenced by effective periodization of training. It is important to be able to recognize the plateau phenomenon and use effective behavioral (see chapter 13 for more information) and FITT strategies to help people remain motivated to maintain or increase their physical activity and exercise behaviors. This is because a lack of improvement may promote relapse. There are numerous ways to apply the periodization principle to physical activity

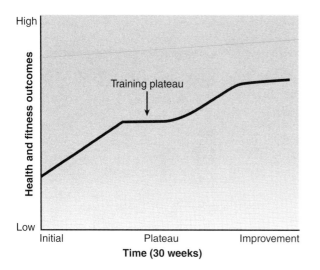

Figure 2.10 Simple periodization example.

and exercise plans; the references and e-media list at the end of the chapter offer more information about FITT variables and seasonal variations associated with periodization.

OVERTRAINING

Overtraining can produce abnormal physical and psychological responses. It is defined as doing too much physical activity or exercise without taking the time to recover appropriately. Overtraining is associated with increased overuse injuries (muscle and skeletal problems) and addictive behaviors that are directly related to relapses and lower exercise compliance levels. Symptoms of overtraining include excessive fatigue, unusual muscle soreness, a lower immune response (high incidence of upper respiratory infections), insomnia, weight loss, increased resting heart rates, and feeling mentally burnt out.

Chronic overtraining that occurs over several weeks or months can lead to burnout and is often associated with permanent physical activity or exercise relapses. You can help individuals or populations avoid overtraining by encouraging them to recognize the signs of addiction to exercise (e.g., the notion that if a little bit is good, more is better) and to take a short break (two or three days) in their physical activity or exercise routine if such signs are evident.

DETRAINING

Detraining refers to the loss of health or fitness following the cessation of a regular program of physical activity or exercise. Health and fitness benefits

often decrease at rates similar to those at which they accrued; however, significant variability can occur between specific factors (e.g., osteoporosis benefits versus functional health benefits; see figure 2.3). Significant detraining does not occur in a day or two, but people who stop all regular physical activity or exercise for two weeks to a month will notice losses in their aerobic and anaerobic abilities.

Bed rest is an extreme example of detraining. Dr. Jere Mitchell and his colleagues from Southwestern Medical School in Dallas, Texas (McGuire et al. 2001), reported in a classic study that 20 days of complete bed rest in 20-year-old males caused a greater deterioration in physical cardiorespiratory capacity than did 30 years of aging. Some research suggests that detraining effects can be minimized by continuing to be physically active even at a reduced volume of training before resuming a regular schedule of participation.

RECOVERY

The rate at which people or populations can recover is influenced by FITT variables, age, past experience with physical activity and exercise, environmental factors such as heat and altitude, average amount of sleep, and the abilities to rehydrate and consume enough energy to meet physical activity and exercise demands. Generally, at least 24 hours are needed to properly recover from high-intensity exercise, especially if someone is active daily. People who are just trying to meet the minimum guidelines for moderate and vigorous physical activity of 2.5 hours per week should not have much trouble recovering unless they have existing musculoskeletal injuries, disabilities, or complicating medical conditions.

COMPLIANCE AND ADHERENCE

Compliance refers to people's ability to continue to participate in regular physical activity or exercise programming. Following are barriers that can negatively affect compliance:

- Lack of time
- Low potential for physical adaptations
- Poor mobility
- Unrealistic physical activity or exercise goals or expectations
- Lack of knowledge about physical activity and exercise

KEY LEADER PROFILE

Gregory W. Heath, DHSc, MPH

Why and how did you get into this line of work? Did any one person have an overriding influence on you?

My initial academic interest in health-related behavior developed during my undergraduate years as a psychology student at Westmont College in Santa Barbara, California. Having an enthusiastic mentor in Dr. Mike Rulon, completing an independent research project, presenting the findings at a major regional meeting, and volunteering at a local county health department motivated me to pursue graduate/professional studies in the health sciences. I received training in epidemiology and completed the requirements for my master of public health (MPH) degree under the late Dr. Roland Phillips at Loma Linda University's School of Public Health. I also concurrently enrolled in a new doctoral program in the health sciences with a concentration in nutrition and physiol-

Courtesy of Gregory W. Heath.

ogy. Both my master's and doctoral training focused on chronic disease prevention and treatment.

I was a postdoctoral fellow at the Washington University School of Medicine in St. Louis, Missouri, where under the guidance of Drs. John Holloszy, Ali Ehsani, and Jim Hagberg I was able to further explore the preventive and therapeutic benefits of physical activity and exercise. During the following years I was engaged in establishing community-based cardiac rehabilitation programs and conducting community-based research addressing cardiovascular disease prevention (i.e., the Heartwatch Program) targeting school children and youth and their parents.

With the success of Heartwatch, my interests turned more and more to the importance of public health approaches to preventing disease and promoting health. The opportunity arose for me to join the CDC as an Epidemic Intelligence Officer under the mentorship of Drs. Ken Powell and Carl Caspersen. I readily immersed myself in the skills of field epidemiology and public health approaches directed toward the prevention and control of chronic diseases through physical activity. This activity has continued in my role as Guerry Professor of Health and Human Performance and Medicine at the University of Tennessee, Chattanooga.

What are your current research interests?

Most of my research efforts have been and continue to be dedicated to using epidemiologic and surveillance data for the purpose of planning, implementing, and evaluating evidence-based community interventions designed to promote physical activity.

What drives you as a researcher and activist?

I am driven by the overwhelming evidence that physical activity promotes health and prevents disease, and the unacceptably low prevalence of an effective dose of physical activity among all segments of the population in the United States and globally. With these facts as a backdrop, coupled with the truth that humans were created to move and the urgent need for scientifically grounded approaches to promoting physical activity among all peoples, I have pursued research and public health practice in physical activity.

What are one or two key issues to be addressed by 2022?

First, understanding and addressing physical activity disparities based on socioeconomic status, race, and ethnicity and second, the integration of active transport and recreational policies and planning into all regional planning organizations throughout the United States.

- Low perceived competency with physical activity
- Injury
- Past negative experiences with physical activity or exercise

Although compliance is presented as the last training consideration, it is one of the most important because, without it, even the best physical activity or exercise plan will fail. By minimizing barriers to compliance the opportunities for developing successful physical activity and exercise plans are significantly enhanced.

HEALTH AND FITNESS BENEFITS OF PHYSICAL ACTIVITY AND EXERCISE

Numerous health, fitness, and performance benefits of physical activity and exercise have been reported in the scientific literature. A complete list is beyond the scope of this chapter, but some specifics are provided in part II of the text. Some of the general benefits gleaned from the *2008 Physical Activity Guidelines for Americans* (USDHHS 2008) are provided in the sidebar Health Benefits Associated With Regular Physical Activity.

As noted in the *Physical Activity Guidelines Advisory Committee Report, 2008* (USDHHS, PAGAC 2008), people who are interested in training programs to increase performance-related fitness (e.g., agility, balance, coordination, speed, power, and reaction time) should seek advice from professionals specializing in sport skill activities, because people with these interests already are more active than the guidelines recommend for health and fitness. People who are interested in performance-related fitness may also be at higher risk for injury related to physical activity and exercise, and as such should consider factors that will help them remain safe and active (see chapter 10 for more information).

HEALTH BENEFITS ASSOCIATED WITH REGULAR PHYSICAL ACTIVITY

Children and Adolescents
Strong Evidence
- Improved cardiorespiratory and muscular fitness
- Improved bone health
- Improved cardiovascular and metabolic health biomarkers
- Favorable body composition

Moderate Evidence
- Reduced symptoms of depression

Adults and Older Adults
Strong Evidence
- Lower risk of early death
- Lower risk of coronary heart disease
- Lower risk of stroke
- Lower risk of high blood pressure
- Lower risk of adverse blood lipid profile
- Lower risk of type 2 diabetes
- Lower risk of metabolic syndrome

- Lower risk of colon cancer
- Lower risk of breast cancer
- Prevention of weight gain
- Weight loss, particularly when combined with reduced calorie intake
- Improved cardiorespiratory and muscular fitness
- Prevention of falls
- Reduced depression
- Better cognitive function (for older adults)

Moderate to Strong Evidence
- Better functional health (for older adults)
- Reduced abdominal obesity

Moderate Evidence
- Lower risk of hip fracture
- Lower risk of lung cancer
- Lower risk of endometrial cancer
- Weight maintenance after weight loss
- Increased bone density
- Improved sleep quality

Note: The Advisory Committee rated the evidence of health benefits of physical activity as strong, moderate, or weak. To do so, the Committee considered the type, number, and quality of studies available, as well as consistency of findings across studies that addressed each outcome. The Committee also considered evidence for causality and dose response in assigning the strength-of-evidence rating.

Reprinted from USDHHS, PAGAC 2008.

CHAPTER WRAP-UP

What You Need to Know

- Definitions for terms like *physical activity, exercise, physical fitness, health-related fitness*, and *skill-related fitness* are all important to understand and to use for effective communication in the field of physical activity and public health.

- The integration of the concepts of kinesiology related to public health is based primarily on the *2008 Physical Activity Guidelines for Americans* (USDHHS 2008) and the *Physical Activity Guidelines Advisory Committee Report, 2008* (USDHHS, PAGAC 2008).

- The study of kinesiology includes the major exercise sciences of exercise physiology, the movement sciences, and sport and exercise psychology.

- An understanding and integration of multiple kinesiology subdisciplines will help practitioners be successful at developing physical activity and exercise plans to promote and achieve public health goals.

- Numerous general and specific health, fitness, and performance benefits of participating in physical activity and exercise have been reported in the scientific literature.

- The *Physical Activity Guidelines Advisory Committee Report, 2008* (USDHHS, PAGAC 2008) includes real-life case studies of children, adolescents, adults, and older adults who have become and remained physically active. By learning how to integrate the physical training concepts presented in the chapter, you can optimize the success of your physical activity and exercise program.

- An understanding and application of training theory can help practitioners understand traditional training models designed for maximizing performance versus new strategies for developing physical activity and exercise plans for promoting positive health outcomes.

- The volume of physical activity and exercise significantly affects dose-response health benefits.

- It is important to understand information from the exercise sciences (data-based research literature outcomes) and apply it (the clinical practice of using training principles based on knowledge and experiences) effectively to improve health, fitness, and performance.

- The intensity of physical activity and exercise when duration is held constant is the most important FITT variable relative to caloric expenditure (volume) per minute, per hour, per day, or per week.

- Intensities are often classified in absolute terms (i.e., energy or work required to do an activity without accounting for the person's physiological capacity) or relative terms (i.e., taking into account the person's exercise capacity, such as a percentage of aerobic capacity).

- Moderate-intensity physical activity or exercise is categorized as working between 3 and 5.9 METs (USDHHS, PAGAC 2008), whereas vigorous-intensity physical activity is defined as working at greater than 6 METs.

- The amount of physical activity and exercise performed can be determined by using the absolute intensity or relative intensity, the time or duration, and the frequency.

- The art of applying physical activity and exercise training principles can help practitioners minimize barriers to meeting public health recommendations for participation in regular physical activity and exercise.

Key Terms

physical activity

exercise

physical fitness

kinesiology

exercise physiology

movement sciences

motor learning

motor control

biomechanics

sport and exercise
  psychology

functional health

frequency

intensity

time or duration

type

practical goal setting

genetics and individual
  variations

motivation

teaching model

fitness evaluation

progressive overload

specificity

modifications

periodization

overtraining

detraining

recovery

compliance

total caloric expenditure

volume

MET

gross energy expenditure

net energy expenditure

accumulation

dose-response

anaerobic activities

resistance training

aerobic activities

combined physical
  movements

interval training

static physical activity
  or exercise

dynamic physical activity
  or exercise

anaerobic power
  (peak power)

anaerobic capacity
  (mean peak power)

aerobic power

aerobic capacity

economy

moderate intensity

vigorous intensity

absolute intensity

relative intensity

absolute strength

relative strength

training plateaus

Study Questions

1. What are the definitions of the terms *physical activity, exercise,* and *physical fitness*?

2. What are five subdisciplines of the field of kinesiology?

3. What is the difference between health-related fitness and skill-related fitness?

4. What three exercise sciences were primarily involved in the development of the traditional exercise training model?

5. What six principles of training can positively influence physical activity and exercise behaviors?

6. How are the terms *dose, accumulation,* and *dose-response* defined with regard to physical activity and exercise?

7. What are the differences between aerobic activities and anaerobic activities?

8. What are the definitions of the following terms: *absolute intensity, relative intensity, moderate intensity,* and *vigorous intensity*?

9. What are six general health and fitness benefits to regular participation in physical activity and exercise?

10. How does Harold, whose case study is in the *Physical Activity Guidelines Advisory Committee Report, 2008* (USDHHS, PAGAC 2008), meet the recommended levels of physical activity and exercise for children?

E-Media

Explore issues related to physical activity, exercise, and public health at the following websites:

Human Kinetics	www.HumanKinetics.com
U.S. Department of Health and Human Services: Physical Activity Guidelines for Americans	www.health.gov/PAGuidelines
World Health Organization	www.who.int
International Society for Physical Activity and Health	www.ispah.org
U.S. Centers for Disease Control and Prevention: Physical Activity	www.cdc.gov/physicalactivity
American College of Sports Medicine	www.acsm.org
President's Council on Fitness, Sports & Nutrition	www.fitness.gov

Bibliography

Ainsworth BE, Haskell WL, Whitt MC, Irwin ML, Swartz AM, Strath SJ, O'Brien WL, Bassett DR, Jr., Schmitz KH, Emplaincourt PO, et al. 2000. Compendium of physical activities: An update of activity codes and MET intensities. *Medicine & Science in Sports & Exercise* 32 (9 Suppl): S498-S504.

American College of Sports Medicine. 2010. *ACSM's Guidelines for Exercise Testing and Prescription*, 8th ed. Philadelphia: Lippincott Williams & Wilkins.

Blair SN, Kohl HW, III, Paffenbarger RS, Jr., Clark DG, Cooper KH, Gibbons LW. 1989. Physical fitness and all-cause mortality. A prospective study of healthy men and women. *Journal of the American Medical Association* 262 (17): 2395-2401.

Caspersen CJ, Powell KE, Christenson GM. 1985. Physical activity, exercise, and physical fitness: Definitions and distinctions for health-related research. *Public Health Reports* 100: 126-131.

Haskell WL, et al. 2007. Physical activity and public health: Updated recommendation for adults from the American College of Sports Medicine and the American Heart Association. *Medicine & Science in Sports & Exercise* 39: 1423-1434.

Kenney, WL, Wilmore, JH, Costill, DL, 2012. *Physiology of Sport and Exercise*, 5th ed. Champaign, IL: Human Kinetics.

McGuire DK, Levine BD, Williamson W, Snell PG, Blomqvist G, Saltin B, Mitchell J. 2001. A 30-year follow-up of the Dallas Bed Rest and Training Study: I, effect of age on the cardiovascular responses to exercise. *Circulation* 104: 1350-1357.

U.S. Department of Health and Human Services. 2008. *2008 Physical Activity Guidelines for Americans*. Washington, DC: U.S. Department of Health and Human Services. www.health.gov/PAGuidelines.

U.S. Department of Health and Human Services, Physical Activity Guidelines Advisory Committee. 2008. *Physical Activity Guidelines Advisory Committee Report, 2008.* Washington, DC: U.S. Department of Health and Human Services. www.health.gov/PAGuidelines.

PHYSICAL ACTIVITY IN PUBLIC HEALTH SPECIALIST

This chapter covers these competency areas as set forth by the National Society of Physical Activity Practitioners in Public Health:

1.2.2, 2.3.3, 2.5.2, 3.1.3, 6.1.3, 6.1.4, 6.2.1, 6.2.2, 6.2.3, 6.2.4, 6.3.3, 6.3.5, 6.4.1, 6.4.2, 6.5.4, 6.5.5

INTEGRATING PUBLIC HEALTH AND PHYSICAL ACTIVITY

OBJECTIVES

After completing this chapter, you should be able to discuss the following:

» The history of physical activity and public health

» How science is translated into practice in physical activity and public health

» How the application of scientific findings differentiates physical activity and public health from other areas such as medicine and exercise physiology

» Knowledge, skills, and aptitudes for careers in physical activity and public health

Opening Questions

» What is the intersection between kinesiology and public health?
» Why is defining this intersection important?
» Is there a career for you in physical activity and public health?

Physical activity and public health—what do you think of when you read that phrase? Elite Olympic athletes who look to be in ideal health? Masters-level swimmers who swim a mile each day—because they enjoy it? Women who work at a factory and spend their lunch hour each day walking 2 miles (3.2 km) together? Construction workers who lift, bend, stoop, and carry loads all day? Kids in a physical education class jumping rope? If you answered *yes* to all of these questions, you are right. Physical activity and public health is a field of study that looks at the health effects and risks of physical activity and ways to help people become active and maintain a healthy level of activity throughout their lives.

Public health is the science and practice of protecting, promoting, and improving the health of populations and communities. An overview of public health concepts was presented in chapter 1. Although individual people are important to health care professionals, public health is primarily interested in the health of communities or groups of people. When we think of public health, notable achievements such as vaccinations against disease, quarantine rules for controlling disease outbreaks, reductions in deaths resulting from motor vehicle accidents, fluoridation of public water supplies to improve dental health, and food safety and restaurant inspections to reduce food-borne illnesses come to mind.

History of Physical Activity and Public Health

Although there is evidence that people understood that exercise was important for health all the way back in ancient Greece, the formal study of the health effects of physical activity is much newer. Even newer is what we now know about how to help people change their behaviors to take advantage of all the health benefits of a physically active way of life.

An understanding of the evolution of the field of physical activity and public health can come from learning how the two fields of exercise science and public health science evolved independently in the 20th century. Exercise science was a developing field in the early 1900s, and much of what we now know about exercise physiology, biomechanics, and sport performance can be traced to early work in exercise science. Fundamental understandings of how oxygen is delivered to working muscles through the cardiovascular system, how carbon dioxide is produced and expelled during exercise, and how glucose (sugar) is metabolized as the body moves resulted in prominent steps forward in the early part of the last century.

The 20th century also witnessed the expansion of personal physical education from having a medical focus in the 1800s to providing sport and games to the masses. Developing fitness for military service was also a prominent role for physical education. The concept of play as a promoter of health also grew during the 20th century.

In the 20th century, life expectancy in the United States increased by 30 years. Even though fundamental practices of personal and community hygiene and quarantine, and the development of vital statistics systems, occurred much earlier, these were modified and updated in the 20th century. Penicillin, the first antibiotic medicine used to control bacterial infections, was discovered in 1928 by Scottish scientist Alexander Fleming. Vaccine development and mass vaccination efforts helped to stem and nearly eliminate serious infectious diseases such as polio and smallpox. Major advances in food safety, including handling and storage, were made in the 20th century, making the food supply safer and reducing the risk of pathogens being carried in food to humans and animals.

Epidemiology, the basic science of public health, is concerned with the study of the causes and consequences of disease and disability in human populations. Initially developed to help people understand how to identify and prevent infectious diseases such as cholera and tuberculosis prior to 1900, epidemiology has evolved to address contemporary health burdens as well, such as heart diseases, cancers, motor vehicle accidents, and air pollution. Major advances in epidemiology also occurred in the 20th century with the emergence of new study designs and observational techniques to assess health outcomes, particularly those related to chronic diseases in populations. Further, our ability to collect, process, and analyze data improved with the advent of computing.

The conditions were ripe in the 20th century for the two fields of exercise science and public health to come together. In fact, such a merger began with the publication of a study conducted in London in 1953. A young epidemiologist named Jeremy N. Morris was interested in heart disease and its causes and consequences. Dr. Morris chose to study workers employed by the London transport system. It was a large group of (mostly) men who worked all day moving people around the city of London. Dr. Morris was particularly interested in the amount of physical activity the men got in the course of their jobs and how that related to their risk of having a heart attack.

A popular form of public transportation in London in the late 1940s and 1950s was the legendary Routemaster (double-decker) bus (see figure 3.1). Hundreds of these buses operated throughout the city. Although similar buses operate in London today, in the 1950s there were no computers, magnetic card readers, or other conveniences on the buses that we know today. Dr. Morris was able to separate the men he was studying into two groups: the bus drivers who were inactive all day long because they were sitting and driving the buses, and the conductors who were walking all day up and down the stairs of the buses taking tickets from riders (Morris et al. 1953; see figure 3.2). Morris found that the physically active conductors had significantly lower rates of coronary heart disease then the less active drivers.

Although these findings may seem fairly straightforward now, at the time they were quite revolutionary. A shift in the paradigm had begun as physical activity for health began to be placed alongside exercise for performance.

The exercise/heart hypothesis, now more of an accepted fact than a hypothesis, essentially was that people who exercised more frequently had

Figure 3.1 The Routemaster bus, a laboratory of early studies of physical activity and health.
Courtesy of Wikimedia Commons/Oxyman

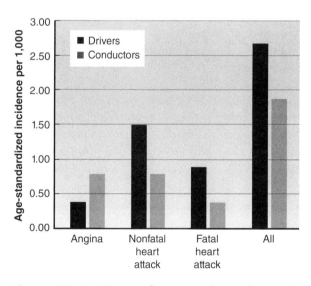

Figure 3.2 Incidence of coronary heart disease in London bus drivers and conductors aged 35 to 64. London Transport Executive 1949-1950.
Data from Morris et al. 1953.

healthier, better functioning circulatory systems than similar people who did not exercise. This physiological benefit resulted in a lower risk of death from heart disease among those who were more active. Dr. Morris went on to publish further studies of occupational physical activity (i.e., physical activity that results from the job one performs) and is widely viewed today as the grandfather of the field of physical activity and health.

Following Dr. Morris, and equally, if not more, effective, was Dr. Ralph S. Paffenbarger Jr. (see figure 3.3). Dr. Paffenbarger was trained in medicine and studied infectious diseases early in his career. He was extremely influenced by Dr. Morris' work on the London Transport workers and set out to improve our understanding of the exercise/heart hypothesis from a public health point of view. As a result of his many influential studies of college alumni and longshoremen, Dr. Paffenbarger was able to more precisely identify the amounts and types of physical activity that were associated with improved health. Thus was born the field of physical activity and public health.

Into the 1970s and 1980s, observational studies of physical activity and health outcomes continued to emerge, as did studies of exercise dose, performance, and the physiological effects of the two. Researchers began to quantify the substantial health benefits of physical activity (and the risks of inactivity). Chapter 2 lists some of these many benefits. Moreover, during this time it became clear that many people throughout the world were not physically active at levels that could reduce their risk of disease, disability, or both. Populations and not just individuals were not as healthy as they could be because of physical inactivity. Suddenly, an exercise and performance problem became a physical activity and public health problem (see figure 3.4).

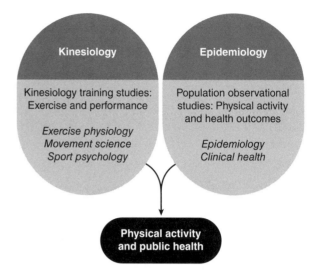

Figure 3.4 The merger of kinesiology and epidemiology to create physical activity and public health.

With the scientific cooperation of exercise science and epidemiology taking hold, additional disciplines began to focus on physical activity and health. Specifically, in the 1990s, behavioral sciences and environmental health sciences were added to the picture. Behavioral sciences began to explore the determinants of physical activity and inactivity and to investigate how inactive people could adopt and maintain healthier behaviors. Environmental health sciences most recently have begun to explore the role that place and the built environment play in encouraging or discouraging physical activity behaviors. Taken together, a new subdiscipline, physical activity and public health, has emerged (see figure 3.5).

ROLE OF PHYSICAL ACTIVITY IN CHRONIC DISEASE DEVELOPMENT

Much of the interest in physical activity and public health arises from the role physical activity plays in the prevention and treatment of chronic diseases. Chronic diseases are conditions and illnesses that occur or develop over a relatively long period of time (months to years), are prolonged, may be preventable, but are rarely completely cured. Examples of chronic diseases are diabetes mellitus (types 1 and 2), cancer, heart disease, pulmonary disease, and osteoporosis.

Figure 3.3 Dr. Ralph Paffenbarger and Dr. Jeremy Morris, trailblazers in physical activity and public health.

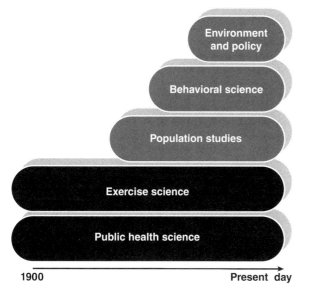

Figure 3.5 Physical activity and public health—the emergence of a subdiscipline.

Sometimes it is good to differentiate chronic diseases (sometimes called *noncommunicable diseases*) from infectious diseases. Infectious diseases are illnesses or conditions that are caused by pathogens such as viruses, bacteria, and fungi. Most infectious diseases can be treated, cured, or controlled with medicines, vaccines, or other measures such as quarantine. Infectious diseases include influenza, tuberculosis, measles, and malaria.

As many countries have improved their public health infrastructures to control the causes of infectious diseases, chronic diseases have become a much more important part of the health burden. In fact, chronic diseases such as cardiovascular diseases (i.e., diseases of the heart and blood vessels), diabetes mellitus, and some cancers present the most urgent threat to public health in developed countries. Around the turn of the 20th century, the three leading causes of death in the United States were infectious in their origins. By 2006, the three leading causes of death were chronic diseases, each with a high degree of association with physical activity. The 10 leading causes of death in 1900 and 2006 are listed in table 3.1.

The most important and powerful health benefits of physical activity are the prevention and treatment of chronic diseases. Through a variety of physiological processes, physically active people are much healthier and much less likely to develop and die from chronic diseases than those who are not. Moreover, physical activity and exercise reduce the risk of dying prematurely (all-cause mortality). You will learn more about these health benefits throughout part II.

FROM SCIENCE TO PRACTICE AND BACK

Public health is characterized by science and action. Epidemiology provides the fundamental science. The action aspect is the implementation of findings

Table 3.1 Ten Leading Causes of Death in the United States—1900 and 2006

Rank	1900	2006
1	Pneumonia (all forms) and influenza	Heart disease
2	Tuberculosis (all forms)	Cancer
3	Diarrhea, enteritis, and ulceration of the intestines	Stroke (cerebrovascular diseases)
4	Diseases of the heart	Chronic lower respiratory diseases
5	Intracranial lesions of vascular origin	Accidents (unintentional injuries)
6	Nephritis (all forms)	Diabetes
7	All accidents	Alzheimer's disease
8	Cancer and other malignant tumors	Influenza and pneumonia
9	Senility	Nephritis, nephrotic syndrome, and nephrosis
10	Diphtheria	Septicemia

Adapted from the U.S. Centers for Disease Control and Prevention (CDC). Available: www.cdc.gov/nchs/FASTATS/lcod.htm and www.cdc.gov/nchs/data/dvs/lead1900_98.pdf.

from scientific studies to improve health, which is what differentiates the field of public health from other basic science and biomedical science disciplines. Public health applies science to practice by addressing three critical areas: surveillance, community interventions, and the development of health guidelines.

SURVEILLANCE

Public health surveillance has been defined as the ongoing, systematic collection, analysis, and interpretation of data (e.g., regarding agent or hazard, risk factor, exposure, or health event) essential to the planning, implementation, and evaluation of public health practice (Thacker 1988). Surveillance is a critical public health function because it helps us understand the extent of a health problem and the types of people and populations that may be at higher risk of that health problem. The number of weekly cases of influenza, the annual death rate due to malaria, the number of heart attacks among older African American adults, and the percentage of high school students who take daily physical education classes are all real-life examples of surveillance data that are routinely collected by public health professionals.

The CDC has developed a variety of surveillance systems to monitor the health of the U.S. population. More information on surveillance for physical activity and public health can be found in chapter 4.

COMMUNITY INTERVENTIONS

Interventions—preferably at the community level—are another cornerstone of public health. The science behind interventions is established during projects called efficacy trials. **Efficacy trials** are studies that are used to establish that a certain intervention or public health program can change a certain condition. For example, efficacy trials have been used in small interventions to test the ability of supplementing a person's diet with iron to cure iron-deficiency anemia, a dangerous condition in which the body does not have enough red blood cells to carry oxygen. A study could be designed that randomly assigns infants with low levels of iron in their blood to one condition in which iron is given in the form of a pill to the mother (and delivered to the infant through breast-feeding) or to another

condition in which the iron is delivered directly to the infant in an infant formula drink. The study would evaluate which of the two methods works better at increasing stores of iron in the bodies of infants.

Effectiveness studies are the other main type of intervention study of interest in public health. In **effectiveness studies**, the main outcome of interest relates to how well a treatment works in practice—or more appropriately, in real life instead of in controlled settings. Taking the preceding iron-deficiency anemia situation, the appropriate effectiveness question becomes, How well does the iron supplementation delivery method work in a community or village in which malnutrition is rampant? Obviously, many questions arise in this kind of study. For example, even if we know that the breast-feeding method appeared to work better in the efficacy study, it may not be particularly effective in a certain area because of cultural or social restrictions on breast-feeding. This presents a clear dilemma for a public health intervention—efficacy data show one route to go, but the real-life (effectiveness) evaluation of the intervention suggests something different. This situation requires additional study and observation to refine the methods to ensure the effectiveness of the intervention.

Similar challenges become apparent in the world of physical activity and public health. From exercise physiology, we know the physiological effects of a prescribed dose of exercise on physiological parameters such as cardiorespiratory fitness, blood pressure, adverse lipid and lipoprotein levels, body weight, and blood glucose response. These results come from carefully controlled laboratory efficacy trials in which the dose of exercise is managed tightly and measured very closely. Exercise has a known efficacy for improving many parameters that are related to poor health in the human body. The effectiveness of such interventions in real life becomes much more difficult and challenging to measure and assess when these studies are taken out of the laboratory and into the community. How do we translate laboratory-based studies into community interventions that will help people increase their physical activity levels and thus become healthier? This *practice* aspect of public health is the main focus of the third part of this textbook.

DEVELOPMENT OF HEALTH GUIDELINES

The development of health guidelines is a third critical function of public health science. Public health guidelines are official policy statements, usually developed by a government body, agency, or other reputable organization, that are based on the best available science. Public health guidelines provide clear recommendations about a course of action to deal with a pressing public health issue. Recommendations for childhood immunization schedules (when and which vaccines and their timing), diabetes treatment (the frequency and method of glucose self-monitoring), nutrient intake (micronutrient daily recommended intakes for health), and annual influenza vaccinations (type and timing) are all examples of public health guidelines that, based on the best available science, give health professionals and the public clear guidance about the most appropriate courses of action for preventing and treating certain health problems.

In 2008 the U.S. Department of Health and Human Services published the first-ever *2008 Physical Activity Guidelines for Americans* (USDHHS 2008, see the following highlight box). These guidelines, based on a summary of scientific literature encompassing more than 600 pages, detail the best

---

2008 PHYSICAL ACTIVITY GUIDELINES FOR AMERICANS

Children and Adolescents (Aged 6-17)

- Children and adolescents should do 1 hour (60 minutes) or more of physical activity every day.
- Most of the 1 hour or more a day should be either moderate- or vigorous-intensity aerobic physical activity.
- As part of their daily physical activity, children and adolescents should do vigorous-intensity activity on at least 3 days per week. They also should do muscle-strengthening and bone-strengthening activity on at least 3 days per week.

Adults (Aged 18-64)

- Adults should do 2 hours and 30 minutes a week of moderate-intensity, or 1 hour and 15 minutes (75 minutes) a week of vigorous-intensity aerobic physical activity, or an equivalent combination of moderate- and vigorous-intensity aerobic physical activity. Aerobic activity should be performed in episodes of at least 10 minutes, preferably spread throughout the week.
- Additional health benefits are provided by increasing to 5 hours (300 minutes) a week of moderate-intensity aerobic physi-

cal activity, or 2 hours and 30 minutes a week of vigorous-intensity physical activity, or an equivalent combination of both.
- Adults should also do muscle-strengthening activities that involve all major muscle groups performed on 2 or more days per week.

Older Adults (Aged 65 and Older)

- Older adults should follow the adult guidelines. If this is not possible due to limiting chronic conditions, older adults should be as physically active as their abilities allow.
- It is important for older adults to avoid inactivity.
- Older adults should do exercises that maintain or improve balance if they are at risk of falling.

For all individuals, some activity is better than none. Physical activity is safe for almost everyone, and the health benefits of physical activity far outweigh the risks. People without diagnosed chronic conditions (e.g., do not have conditions such as diabetes, heart disease, or osteoarthritis) and who do not have symptoms (e.g., chest pain or pressure, dizziness, or joint pain) do not need to consult with a health care provider about physical activity.

science-based recommendation for the weekly amount of physical activity necessary to prevent disease and promote positive health outcomes. The physical activity guidelines were developed for children and adolescents, adults, and older adults. These guidelines should be used as targets for physical activity participation.

Several other countries, as well as the World Health Organization, have published guidelines similar to the *Physical Activity Guidelines for Americans*. Japan, Canada, Australia, and England all have taken leadership roles in establishing physical activity as a public health priority by setting physical activity guidelines and recommendations.

Promoting Physical Activity for Health

As the field of physical activity and public health has emerged, it has become clear that many factors at many levels influence physical activity behaviors. Many investigators have used the social ecological model as a guiding framework to explain these multiple levels. They are illustrated in figure 3.6 and form the basis for the third section of this textbook.

At the center of the target in the social ecological model for physical activity behaviors are individual factors. These are the factors that are innate to each person and that differ among people. A person's genetic makeup, early life experiences (e.g., youth

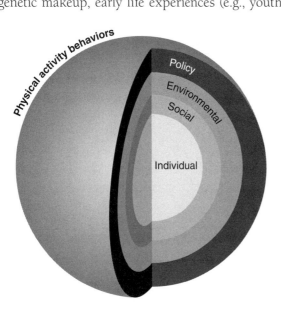

Figure 3.6 Multiple levels of influence on physical activity behaviors: The social ecological model.

sport participation), self-efficacy, and other factors, such as sex, disability, growth and development, and socioeconomic status, may all be important determinants of health behaviors such as physical activity.

Moving out from the center, social influences on physical activity are important. Research has now shown that individual factors are not enough to explain physical activity behavior. Determinants at the social influences level aren't characteristics of the person per se, but are, rather, characteristics of how the person interacts with society or the culture. Influences at this level can include peers, medical care organizations (doctors), family members, and organizations (schools, places of worship, worksites).

The third level in the social ecological model represents environmental influences. These influences that may enhance or restrict physical activity behaviors are external to the person but common across societies and cultures. Research on the effects of the physical environment on physical activity participation has exploded as the field of physical activity and public health has emerged. The ability to influence the physical activity of vast numbers of people (instead of one person at a time) by making a single change makes environmental influences on physical activity a particularly interesting area of research. Aspects of the built environment such as the availability of places to be active (e.g., trails, sidewalks, fitness facilities, bicycle lanes, community and neighborhood design elements) are examples of environmental influences on physical activity.

Finally, the outer level represents policy influences on physical activity. Influences in this sphere include written or unwritten rules, codes, and norms that influence environmental or social determinants of physical activity. As with the physical environment, determinants at this level are particularly attractive because of their potential to influence many people. Examples of policy influences on physical activity include policies allowing increased access to places to be physically active (making it easier to be physically active), educational policies (e.g., mandating high-quality daily physical education for schoolchildren), and transportation-related policies (e.g., making it easier to walk or bicycle for transportation).

By combining knowledge, skills, and abilities related to the basic exercise sciences and public health, you can better explain and discuss professionally the health benefits and risks of exercise and physical activity (see part II, chapters 5 through 10) to your peers, colleagues, and the communities you serve. An understanding of the specific challenges that affect physical activity and exercise in professional exercise science jobs can clarify how the exercise sciences affect public health, and vice versa. The highlight box Careers Combining Exercise Science and Public Health lists careers in exercise physiology, biomechanics, and sport and exercise psychology that are commonly seen in the professional areas of health and fitness, preventive medicine, athletic performance, and rehabilitation.

PRACTITIONERS OF PHYSICAL ACTIVITY IN PUBLIC HEALTH

Physical activity and public health is an emerging discipline with many opportunities. People with training in this area, particularly at the master's level, are employed in state and local health departments working on public health programming for physical activity promotion. Many universities have research opportunities for people interested in studying physical activity and public health. Finally, private foundations and nongovernmental organizations are interested in people with training and interests in physical activity and public health.

In 2006, a new professional organization was created in the United States. The National Society of Physical Activity Practitioners in Public Health (NSPAPPH) is now called the National Physical Activity Society (NPAS). The NPAS is a dedicated group of professionals interested in advancing the capacity of professionals in physical activity and public health in the United States. The group has taken a major leadership role in developing a definition of what it means to be a professional in the field of physical activity and public health and what core competencies are necessary for leadership in community-level interventions to promote and evaluate the effectiveness of those interventions. Many of the core competencies put forth by NPAS are addressed in this textbook and are found at the end of each chapter as a cross-reference.

NPAS has developed resources, trainings, and a certification (Physical Activity and Public Health Specialist) to create professionals in physical activity and public health. The certification is a voluntary credential that is a U.S. national standard for professionals working in this area. Important knowledge, skills, and abilities in partnership development, planning and evaluation, exercise science, development of effective interventions, and evaluation of scientific data are part of NPAS' training and certification procedure. Knowledge and skills in each of these areas are critical for advancing physical activity and public health and developing a successful career in the field.

CAREERS COMBINING EXERCISE SCIENCE AND PUBLIC HEALTH

Physical education teacher	Coach
Physical activity specialist	Sporting goods representative
Firefighter	Biomechanist
Police or military	Sport physiologist
Personal trainer	Health or fitness facility owner
Physician	Physical therapist
Nurse	Cardiac rehabilitation specialist
Researcher	Occupational therapist
Movement specialist	Diabetes or obesity prevention specialist
Wellness coach	Athletic trainer
Clinical exercise physiologist	Consultant
Translational exercise physiologist	Translational biomechanist

KEY LEADER PROFILE

Steven N. Blair, PED

Courtesy of Steven N. Blair.

Why and how did you get into this field? Did any one person have an overriding influence on you?

My football career at Kansas Wesleyan University was ended by an injury after two years; but my coach, Gene Bissell, invited me to serve as an assistant coach. He also got me interested in research, and I did a simple research project as an independent study in my senior year.

Professors Karl Bookwalter, George Cousins, and Arthur Slater-Hammel at Indiana University encouraged my academic interests as a doctoral student. Professor Sid Robinson was an early leader in the field of exercise physiology, and his professor, D.B. Dill, had retired and moved to Indiana to work with him. Thus, I was exposed to two giants in the field who helped develop my interest in the effects of exercise on the body.

My first job after graduate school was at the University of South Carolina, where I was asked to develop a clinical exercise physiology laboratory; this was relatively rare in U.S. universities in 1966. As I taught exercise physiology and began doing research in the area, I also began to work with the South Carolina chapter of the American Heart Association, and helped plan a major national conference on physical activity and cardiovascular disease.

My research interests began to shift to topics related to preventive cardiology and epidemiology. At Stanford University during 1978-1980 I had the opportunity to work with many leaders in exercise and heart disease prevention, notably Peter Wood and Bill Haskell. Most important for my future career, however, was the opportunity to develop a relationship with Professor Ralph S. Paffenbarger, Jr (Paff). Paff was on the Scientific Advisory Board at the Cooper Institute in Dallas, and after returning to South Carolina I worked with the Institute on a research grant. I did this for four years, and then moved full-time to the Institute in 1984.

My first employee at the Institute was one author of this textbook, Professor Bill Kohl. Paff served as a major mentor and colleague with us during our early years at the Institute. He helped us develop the database by doing mortality surveillance and conducting mail-back surveys to monitor health and behaviors. This work launched research on physical activity, fitness, and health that is still ongoing. We have published over 200 papers from the Aerobics Center Longitudinal Study (ACLS) database, and believe that this work has shaped current ideas and recommendations in the area of physical activity and public health. I will be forever grateful to Paff for helping launch this work, and for his guidance and mentoring over many years.

What are your current research interests?

I continue to be interested in the role of physical activity in the prevention, treatment, and rehabilitation of chronic diseases. Over the past several years I also have become interested in randomized controlled trials of physical activity and a variety of outcomes. One of my current major interests is to learn how to use modern technology such as iPhones, PDAs, and the Internet to deliver interventions to help more people become and stay more physically active. Another current project is the Energy Balance Study. The reason for this study is that although it is clear that we have been experiencing an obesity epidemic around the world for the past few decades, I believe the causes are not clear.

What are one or two key issues to be addressed by 2022?

We have been engineering energy expenditure out of life for the past several decades. For example, occupational energy expenditure has declined more than 100 kcal/day over the past 50 years. Of course, we also have made it easier to do housework, gardening, and transport ourselves. We need to implement the U.S. National Physical Activity Plan (www.physicalactivityplan.org). We must all get behind this effort and create a more physically activity society.

CHAPTER WRAP-UP

What You Need to Know

- Physical activity and public health is an emerging discipline that combines science and practice from the exercise science disciplines and from public health and epidemiology.
- Physical activity, exercise, and physical fitness are three concepts that are relevant to physical activity and public health.
- Surveillance, community interventions, and the development of health guidelines are three key pillars of public health.
- A variety of careers are available to people with an interest and training in physical activity and public health.

Key Terms

public health surveillance

interventions

efficacy trials

effectiveness studies

Study Questions

1. What defines physical activity and public health as a field?

2. What was the unique way Dr. Morris combined the areas of physical activity and public health?

3. What populations did Dr. Paffenbarger study with regard to physical activity and public health?

4. What is the exercise/heart hypothesis?

5. What are five chronic disease conditions that might be positively influenced by participation in regular physical activity? Discuss each.

6. What is public health surveillance? Give an example.

7. What are examples (one each) of physical activity interventions related to efficacy and effectiveness?

8. What are the *2008 Physical Activity Guidelines for Americans*?

9. What is the social ecological model? What levels are associated with the model?

10. What is the National Society of Physical Activity Practitioners in Public Health, and how might joining this organization benefit professionals in the field of physical activity and public health?

E-Media

Explore issues related to physical activity, exercise, and public health at the following websites:

Human Kinetics	www.HumanKinetics.com
U.S. Department of Health and Human Services: Physical Activity Guidelines for Americans	www.health.gov/PAGuidelines
International Society for Physical Activity and Health	www.ispah.org
U.S. Centers for Disease Control and Prevention: Physical Activity	www.cdc.gov/physicalactivity/
U.S. National Society of Physical Activity Practitioners in Public Health	www.nspapph.org
American College of Sports Medicine	www.acsm.org
President's Council on Fitness, Sports & Nutrition	www.fitness.gov
World Health Organization: Global Recommendations for Physical Activity	www.who.int/dietphysicalactivity/factsheet_recommendations/en/index.html
Australia's Physical Activity Recommendations	www.health.gov.au/internet/main/publishing.nsf/content/health-pubhlth-strateg-active-recommend.htm
Public Health Agency of Canada: Physical Activity Guidelines	www.phac-aspc.gc.ca/hp-ps/hl-mvs/pa-ap/index-eng.php
UK Department of Health	www.dh.gov.uk/en/Publichealth/index.htm

Bibliography

Morris JN, Heady JA, Raffle PAB, Roberts CG, Parks JW. 1953. Coronary heart disease and physical activity of work. *Lancet* 262: 1053-1108.

Thacker SB, Berkelman RL. 1988. Public health surveillance in the United States. *Epidemiology Review* 10:164-190.

U.S. Department of Health and Human Services. 2008. *2008 Physical Activity Guidelines for Americans.* Washington, DC: U.S. Department of Health and Human Services. www.health.gov/PAGuidelines. Accessed 20 June 2011.

Physical Activity in Public Health Specialist

This chapter covers these competency areas as set forth by the National Society of Physical Activity Practitioners in Public Health:

1.4.1, 2.1.1, 2.1.3, 2.2.1, 2.2.2, 2.2.3, 4.5.1, 4.5.2, 4.5.3, 4.5.4, 5.1.1, 5.1.2, 5.4, 5.4.1

CHAPTER 4

MEASURING PHYSICAL ACTIVITY

OBJECTIVES

After completing this chapter, you should be able to discuss the following:

» Techniques to measure individuals' physical activity and the strengths and weaknesses of these techniques

» Fundamentals of physical activity surveillance

» Sources of public health information on physical activity

OPENING QUESTIONS

» What are the best ways to measure physical activity in individuals?

» Can people accurately remember their exercise patterns for weeks or months?

» Can technology improve our efforts to assess physical activity?

As you learned in chapter 2, *physical activity* is any skeletal movement that results in energy expenditure. This energy expenditure can be quantified in units called *kilocalories,* which are often referred to as just *calories.* Anytime you move your body, from getting out of bed in the morning, to walking to the bus stop, to playing basketball, to walking down the hallway, you are being physically active and burning calories.

Likewise, *exercise* is a subset or type of physical activity that is planned, structured, repetitive, and designed to increase or maintain physical fitness. Because it must be planned and repetitive, exercise is a somewhat clearer concept than physical activity. Taking a 30-minute walk, jogging, playing basketball or tennis, and hiking and mountain climbing are all examples of exercises that increase or maintain physical fitness.

Understanding measurement and assessment techniques for physical activity, exercise, and physical fitness; the strengths and weaknesses of these techniques; and how to use them is critical for understanding and conducting studies of physical activity and health. This understanding is also important when developing, implementing, and evaluating programs for increasing physical activity in individuals and populations. This chapter gives an overview of current techniques in this area.

IMPORTANCE OF FITNESS ASSESSMENTS

As highlighted in chapter 2, physical activity, exercise, and physical fitness are unique constructs. Physical activity and exercise are behaviors, and physical fitness is a physiological state. Although this chapter focuses on the measurement of physical activity and exercise, there are also a wide variety of fitness assessment and measurement strategies and tests. These tests have been developed following

years of research about the human body's adaptation to exercise.

We know that physical activity and exercise are good. Since the early studies in the 1950s and 1960s, we have learned more about the importance of the intensity, frequency, and duration—as well as of the total volume—of physical activity as they relate to health. This new knowledge has its roots in exercise science and exercise training studies.

The best science-based guidelines state that adults need at least 150 minutes per week of moderate-intensity physical activity, at least 75 minutes per week of vigorous-intensity physical activity, or some equivalent combination of intensities (U.S. Department of Health and Human Services [USDHHS] 2008). We also know that adults should do muscle-strengthening activities that are moderate or high intensity and involve all major muscle groups on two or more days a week. Finally, we know that children need at least 60 minutes of moderate-intensity physical activity each day.

With these guidelines in mind, we consider these questions: What are the best ways to measure physical activity and exercise? How well do these techniques measure what we want them to measure? Are they repeatable? What is needed to measure physical activity most accurately among individuals versus small or large groups of people? What methods should we use in controlled laboratory settings and in free-living populations?

LABORATORY MEASURES OF ENERGY EXPENDITURE

Laboratory-based techniques that measure physical activity are based largely on the desire to assess caloric expenditure, or the amount of energy it takes for a person to be physically active, to breathe, to circulate blood, and to digest food. **Total energy** expenditure (TEE) is the sum of all of these components. **Physical activity energy expenditure (PAEE)** is the energy expenditure that is specifically the result of physical activity. The **thermic effect of food (TEF)** is the amount of energy that is used to digest and metabolize energy that is ingested (food and drink). Finally, **basal metabolic energy expenditure (BMEE)** is the energy expended to maintain breathing and circulation at rest.

Roughly 60 to 70% of a person's TEE is composed of BMEE. Approximately 10% of TEE is used to digest food. The remaining expenditure (20 to 30%) is for physical activity. Obviously, people who are more active and those who exercise use a higher proportion of their TEE for physical activity than people who are largely sedentary throughout the day. Because PAEE is the only component of TEE that is very changeable, we need to know how to measure that 20 to 30% very well. See figure 4.1 for more details about the determinants of TEE.

Figure 4.1 Total energy expenditure = basal metabolic energy expenditure + physical activity energy expenditure + thermic effect of food.

INDIRECT CALORIMETRY

In the laboratory, the process of measuring the components of energy expenditure is fairly straightforward. Indirect calorimetry is the technique that is used most often. With indirect calorimetry, the amount of oxygen used and carbon dioxide expelled is used to estimate energy expenditure. This can be done using a facemask and gas analysis system. Another method is when a person lives in a controlled room or setting for a period of time. Energy expenditure is assessed from air samples that are collected while the person is in the room. Calibrated gas collection devices are used to measure and analyze oxygen and carbon dioxide changes in the room.

Because carbon dioxide is a by-product of energy metabolism, it is a very good indicator of total energy expenditure in this controlled setting. Someone who burns more energy in this setting (e.g., through exercise) is easily differentiated from someone who is sedentary and not burning any additional calories. By having the subject keep a diary, or by having study personnel observe the subject, investigators can quantify the amount (frequency, intensity, and duration) and type of physical activity performed during an indirect calorimetry study. Thus, PAEE can be measured as well. This, coupled with careful measurements of the food eaten and waste produced, can give a very accurate measure of TEE.

Although indirect calorimetry is very accurate, it has limitations. First, it is relatively expensive. A special room must be built, gas analyzers must be installed, and study personnel must be hired to maintain the facility, among other costs. Second, only one person in a room (two if there are two rooms) can be measured at a time. This makes larger studies more tedious to conduct and quite lengthy to carry out. Another limitation is that the equipment needed to do these analyses can break down and usually must be calibrated frequently so that the investigator has confidence in the measurements. A final limitation is that this type of protocol does not capture TEE or its components in real life. That is, the artificial laboratory settings do not allow people to live as they would normally. This is of obvious importance in the public health world.

Doubly Labeled Water Technique

A second laboratory technique that has been shown to be very accurate in measuring TEE is the doubly labeled water technique. This technique uses the principles of indirect calorimetry in that turnover rates of oxygen and hydrogen are measured. The higher the TEE is, the higher the turnover rates are (because of higher metabolic activity).

Doubly labeled water study participants are given a prespecified dose of stable radioisotope-labeled water. After taking some baseline measurements, investigators measure the excretion of the isotopes over the course of one to three weeks (depending on the study question and protocol) in collected urine samples. The difference in elimination rates between the labeled oxygen and hydrogen is used to estimate carbon dioxide production, which, as in indirect calorimetry, can be used to estimate TEE.

Doubly labeled water is a very safe procedure that allows people to live their lives normally (i.e., they are not confined to a calorimetry room). It has been shown to be very precise. Participants must collect and store urine samples frequently, which can be a burden, and the radiolabeled hydrogen and oxygen are fairly expensive ($350 to $400 in USD in 2011). Moreover, substantial equipment (a mass spectrometer) and assay tools are needed for analyzing the samples. Another important limitation is that the assay allows only the estimation of TEE. Estimation of PAEE is unattainable using the doubly labeled water method unless the investigator also has a measure of BMEE from indirect calorimetry (remember that TEE = BMEE + PAEE + TEF). This can be a significant limitation if assessment of PAEE is the desired outcome. Further, the method tells us nothing (alone) about how the physical activity was performed (e.g., walking for recreation or walking for transportation). These types of data are usually helpful if changing physical activity behaviors is of interest.

Electronic Measurements to Estimate Energy Expenditure

Self-report measures of physical activity have important limitations—namely, they rely on partici-

pants to accurately remember their physical activity behaviors. Advances in technology, however, have resulted in the development of devices that can objectively measure movement.

Accelerometers

Accelerometers can be used to measure the magnitude and direction of acceleration. Accelerometers of many varieties are used for applications as varied as oil and gas drilling, navigation, and transportation. For human movement and physical activity, accelerometers have proven very useful in determining total physical activity and in estimating energy expenditure. However, although accelerometers have advanced our ability to measure physical activity, they provide no information regarding the type of activity being performed. For example, a period of brisk walking is indistinguishable from a tennis match to an accelerometer.

Accelerometers (see figure 4.2) measure movement and physical activity by measuring acceleration forces in one, two, or three planes (or axes) as a result of a change in the velocity of the body. Semiconductors translate the processing of movement to acceleration. The following simple equation shows the mathematical relationship between acceleration (a), change in velocity (Δv), and change in time (Δt). As velocity increases for a given time, so does acceleration. When the amount of time decreases (at a constant velocity), acceleration also increases.

$$a = \Delta v \,/\, \Delta t$$

Figure 4.2 Accelerometers have greatly increased the ability to measure human physical activity.
3DNX™ instrument photos courtesy of BioTel Ltd.

Using this relationship, an accelerometer attached to a person's body measures changes in velocity and time. These measurements can then be converted into estimates of energy expenditure.

Accelerometers are very useful for measuring physical activity because they take human recall out of the equation. Participants can attach an accelerometer unit to a waistband and forget it throughout the course of the day—they are free to go about their usual activities of daily living without having to remember anything. More advanced units can estimate time spent in a certain intensity of physical activity (e.g., moderate or vigorous) and time spent in periods of physical activity above a certain time threshold (e.g., 10-minute bouts). Continued technological advances allow for multiday data storage (sometimes more than 30 days) for long-term behavior monitoring.

Accelerometers can be relatively expensive (the better ones exceeded $350 USD in 2011) and may not be accurate for all kinds of activities. Accelerometers are not waterproof; thus, they cannot be used to measure physical activity in the pool, ocean, or anywhere where it is wet.

PEDOMETERS

Pedometers (or step counters) are another kind of electronic monitoring device that can be used to take the recall bias out of physical activity assessment. They are usually most useful for measuring walking, running or jogging, or any other type of physical activity that involves the lower body. Many kinds of pedometers exist, using several types of mechanisms, although they all fundamentally measure total steps (see figure 4.3). Some rely on a spring or a spring lever to record the movement, others use a strain gauge, and still others use a magnetic switch.

A key strength of pedometers is that they are fairly inexpensive and thus can be used by many people. Good pedometers could be purchased for $15 to $20 USD in 2011. They are fairly simple and straightforward to use and seem to accurately measure the number of steps taken. Studies have shown a range of accuracies among various brands, with the less expensive pedometers generally being less accurate than more expensive ones. Pedometers help people become more active by reporting the total steps they have taken (e.g., toward a preset goal), thereby reminding them to be active.

Figure 4.3 Pedometers are inexpensive and easy ways to track the number of steps taken in a period of time. They can also be useful reminders and help people set goals for steps per day.

A substantial drawback to using pedometers for measuring physical activity is that they do not directly measure velocity or time and thus cannot estimate acceleration. The least common denominator for pedometers is steps taken during the observation period. With additional data collection, such as a diary or questionnaire, time and velocity may be estimated.

The *2008 Physical Activity Guidelines for Americans* (USDHHS 2008) report the amount of physical activity related to improving health status. The inability of pedometers to reveal anything about time or velocity means that they cannot differentiate between a person who took 9,500 steps in 30 minutes while playing soccer and another person who took the same number of steps in 24 hours. Clearly, the dose of physical activity between the two situations is different, but a pedometer would not be able to tell us that without additional data collection.

COMBINING MEASUREMENT APPROACHES

Many ways of assessing physical activity exist, and none of them can be recommended for every situation. Large studies that follow many people for long periods in order to track health problems are likely to need self-report questionnaires to assess physical activity. Smaller studies of shorter duration that are interested in short-term behavior changes may be best suited for relying on accelerometers as the primary physical activity assessment method. Characteristics of the environment and population under study all must be factored in to the decision of which physical activity assessment technique to use. Each type has strengths and weaknesses that must be thoroughly understood.

Ideally, combined approaches should be used to arrive at the most thorough estimate of physical activity. Strategies that combine self-report, direct observation, and questionnaires have rarely been used. This future of the field remains to be written.

DIRECT OBSERVATION TECHNIQUES

A popular way to measure physical activity behavior (and by extension to estimate PAEE) is the direct observation of the people or places of interest. Using trained observers, researchers can get a standardized view of physical activity participation for individual people (e.g., the amount of time a child is inactive during a physical education class). These observers, who are trained to recognize certain characteristics of physical activity, record what they see during a period of time. These observations can then be converted to estimates of energy expenditure based on the intensity, frequency, and duration of the physical activity or exercise observed.

Direct observation techniques can be very useful in many situations. They are particularly helpful in identifying how people use defined spaces (e.g., parks, playgrounds, and neighborhoods) for physical activity. Direct observation techniques, however, do have drawbacks. First, they are usually fairly expensive, given the cost of training and employing observers. Second, the observer training is a very important aspect of this technique. The investigator needs to be sure that what one observer calls moderate-intensity physical activity is the same as what all the other observers are identifying as such. Further, that definition cannot drift over time—that is, an exercise that is classified as vigorous intensity should be classified as such throughout the study. Direct observation techniques are difficult to use with hundreds of people in multiple situations.

Finally, people's behaviors often change simply because they know they are being observed (see figure 4.4). If you knew that your physical activity and exercise behavior were being monitored, do you think you might alter what you do and how long or

SOPLAY AND SOPARC

SOPLAY (system for observing play and leisure activity in youth) and **SOPARC** (system for observing play and recreation in communities) are two excellent examples of direct observation techniques to assess physical activity among children and adolescents in defined areas. Both measures, developed at San Diego State University by Dr. Thom McKenzie and colleagues, have helped us get objective measures of physical activity in youth. SOPLAY is used to assess the amounts of physical activity during free-play settings, and SOPARC is used to measure physical activity and associated environmental characteristics in park and recreation settings. More information on these two techniques can be found at www.activelivingresearch.org/node/10643.

Figure 4.4 Would you do things a bit differently if you knew someone was observing or recording your physical activity behavior?

how intensely you participate? This phenomenon is referred to as **reactivity** and can be very difficult to control in studies of physical activity behavior. Most often, investigators try to observe physical activity in public areas where observers aren't identifiable. This anonymity minimizes reactivity and results in a better measure of physical activity behavior.

SELF-REPORT INSTRUMENTS

Historically, the most frequently used techniques for physical activity assessment, particularly in studies of how physical activity and exercise influence health outcomes, have been based on self-report. Using self-report instruments, science has been able to not only show that physical activity greatly decreases (and physical inactivity greatly increases) a person's risk of a variety of chronic diseases, but also demonstrate the dose-response effect. Higher

levels of self-reported physical activity are associated with lower risk of disease outcomes. Part II of this textbook details some of this evidence.

With self-report, study participants are asked to tell the investigators, either in interviews or via questionnaire or diary, about their participation in physical activity, usually over a defined period. This period can be short (24 hours), a bit longer (the last 7 or 30 days), or much longer (sometime in the distant past). These responses are then classified across the study population, and people are categorized based on their reported physical activity. Analysis of such data can be done in a variety of ways, but usually is set up to examine exposure to physical activity (low amount versus high amount, or low, medium, or high). This exposure, which is meant to be an estimate of PAEE, can then be related to an outcome such as risk of heart attack or osteoporosis.

DIARIES

Physical activity diaries are essentially the same as any other type of diary. Study participants are asked to record their physical activities at various time points during the study. The idea is to develop a protocol that maximizes recall and minimizes error—such as asking participants to record their physical activities as they are happening or even at bedtime. In this way, they are less likely to forget meaningful physical activities and their intensity, frequency, and duration. Personal trainers and coaches frequently use this technique to help exercisers and athletes understand and adhere to their training regimens.

Diaries can be particularly helpful in understanding the context and type of physical activity. Walking or bicycling to school, for example, is often an important type of physical activity behavior to measure, and people—even children—can remember these events and record them in a diary. Beginning and ending times can be recorded in a diary to give the investigator an idea of the duration (and therefore the intensity). Diaries may also help investigators understand the behavior pattern (day-to-day variability) of people in a defined time period. The data gathered in physical activity diaries can be hard to summarize, however, particularly when study participants keep diaries for extended periods.

Interviews

Interviews have been used to measure physical activity in a variety of settings. The person conducting the interview, after appropriate training, most often follows a predetermined interview protocol to learn details about study participants' physical activity behaviors. Memory cues and prompts are often used to aid in accurate reporting. For example, asking a person what time she went to sleep and what time the next day she awoke provides an excellent estimate of sleep time, or energy expenditure at or near BMEE. A good interviewer can prompt the participant to provide information that otherwise may go unreported.

Interviews have several advantages. They are conducted by trained personnel who can probe for items that participants may not readily recall, which enhances the overall accuracy of the measurement. They can be structured so that the context as well as intensity, frequency, and duration of the physical activities are reported. They also are not subject to the reactivity problem because people are asked to recall the past—and the past cannot be altered in any way that we know of!

Interviews can be expensive and have a substantial participant burden, however. Some interviews can take 30 to 45 minutes, and people sometimes struggle to remember even the broadest details about days in the past. Studies have shown that recalls of moderate-intensity activities are much more prone to reporting errors than are those of vigorous-intensity activities or rest (Pettee-Gabriel et al. 2010). Finally, interviewer training (or lack of it) can be a source of substantial error. A good interviewer is essential to help a participant remember details. A bad or inconsistent interviewer does not elicit the same response.

Questionnaires

The final self-report method for physical activity assessment is the use of questionnaires. Questionnaires can be administered directly to participants (in person or on the internet) or over the telephone as part of a telephone survey. The difference between a self-administered questionnaire and one administered in a telephone survey or an interview, as discussed earlier, is that the telephone survey administrator most typically follows a predetermined script. People who facilitate telephone surveys rarely, if ever, have the latitude to probe and deviate from the script.

Questionnaires have evolved over the years as our understanding of the effects of physical activity on health outcomes have evolved. The earliest questionnaires asked respondents to report occupational physical activity. Job classifications and estimated energy expenditure in broad job categorizations were used to estimate PAEE. Very quickly, though, it became clear that an understanding of other domains of physical activity (including transportation, recreation, and household) were important, making an understanding of occupational physical activity less of a priority than an understanding of the full range of physical activity.

The 7-Day Physical Activity Recall

As with direct observation and diary techniques, physical activity interviews work best when the period of time is defined. The 7-Day Physical Activity Recall (PAR) is an interviewer-administered questionnaire that has been used for many years in physical activity research (Blair et al. 1985). The participant is asked first about the most distal day (i.e., seven days ago). The interviewer asks about sleep time, sitting time, vigorous-intensity activities (e.g., jogging and cycling), and moderate-intensity activities (e.g., walking). Each activity recalled is probed for duration of participation. With the help of the interviewer, the participant reports all activities in each of these categories. Low-intensity activities (i.e., between 1.0 MET and 2.99 METs) are inferred once the other intensities have been assessed. The interviewer then proceeds to the next day and repeats the process. The protocol continues until the most recent (proximal) day. The data are then summarized and an estimate of PAEE, as well as TEE, is calculated based on the reported activities, their duration, and their frequency. These interviews often can take more than 45 minutes to complete!

SELF-REPORT IN CHILDREN

Physical activity among children and adolescents is an important concern for health and growth and development. Children who are inactive are more likely to grow up to be inactive than their active peers. Many attempts to assess physical activity in youth using questionnaires have been made over the years. Because of a child's lack of awareness and inability to recall, however, questionnaires are generally *not* recommended for children until the age of 12.

Questionnaires are usually developed to assess physical activity exposure(s) of interest. For example, participants in a study of the effectiveness of a training program could be asked about their exercise habits (i.e., frequency, intensity, and duration) before beginning the program. The data from the questionnaire would be summarized and used in planning the training protocol of interest.

The time frame of interest is usually a critical component of questionnaire assessments of physical activity. This component also presents many difficulties in comparing results across studies because many different time frames have been used, such as the following:

- Past three days
- Past week
- Past 30 days
- Usual week
- Usual month

Questionnaires have also been used to quantify physical activity in the distant past. This technique is particularly appealing for studies of chronic diseases such as heart disease and cancers because it attempts to assess physical activity prior to the clinical manifestation of the disease. Because the biological influences of physical activity on disease presumably occur over a long period of time, historical recall of physical activity can be very helpful.

Questionnaires and telephone surveys are cost-effective ways of obtaining a substantial amount of physical activity data on a large group of people and to track them over time. Questionnaires can be mailed, sent electronically, or administered in a group format. The emergence of handheld technologies (smartphones and tablet computers) has opened up a new set of opportunities for physical activity assessment using questionnaires and surveys. Questionnaires can be written and implemented in multiple languages and can be repeated fairly easily over the course of a study. Questionnaires can also be tailored for the specific purpose of the study and population of interest (e.g., older adults, children, cardiac rehabilitation patients).

Questionnaires for physical activity assessment do have some substantial drawbacks, however. **Recall bias**—the inability to accurately recall or the selective recall of only certain activities—can substantially influence respondents' answers to questions. The inability to probe respondents for more complete answers may result in many physical activities not being reported. Also, the validity of the responses is always a question. Are respondents overestimating (or underestimating) their behaviors? Despite these challenges, questionnaires have taught us a lot about the relationship between reported physical activity and health and disease.

SURVEILLANCE IN POPULATIONS

Disease surveillance has been a fundamental pillar of public health at least since the era of the Black Death in Europe in the 14th century. During this time, fundamental health interventions (quarantine, determining an outbreak) were developed based on counting the number of people affected by the deadly disease. Other epidemics in the centuries since led to more systematic data collection of births and deaths and the emergence of vital statistics systems. A plague in London in the 16th century led to the development and routine dissemination of the Bills of Mortality, the first known systematic system for collecting death information. These weekly summaries were used to map and monitor the extent of the plague in the city. In the next century, John Graunt famously converted these simple counts of dead citizens to useful and interpretable surveillance techniques in his work *Natural and Political Observations Made upon the Bills of Mortality*.

Public Health Surveillance

Public health surveillance has been defined in the modern era as the ongoing, systematic collection, analysis, and interpretation of health-related data. These data are meant to be used in the planning, implementation, and evaluation of public health practices, and their interpretation is meant to be disseminated to those responsible for prevention and control. Thus, surveillance is not merely the collection and analysis of data; it also involves action on the part of public health officials.

As population disease burdens have expanded over the years beyond infectious diseases to include noncommunicable diseases and their precursors, techniques and strategies for surveillance have evolved to address these health issues. Public health surveillance expanded from counting deaths and cases of a certain disease to monitoring those who may have been exposed to a certain disease and the trends and patterns of the disease in populations. It now also involves monitoring behavior and environmental exposures regardless of disease status and policy and environmental correlates of the risk of disease (USDHHS 2011a). New techniques have also emerged as technology has advanced. Given this evolution and expansion of public health surveillance, Declich and Carter (1994) proposed that it has emerged as a new subdiscipline separate from epidemiology.

Physical Activity Surveillance

Surveillance of physical activity is in its infancy relative to the surveillance conducted on other important public health problems. Its recent emergence is likely partially due to the fairly recent union of the fields of kinesiology and public health to create the subdiscipline of physical activity and public health (Kohl et al. 2006). It is also partially due to the recent realization that a lack of physical activity is a major determinant of multiple noncommunicable diseases. This acknowledgment of the importance of physical activity to public health has resulted in the development and evolution of surveillance techniques as our understanding of the dose and types of physical activity that promote health and prevent disease and disability becomes clearer. Also of interest are determinants, individual and

environmental, that promote physical activity (Kohl and Kimsey 2009).

Different tools have been used in different countries and regions to obtain country-specific surveillance data. Although these tools have been helpful for the countries using them, the situation limited the ability to compare and contrast data among and between countries and populations. Happily, major advances have been made toward a standardized tool for use around the world, in developed as well as developing countries.

Internationally, two major developments have greatly assisted with the population surveillance of physical activity behaviors. The **International Physical Activity Questionnaire** (IPAQ; Craig et al. 2003) and the **Global Physical Activity Questionnaire** (GPAQ; Bull et al. 2009) represent two responses to this question: Can a physical activity surveillance system be developed to allow consistent measures, within and between countries, of physical activity participation? The IPAQ, first developed in the 1990s, was the first to allow such consistent measures across countries; it has since been used in more than 100 countries.

What proportion of adults in a population are active at recommended levels? How many report regular walking? Do these proportions vary by age, sex, or race or ethnicity? What are long-term trends in these values over time? Is the population becoming more active? Less active? Remaining the same? These are the kinds of questions physical activity surveillance systems can answer. These data are used to inform public health professionals and others who can plan strategies to make changes to improve the health of populations.

Population Indicators of Physical Activity

Surveillance data on physical activity provide a wealth of information—sometimes too much. The challenge of using these kinds of data is to find the correct indicator or indicators that best suit the purpose. Frequently used measures of interest from physical activity surveillance are those associated with health outcomes.

Most physical activity surveillance systems provide data such as the number of people in a population who participate in moderate-intensity and vigorous-intensity physical activity, the number of people who meet physical activity guidelines, the prevalence of specific physical activities such as walking, and the prevalence of sedentary behavior. Further, domain-specific surveillance becomes important for understanding the context associated with the physical activity. Recreational (or discretionary) time, occupational physical activity, transport-related activity, and domestic (or household) activity are the four common domains of activity that are typically assessed. Each of these domains, and the physical activity behaviors performed in them, are different for different people. The recent emergence of evidence-based physical activity guidelines has greatly helped define and standardize key outcome measures to use in physical activity surveillance systems.

Attention has recently turned to the role environment and policy indicators can play in physical activity surveillance. Characteristics of the built environment, as well as policies that support or inhibit physical activity, appear to be related to physical activity behavior. Policies can include anything from state laws to support the training of physical education teachers (to increase capacity for physical education) to local government actions to construct bicycle-friendly pathways that encourage active commuting. Chapter 15 contains more of a discussion on physical activity policies, but questions that can be asked include the following: What role does increased access to places in which to be active play? and Are there measurable changes over time in the prevalence of these places? Examples of environmental indicators include the location of and access to parks, trails and other green spaces, recreation and fitness centers, and other places to be active. Legislative policies include mandates for school-based physical education and active-transport-to-school programs.

SOURCES OF PHYSICAL ACTIVITY DATA

In the United States, the ongoing surveillance of physical activity behaviors of adults occurs in the CDC-run National Health Interview Survey (NHIS; USDHHS 2010) and in the Behavioral Risk Factor Surveillance System (BRFSS; USDHHS 2011a). Periodic surveys such as the National Health and Nutrition Examination Survey (NHANES; USDHHS 2011b) supplement our understanding. The Youth Risk Behavior Surveillance System (YRBSS; USDHHS 2011c) is used to monitor physical activity levels in high school students. No physical activity surveillance data are available for children younger than high school. Table 4.1, adapted from Carlson and colleagues (2009), highlights key data available from each of the three major sources of adult physical activity surveillance in the United States. Other countries (Australia, Canada, and Brazil, among others) also have substantial physical activity surveillance systems that address the prevalence of physical activity and inactivity in those countries.

Table 4.1 Characteristics of Physical Activity Assessments and Physical Activity Levels

Category	NHIS	NHANES	BRFSS
Survey years physical activity data were collected[a]	1998-2007	1999-2006	2001, 2003, 2005, 2007
Recall period	Respondent selects recall period[b]	Past 30 days[c]	Usual week[d]
Self-reported	Yes	Yes	Yes
List of specific activities	No	Yes	No
Assesses moderate-intensity physical activity	Yes, but includes light intensity	Yes	Yes
Assesses vigorous-intensity physical activity	Yes	Yes	Yes
Which intensity level is asked about first?	Vigorous	Vigorous	Moderate
Definition of moderate-intensity physical activity	Light sweating or a slight to moderate increase in breathing or heart rate	Light sweating or a slight to moderate increase in breathing or heart rate	Small increases in breathing or heart rate
Definition of vigorous-intensity physical activity	Heavy sweating or large increases in breathing or heart rate	Heavy sweating or large increases in breathing or heart rate	Large increases in breathing
Definition of active based on the *Healthy People 2010* objective	Light or moderate activity (≥5 times/week and ≥30 min/time)—or—vigorous activity (≥3 times/week and ≥20 min/time)	Moderate activity (≥20 days/month and ≥600 min/month—or—vigorous activity (≥12 days/month and ≥240 min/month)	Moderate activity (≥5 days/week and ≥30 min/day—or—vigorous activity (≥3 days/week and ≥20 min/day)
Definition of inactive based on the *Healthy People 2010* objective	No reported light- to moderate- or vigorous-intensity activity for at least 10 minutes	No reported moderate- or vigorous-intensity activity for at least 10 minutes	No reported moderate- or vigorous-intensity activity for at least 10 minutes

Abbreviations: BRFSS, Behavioral Risk Factor Surveillance System; NHIS, National Health Interview Survey; NHANES, National Health and Nutrition Examination Survey.

[a]Includes years in which the same physical activity question was asked of respondents. NHIS asked a slightly different physical activity question in the first half of 1997, which included a minimum duration of "at least 20 minutes." This changed midyear in 1997 to "at least 10 minutes" and has remained unchanged ever since.

[b]NHIS physical activity questions allow respondents to select the recall period. To define physical activity levels, the average number of times per week (rounded to the nearest time) was calculated for those respondents who selected monthly or yearly time periods.

[c]NHANES has separate questions about active transportation and moderate household activities that are not included as part of this analysis.

[d]BRFSS has a separate question about monthly participation (yes or no) in any physical activities or exercises such as running, calisthenics, golf, gardening, or walking for exercise that was not included as part of this analysis.

Reprinted, by permission, from S.A. Carlson et al., 2009, "Differences in physical activity prevalence and trends from 3 US surveillance systems," *Journal of Physical Activity and Health* 6 (Suppl 1): S18-S27.

KEY LEADER PROFILE

Wendy J. Brown, PhD

Why and how did you get into this field?

My mother tells me that when I was a child I never sat still. I think I must have an activity gene, which encouraged me to take every opportunity to move and stimulated my interest in studying many aspects of human movement. During my undergraduate studies in Birmingham, UK, I specialized in human physiology and was excited to have my first paper published in the peer-reviewed *Journal of Physiology.* Because I was still suffering from hyperactivity and didn't really enjoy working with animal models, I went on to train as a PE teacher. After teaching for a few years, I went back to complete a master's degree in human biology—with a strong focus on what was then called work physiology.

Courtesy of Wendy J. Brown.

After a stint on the lecturing staff at Loughborough University, I moved to Australia and completed my PhD—looking at sex differences in substrate metabolism during endurance exercise. Interestingly, having shown that women's metabolism was quite well suited to distance events, I graduated on the day women were first allowed to compete in the Olympic Marathon.

As a working mother, I then juggled part-time jobs while caring for two small boys. This period saw me developing training programs for fitness leaders and working in a large private hospital with cardiac patients. It was only then that I realized we should be paying much more attention to primary prevention and to activating the population. Everyone I worked with during these formative years probably influenced my gradual move from exercise science toward my love of behavioral science and epidemiology.

What are your current research interests?

I am passionate about the prevention of chronic disease through the promotion of physical activity. My work with the Australian Longitudinal Study on Women's Health has provided many opportunities to examine novel determinants of physical activity and of changes in physical activity in women across the adult life span. We also have a wonderful dataset (courtesy of the fantastic women who are still providing us with data after 16 years) with which to examine the health outcomes of various long-term activity patterns. I also love to be involved with innovative intervention efforts such as 10,000 Steps Rockhampton, in which we tried to activate the entire adult population of an Australian city. Currently, my research is focused more on the relatively new field of understanding, measuring, and influencing sitting time. Ironically, I often sit for more than 12 hours a day to do my work!

What drives you as a researcher and activist?

I really believe that research can inform interventions that will make a difference to population health.

What are one or two key issues to be addressed by 2022?

Issue 1: We need to understand much more about the dose of physical activity required for the prevention of health problems and for healthy aging, and about the contribution of all forms of activity (e.g., housework, gardening, active transport, leisure walking, occupational and unpaid physical work, and sport participation) to this dose, in people of all ages, right across the life span. Issue 2: We need to continue to work out how to get people to move more, in the context of living and working in situations that are not conducive to movement.

Chapter Wrap-Up

What You Need to Know

- There are many ways to assess physical activity and exercise behavior. None are perfect, and all are subject to substantial errors in measurement.
- The total amount of energy expenditure due to physical activity is a small portion of the total daily energy expenditure.
- Domain-specific physical activity participation is important to measure so that the context of the behavior can be understood.
- Surveillance of physical activity is important for understanding trends in participation over time in populations.
- Several sources of surveillance data exist in the United States and throughout the world.

Key Terms

total energy expenditure (TEE)

physical activity energy expenditure (PAEE)

thermic effect of food (TEF)

basal metabolic energy expenditure (BMEE)

SOPLAY

SOPARC

reactivity

recall bias

International Physical Activity Questionnaire (IPAQ)

Global Physical Activity Questionnaire (GPAQ)

Study Questions

1. What is the definition of *total energy expenditure*?

2. How much of daily total energy expenditure does physical activity account for?

3. How much of daily total energy expenditure does basal metabolic energy expenditure (BMEE) account for?

4. What are the basic scientific concepts supporting indirect calorimetry and the doubly labeled water technique as measures of physical activity?

5. What two direct observation techniques are commonly used for measurement of physical activity?

6. What are the pros and cons of collecting self-report data about physical activity?

7. Why are questionnaires and surveys regularly used to collect data about physical activity behaviors?

8. What are the major differences between physical activity measurements collected with accelerometers and those collected with pedometers?

9. What is public health surveillance, and how does it apply to physical activity measures?

10. What are three sources of public health surveillance data related to physical activity behaviors?

E-Media

Explore issues related to physical activity, exercise, and public health at the following websites:

Human Kinetics	www.HumanKinetics.com
U.S. Department of Health and Human Services: Physical Activity Guidelines for Americans	www.health.gov/PAGuidelines
World Health Organization	www.who.int
International Society for Physical Activity and Health	www.ispah.org
U.S. Centers for Disease Control and Prevention: Physical Activity	www.cdc.gov/physicalactivity/
American College of Sports Medicine	www.acsm.org
President's Council on Fitness, Sports & Nutrition	www.fitness.gov
University of Pittsburgh Physical Activity Resource Center for Public Health	http://dev.edc.pitt.edu/about.aspx

Bibliography

Blair SN, Haskell WL, Ho P, Paffenbarger RS, Vranizan KM, Farquhar JW, Wood PD. 1985. Assessment of habitual physical activity by a seven-day recall in a community survey and controlled experiments. *American Journal of Epidemiology* 122:794-804.

Bull FC, Maslin TS, Armstrong T. 2009. Global Physical Activity Questionnaire (GPAQ): Nine country reliability and validity study. *Journal of Physical Activity and Health* 6: 790-804.

Carlson SA, Densmore D, Fulton JE, Yore MM, Kohl HW III. 2009. Differences in physical activity prevalence and trends from 3 US surveillance systems. *Journal of Physical Activity and Health* 6 (Suppl 1): S18-S27.

Craig CL, Marshall AL, Sjöström M, Bauman AE, Booth ML, Ainsworth BE, Pratt M, Ekelund U, Yngve A, Sallis JF, Oja P. 2003. International Physical Activity Questionnaire: 12-country reliability and validity. *Medicine & Science in Sports & Exercise* 35 (8): 1381-1395.

Declich S, Carter AO. 1994. Public health surveillance: Historical origins, methods and evaluation. *Bulletin of the World Health Organization* 72 (2): 285-304.

Kohl HW III, Kimsey CD Jr. 2009. Physical activity surveillance. In Lee, I-M, ed. *Physical Activity Epidemiology.* New York: Oxford University Press.

Kohl HW III, Lee I-M, Vuori IM, Wheeler FC, Bauman A, Sallis JF. 2006. Physical activity and public health: The emergence of a subdiscipline. *Journal of Physical Activity and Health* 3: 344-364.

Pettee Gabriel K, James J, McClain JJ, Schmid KK, Kristi L, Storti KL, Ainsworth BE. 2010. Reliability and convergent validity of the past-week Modifiable Activity Questionnaire. *Public Health Nutrition* 14 (3): 435-442.

U.S. Department of Health and Human Services. 2008. *2008 Physical Activity Guidelines for Americans.* www.health.gov/PAGuidelines. Accessed 20 June 2009.

U.S. Department of Health and Human Services. 2010. National Health Interview Survey (NHIS). www.cdc.gov/nchs/nhis.htm. Accessed 10 May 2011.

U.S. Department of Health and Human Services. 2011a. Behavioral Risk Factor Surveillance System (BRFSS). www.cdc.gov/BRFSS. Accessed 10 May 2011.

U.S. Department of Health and Human Services. 2011b. National Health and Nutrition Examination Survey (NHANES). www.cdc.gov/nchs/nhanes.htm. Accessed 10 May 2011.

U.S. Department of Health and Human Services. 2011c. Youth Risk Behavior Surveillance System (YRBSS). www.cdc.gov/HealthyYouth/yrbs/index.htm. Accessed 10 May 2011.

PHYSICAL ACTIVITY IN PUBLIC HEALTH SPECIALIST

This chapter covers these competency areas as set forth by the National Society of Physical Activity Practitioners in Public Health:

1.1.3, 1.1.5, 2.1.2, 2.1.3, 2.2.3, 2.3.2, 2.3.3, 2.5.2, 2.6.1, 2.6.2, 2.6.3, 3.2.1, 3.2.2, 3.8.1, 3.8.2, 4.1.3, 4.1.4, 4.2.1, 5.5.5, 5.5.6, 5.5.7, 6.2.1, 6.3.1, 6.3.2, 6.4.1, 6.4.2

© EastWest Imaging

PART II

HEALTH EFFECTS OF EXERCISE AND PHYSICAL ACTIVITY

CHAPTER 5

CARDIORESPIRATORY AND METABOLIC HEALTH

OBJECTIVES

After completing this chapter, you should be able to discuss the following:

- » The prevalence of, economic costs of, and risk factors for cardiovascular disease
- » The evidence of a correlation between physical activity and cardiorespiratory health
- » The physical activity and exercise recommendations for cardiorespiratory health
- » What metabolic diseases are, how prevalent they are, and their risk factors
- » The evidence of a correlation between physical activity and metabolic health
- » The physical activity and exercise recommendations for metabolic health
- » Testing methodologies used to predict and diagnose cardiovascular and metabolic diseases

OPENING QUESTIONS

» What is the leading cause of death worldwide?

» How do metabolic diseases relate to the risk for developing cardiovascular disease?

» How much physical activity and exercise do people need to decrease their cardiovascular and metabolic health risks?

Part II of this textbook provides an overview of the scientific evidence that supports the health benefits of participation in regular physical activity and exercise. It also reviews the potential health risks associated with a lack of physical activity and exercise.

A primary benefit of engaging in physical activity and exercise is that doing so significantly reduces the risk of premature death (i.e., dying earlier than the average age of death for a specific population group) from any cause, or **all-cause mortality,** as compared to being inactive (U.S. Department of Health and Human Services [USDHHS], Physical Activity Guidelines Advisory Committee [PAGAC] 2008). As noted in the *2008 Physical Activity Guidelines for Americans* (USDHHS 2008), the effects of physical activity and exercise on all-cause mortality are remarkable for two reasons:

• Only a few lifestyle choices have as large an effect on mortality as physical activity. It has been estimated that people who are physically active for approximately 7 hours a week have a 30 to 40% lower risk of dying early than those who are active less than 30 minutes per week.

• It is not necessary to do high amounts of activity or even vigorous-intensity activity to reduce the risk of premature death. Studies show a substantially lower risk of mortality when people do at least 150 minutes of moderate-intensity aerobic physical activity a week.

The main messages to share are that (1) research clearly demonstrates the importance of avoiding inactivity and (2) some physical activity is much better than none at all, because there is a dose-response relation between the amount of physical activity and risk of poor health. Figure 5.1 illustrates the relationship between the risk of all-cause mortality, perhaps the ultimate indicator of poor health, and the minutes per week of moderate- or

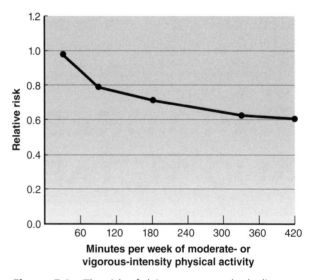

Figure 5.1 The risk of dying prematurely declines as people become physically active.
Reprinted from USDHHS 2008.

vigorous-intensity physical activity and exercise. As you can see, the risk of dying prematurely is lower when one is physically active for 1.5 to 2.5 hours a week versus for only 30 minutes a week. Moreover, the risk continues to decline with higher amounts of physical activity. Thus, people who are more active are better off than those who are inactive or somewhat active.

The health benefits of physical activity and exercise (i.e., lowering the risk of all-cause mortality) apply equally for men and women, adults of all ages, active people of all body weights (normal, overweight, and obese), and there is no evidence for racial or ethnic differences when adjusted for total volume of physical activity. This chapter addresses the specifics of how participation in physical activity and exercise has positive health effects related to cardiovascular health and metabolic health.

According to the *Physical Activity Guidelines Advisory Committee Report, 2008* (USDHHS, PAGAC 2008), **cardiovascular disease (CVD)** and the underlying metabolic disorders (e.g., metabolic syndrome and diabetes mellitus) can be prevented and treated with physical activity. Although people can be independently diagnosed with CVD, metabolic syndrome, or diabetes, all of these health challenges can, and commonly do, occur together. People with these diseases commonly have other chronic health issues and have not been physically active.

The cardiovascular diseases (CVDs) are a group of disorders of the heart and blood vessels that include the following:

- Coronary heart disease (CHD or ischemic heart disease, heart attacks)
- Cerebrovascular disease (stroke)
- Elevated blood pressure (hypertension)
- Peripheral artery disease
- Rheumatic heart disease
- Congenital heart disease
- Heart failure

Numerous **risk factors** (lifestyle or genetic variables that can predict the occurrence of disease) are known to contribute to the development of CVD. In turn, many of these are modifiable with regular physical activity and exercise. Elevated blood pressure, poor lipid and lipoprotein profiles, low cardiorespiratory fitness, and high body fat levels are but a few of these risk factors for CVD. To state this another way, physical activity and exercise are thought to reduce the risk of CVD in two ways: (1) by reducing other risk factors for the disease and (2) by playing a direct role in the physiological changes at the cellular level and in the blood vessels, the impairment of which contributes to CVD. Common respiratory disorders such as chronic obstructive pulmonary disease can also contribute to increased CVD risk and decreased cardiorespiratory function, but are most modifiable by other interventions such as smoking cessation. As noted in chapter 2, regular physical activity and exercise can also improve cardiorespiratory fitness, which can reduce CVD risk.

Metabolic risk factors contribute to a variety of conditions that increase the risk of CVD development. **Metabolic syndrome** is actually a cluster of clinical characteristics that has been defined differently by several organizations (e.g., the U.S.

National Cholesterol Education Program and the World Health Organization [WHO]), but it has the following profile in adults and adolescents:

- Abnormal levels of lipids and lipoproteins (low high-density lipoprotein [HDL] levels and high triglyceride levels with small, dense low-density lipoprotein [LDL], or **atherogenic dyslipidemia**)
- Elevated glucose or insulin levels
- Hypertension
- Excess abdominal obesity (USDHHS, PAGAC 2008)

Diabetes mellitus (or diabetes) is a syndrome associated with low insulin secretion, a limited ability of insulin to act on target tissues to maintain glucose homeostasis, or both of these conditions. Glucose is a main source of energy for the human body, and glucose levels must be kept in equilibrium so the body's metabolic processes can continue. Metabolic dysfunction (including some, if not all, of the conditions described for metabolic syndrome) is commonly seen in people diagnosed with diabetes. As discussed later in the chapter, CVD, metabolic syndrome, and type 2 diabetes can all be prevented or managed by engaging in appropriate, regular physical activity and exercise.

Diabetes can be further defined as type 1 and type 2. In people with **type 1 diabetes,** the immune system attacks and destroys the insulin-producing beta cells of the islets of Langerhans in the pancreas. Type 1 diabetes is treatable with insulin. Type 1 diabetes (the more rare of the two) usually affects children and adolescents (it is also called *juvenile diabetes or insulin-dependent diabetes*), but it can occur in adults after a viral infection or as postgestational diabetes in women who become pregnant after age 35.

Type 2 diabetes (adult-onset, non-insulin-dependent diabetes mellitus, or NIDDM) is related to overweight and obesity and insulin resistance (IR, impaired glucose homeostasis). In people with this condition, the pancreas cannot secrete enough insulin to compensate for the IR, which results in high blood glucose and high lipid levels. People with type 2 diabetes usually have the risk factors associated with metabolic syndrome, and they are at a higher risk for developing CVD.

PREVALENCE OF CARDIOVASCULAR DISEASE

As a whole, CVD is the number one cause of death globally. The WHO (2011, please see table 1.1) reported that over 23% of all global deaths; of these, 7.25 million were associated with CHD and 6.15 million were associated with stroke (see table 1.1). The leading causes of death (heart disease, some cancers, stroke, and diabetes) have remained fairly stable since 2008, and participating in regular physical activity and exercise can decrease the risk of dying from all four.

Looking to the future, the WHO (2011) reported that the global burden of CVD is most likely to continue. If the current global trends continue, by 2015 an estimated 20 million people would die every year from CVD. This represents a monumental shift, as mentioned in chapter 1, away from infectious disease and toward chronic (noncommunicable) diseases as the leading causes of death. This also presents a terrific opportunity, given the positive influence that physical activity has on many chronic diseases.

In addition to the health costs of CVD, there are economic costs to consider: individual and family health care, time lost from work, costs to government and industry for health care, and health care costs to countries due to lost productivity. The American Heart Association (2009) estimated the direct costs (e.g., physician care and medications) and indirect costs (e.g., mortality and morbidity) of CVD and stroke in the United States at $475.3 billion US dollars in 2008. As a comparison, the cost of cancer and benign neoplasms in the United States was $228 billion in 2008.

RISK FACTORS FOR CARDIOVASCULAR DISEASE

Numerous CVD risk factors have been identified in the scientific literature, many of which can be reduced by participating in regular physical activity and exercise. The major risk factors for CVD are divided between those that are modifiable (i.e.,

something can be done about them) and those that are nonmodifiable (i.e., not changeable).

Modifiable Risk Factors for CVD

- Hypertension
- Atherogenic dyslipidemia
- Tobacco use
- Physical inactivity
- Obesity
- Metabolic syndrome
- Diabetes mellitus
- Elevated inflammation biomarkers (e.g., C-reactive protein)

Nonmodifiable or Less Modifiable Risk Factors for CVD

- Age
- Sex
- Heredity (genetics)
- Ethnicity or race

The modifiable risk factors often can be significantly altered with lifestyle changes or pharmaceutical interventions. Nonmodifiable risk factors are often regulated by behavioral interventions (e.g., physical activity and exercise) that influence molecular and cellular changes based on individual or clusters of genes.

Even though these risk factors have been traditionally thought of as binary (modifiable or non-modifiable but not both), recent research suggests that this may not always be the case—that is, a person's genetic code for disease may actually be influenced by external factors such as physical activity and exercise. For example, engaging in regular physical activity and exercise appears to affect epigenetic markers (i.e., regulators of gene function) that cause the down-regulation (repression) or up-regulation (enhanced expression) of specific genes. This in turn may influence the disease processes. In some cases a down-regulation of a gene or series of genes would reduce risk, whereas in other cases an up-regulation would be considered a positive effect. For example, if a person is at risk for developing hypertension because both parents have the disease, by participating in regular physical activity and exercise, that person can lower her or his hypertension and CVD risk, despite any genetic predisposition.

Following are descriptions of the modifiable and nonmodifiable risk factors for CVD:

- *Hypertension.* Hypertension is a major risk factor for stroke and other CVDs, especially if blood pressure is uncontrolled and >140/90 mmHg. Even somewhat elevated blood pressure (between 120/80 and 139/89—also known as *prehypertension*) can indicate an increased risk of CVD. Hypertension exhibits few symptoms, but it is associated with heredity, aging, physical activity, diet, obesity, and alcohol consumption.

- *Atherogenic dyslipidemia.* Elevated total cholesterol (>200 mg/dl), high levels of LDL cholesterol (>100-130 mg/dl), low levels of HDL cholesterol (<40 mg/dl for men and <50 mg/dl for women), and high levels of triglycerides are associated with greater risk for CVD.

- *Tobacco use.* Smokers have two to three times the risk for CVD that nonsmokers have, and they tend to be less physically active than nonsmokers. Quitting smoking reduces CVD risk and may facilitate greater participation in regular physical activity and exercise.

- *Physical inactivity.* Physical inactivity is a risk factor for CVD independent of other risk factors.

- *Obesity.* A body mass index (BMI) >25 kg/m^2 (overweight), or >30 kg/m^2 (obese) increases CVD risk and is highly correlated to metabolic disorders.

- *Diabetes mellitus.* Diabetes mellitus doubles the risk of developing CVD compared to those without the disease. Blindness, limb amputation, and renal nephropathy are health problems that arise when diabetes (type 1 or 2) is not controlled.

- *Metabolic syndrome.* Metabolic syndrome is identified as a cluster of abnormal characteristics (see the preceding definition of metabolic syndrome) associated with prolonged sitting, poor diet, and sedentary behaviors.

- *Elevated inflammation biomarkers.* C-reactive protein (CRP) is one of several biomarkers and responses to internal systemic inflammation that have been found to be associated with the development of atherogenic plaques, plaque rupture, or both (i.e., increased CVD risk).

- *Age.* Advancing age (men >40, women >50) is associated with increased CVD risk because of changes in vascular health (vasodilation versus vasoconstriction with or without artery narrowing) due to vascular stiffening.

- *Sex.* Men are at a higher risk for CVD than women at an earlier age; however, women's risk for CVD increases significantly postmenopause.

- *Heredity (genetics).* Genetics can account for 20 to 50% or more of the variability in people's CVD risk, which can predispose them to a lower or higher overall CVD risk. However, individual lifestyle and health behaviors can significantly reduce CVD risk.

- *Ethnicity or race.* Evidence supports the fact that some groups have higher rates of CVD than others (e.g., African Americans have higher stroke rates than other Americans). Contributing factors to ethnicity and race may include socioeconomic status and stress.

Based on information from *Physical Activity Guidelines Advisory Committee Report, 2008* (USDHHS, PAGAC, 2008), American Heart Association, American Diabetes Association, and the CDC.

KINESIOLOGY AND CARDIORESPIRATORY HEALTH

As noted in chapter 2, a general understanding of several of the exercise sciences is helpful for understanding the effects of physical activity and exercise on health outcomes such as cardiorespiratory health. Recall from chapter 2 that we are focusing on three key domains of kinesiology that guide the health aspects of physical activity: exercise physiology, the movement sciences, and the behavioral sciences.

The highlight box Cardiorespiratory Health Benefits From Physical Activity and Exercise Programming contains some of the desired exercise-related adaptations in each of these three areas that positively influence cardiorespiratory health. Practitioners should be able to explain the common changes that can occur following participation in regular personalized or targeted population physical activity and exercise programs (refer to the Scientific Evidence and Guidelines sections later in the chapter).

Cardiorespiratory Health Benefits From Physical Activity and Exercise

Physiological
- Lower resting heart rate
- Greater stroke volume
- Increased $\dot{V}O_2max$
- Increased ventilatory fatigue
- Increased arteriovenous oxygen difference (AV O_2 diff) (max)
- Lower submaximal blood pressure
- Increased lactate threshold (max)
- Improved functioning of autonomic nervous system
- Improved endothelial function
- Reduced inflammation due to oxidative stress
- Increased total energy expenditure
- Increase in oxidative enzymes
- Increase in anaerobic enzymes
- Improved glucose homeostasis
- Reduced body fat
- Reduced waist girth

- Increased muscular strength
- Increased muscular endurance
- Lower total cholesterol level
- Increased HDL cholesterol level
- Lower triglyceride level

Biomechanical
- Improved economy
- Increased motor skill and confidence to engage further in physical activity and exercise
- Improved proprioception, which helps coordination system response and balance

Behavioral
- Increased self-confidence
- Improved self-efficacy
- Decreased depression and anxiety
- Experience with behavioral change
- Improved stress management
- Improved sleep patterns

Physical activity and exercise improve cardiovascular function by lowering resting heart rate and increasing stroke volume during submaximal workloads. During moderate-intensity physical activity, most people have lower heart rates and blood pressures for the same amount of work (comparing pretraining to posttraining workloads). The muscles used for breathing (intercostals and abdominals) become more resistant to fatigue, and people can work for longer periods. The maximal cardiorespiratory endurance ($\dot{V}O_2max$) increases along with the ability to extract oxygen for muscle activity (AV O_2 diff).

The Fick equation, shown next, shows the central (pump function) and peripheral (muscle function) components of the cardiorespiratory system, and both parts can adapt specifically to provide increases in cardiorespiratory capacity after conditioning. Maximal heart rate (MHR) and

stroke volume (the squeeze of the heart with each beat, SV) make up the pump function components, whereas AV O_2 diff represents muscle adaptations of increasing oxygen extraction:

$$\dot{V}O_2max = (MHR \times SV) \times AV\ O_2\ diff$$

MHR typically stays the same or decreases slightly with training (as a result of an increase in SV, the heart needs more time to fill); AV O_2 diff also increases. The lactate threshold (LT) (i.e., point of increased lactic acid accumulation, which is associated with recruiting more fast-twitch muscle fibers) typically improves from ~50% of $\dot{V}O_2max$ to ~75% of $\dot{V}O_2max$.

Reductions in the percentage of body fat are often observed over time during physical activity and exercise training along with reductions in waist girth. The muscles used during aerobic activities show improvements in strength and endurance

WHY IS THE FICK EQUATION IMPORTANT?

Different types of physical activity affect different parts of the circulatory system. For example, some people (e.g., those with heart disease) may not be able to strengthen their heart as much as they can improve their peripheral circulation (and therefore improve their health). However, a healthy person should expect to see improved pump function (due to increases in SV) and improved peripheral function (due to increases in AV O$_2$ diff).

reduced inflammation related to oxidative stress are also common physiological adaptations observed with regular physical activity and exercise. Figures 5.2 through 5.4 illustrate common changes in cardiorespiratory function after physical activity or exercise training.

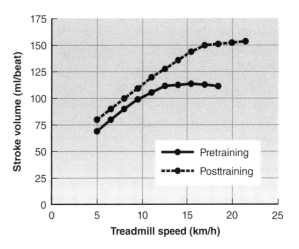

Figure 5.3 Significant changes in stroke volume after exercise training encompassing walking, jogging, and running.

Reprinted, by permission, from L. Kenney, J. Wilmore, and D. Costill, 2012, *Physiology of sport and exercise,* 5th ed. (Champaign, IL: Human Kinetics), 253.

(as compared to baseline). Regular participation in physical activity and exercise is associated with an improved atherogenic dyslipidemia profile that includes lower total cholesterol, increased HDL cholesterol, and lower triglyceride levels. Improvements in autonomic nerve function, increases in aerobic and anaerobic enzymes, improved glucose homeostasis, improved endothelial function, and

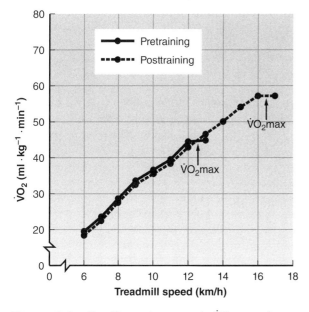

Figure 5.2 Significant increase in VO$_2$max shown after one year of intensive exercise training in an untrained person.

Reprinted, by permission, from L. Kenney, J. Wilmore, and D. Costill, 2012, *Physiology of sport and exercise,* 5th ed. (Champaign, IL: Human Kinetics), 249.

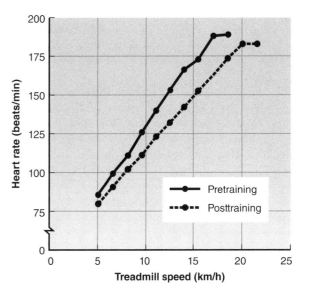

Figure 5.4 Significant changes in heart rate after exercise training encompassing walking, jogging, and running.

Reprinted, by permission, from L. Kenney, J. Wilmore, and D. Costill, 2012, *Physiology of sport and exercise,* 5th ed. (Champaign, IL: Human Kinetics), 254.

Does being physically active or participating in regular exercise help in the secondary prevention (like surviving a myocardial infarction [MI], or heart attack) of CHD and other atherosclerotic diseases? According to research reported by the American Heart Association (AHA) and the American College of Cardiology (ACC) the answer is, yes! The clinical practice of having cardiac patients participate in multidisciplinary (including physical activity and exercise) cardiac rehabilitation programs has been shown to stabilize, slow, or even reverse the natural progression of the underlying atherosclerotic process. Participation in exercise-based cardiac rehabilitation programs has been shown to reduce total mortality for patients by 20% within the first 6 months of a cardiac event, and by 26% with regards to cardiac mortality.

Comprehensive cardiac rehabilitation includes baseline patient assessments, risk factor interventions (stop smoking, hypertension management, lipid management, diabetes management, and weight control), physical activity counseling and exercise training, nutritional counseling, psychosocial counseling, and vocational counseling. Cardiac rehabilitation programs have traditionally included three phases with the following objectives related to physical activity and exercise interventions:

- Phase 1 involves inpatient (2-3 days for an uncomplicated MI) with education, baseline data collection, identification of physical limitations, and physical activity to avoid the negative affects of prolonged bed rest and inactivity.

- Phase II involves outpatient programming (4 weeks or 12 visits) that is often supervised in the clinical setting, but can also be monitored via new technologies from the patient's home. Prior to exercise programming patients should be evaluated with diagnostic graded exercise testing with physician supervision and an exercise prescription that may include clinical supervision. Most patients are then encouraged to accumulate 30 to 60 minutes of moderate intensity aerobic activity, preferably all days of the week, and they are encouraged to acquire two days a week of resistance training.

- Phase III usually involves voluntary patient participation in outpatient programs like those at local hospital-based fitness centers or YMCAs. In Phase III cardiac rehabilitation programming (>12 weeks post hospital discharge), patients are encouraged to at least maintain Phase II levels of physical activity and exercise.

Phases I and II of cardiac rehabilitation programs are usually covered by personal insurance for a specified period of time or number of visits (like 12 visits in Phase II cardiac rehabilitation). However, Phase III cardiac rehabilitation costs are covered directly by patients. The safety of participating in medically supervised cardiac rehabilitation programs is well documented and has been shown to be safe and effective with only 2 fatalities reported in 1.5 million patient-hours of exercise. Unfortunately, only about 10 to 20% of cardiac patients (approximately >2 million) participate in cardiac rehabilitation programs yearly, which means that many individuals do not regain or achieve basic functional health levels associated with higher levels of quality of life.

Some additional health benefits of participation in regular physical activity and exercise that have been reported for cardiac rehabilitation patients include: greater exercise capacity, improved success when returning to work, improved cardioprotective mechanisms, reduced anxiety, improved self-esteem, and better overall quality of life.

Adapted from AHA 2005 and AHA and ACC 2006.

Movement-wise, people who become physically active experience improved economy (i.e., a reduced oxygen–energy cost at a given speed or workload), which is a function of improved efficiency (this is complicated to actually measure). They also develop motor skills and most likely gain more confidence to engage in future physical activity and exercise activities. Peripheral proprioception (i.e., the sense of position and movement) response and balance often improves as well.

From a behavioral science standpoint, people undertaking physical activity or exercise training likely experience several of the steps of behavioral change (contemplation, preparation, action, maintenance, relapse) and learn to cope with the unique challenges of each step. By experiencing the stages of behavioral change, people acquire coping skills and strategies that help them remain in the maintenance stage for activity; they also avoid long periods of relapse as a result of overuse injuries, boredom, or lifestyle schedule changes.

Some people report that they feel better and have more self-confidence after several weeks of participation in physical activity and exercise. Many report experiencing lower levels of depression and anxiety and higher self-efficacy (i.e., personal accomplishment and well-being) levels. Participation in physical activity and exercise has been shown to be a very useful stress management tool and effective for improving sleep patterns.

CARDIORESPIRATORY FITNESS ASSESSMENTS

Many tests that require maximal or submaximal work levels have been developed and used by exercise physiologists and medical specialists to measure changes in cardiorespiratory fitness or function (CRF). The CRF measures and tests described here promote cardiorespiratory health and the prevention of CVDs. People with diagnosed CVD should seek clinical and rehabilitation advice before undertaking physical activity and exercise programs.

When evaluating CRF, two helpful clinical measures to acquire are $\dot{V}O_2$max and $\dot{V}O_2$peak (the highest O_2 uptake value or workload obtained by a person without achieving true maximal criteria),

which represent valid measures or estimates of aerobic power (see chapter 2 to review aerobic power and $\dot{V}O_2$max). The percentage of $\dot{V}O_2$max is also important clinically and can be used to determine intensity over an extended time, or for bouts of physical activity and exercise. Exercise physiologists consider the measurement of $\dot{V}O_2$max to be one of the single best predictors of overall CRF.

Graded exercise testing (GXT) is most commonly used to measure or estimate $\dot{V}O_2$max (see www.acsm.org and figure 5.5 for more on GXT testing). The GXT protocol selection should be based on target goals, physical abilities, and clinical considerations. By determining the person's $\dot{V}O_2$max, a practitioner can calculate her maximal working capacity or absolute intensity (see chapter 2) and classify physical activity and exercise workloads accordingly (see figure 2.4).

Figure 5.6 illustrates the expected change in CRF (as measured by $\dot{V}O_2$peak) with increasing doses or volumes of physical activity and exercise. Of course, all changes depend on factors such as baseline fitness, sex, age, BMI, and genetics.

Having determined the percentage of $\dot{V}O_2$max that people can work at for several minutes or for multiple short bouts throughout a day, practitioners can calculate their relative intensity as described in chapter 2. They can measure or estimate the

Figure 5.5 Graded exercise testing.

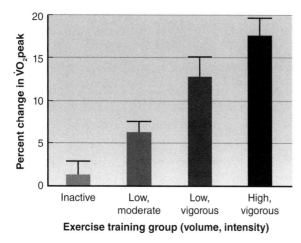

Figure 5.6 Changes in $\dot{V}O_2$peak by exercise group. Reprinted from USDHHS 2008.

percentage of $\dot{V}O_2$max by using any protocol that requires self-selected physical activity or exercise at moderate- to vigorous-intensity levels for 20 to 30 minutes, or for three bouts of ~10 minutes per day.

Can CRF be estimated without a laboratory and a treadmill or without doing maximal testing? Yes it can, because significant positive relationships exist among heart rate responses, speed of walking or running, and increasing workloads (resistance) with increasing oxygen uptake and $\dot{V}O_2$max. Several field assessments can be used as maximal or submaximal tests. Typically, maximal testing (like running on a treadmill to exhaustion) works best with low risk, younger, healthy populations. Submaximal tests can be used with all populations except those with diagnosed CVD, those at high risk for CVD, or those with orthopedic challenges who often require medical supervision for CRF evaluation (see ACSM [2010] for more on supervised exercise testing).

Distance run tests (1-mile run, 1.5-mile run, 12-minute run), the 1-mile walk, cycle tests, step tests lasting 3 to 5 minutes, and nonexercise prediction equations are all commonly used to predict $\dot{V}O_2$max and have acceptable errors of <1 MET. It is also possible to estimate $\dot{V}O_2$max using nonexercise protocols (e.g., questionnaires). The following commercial websites have listings of simple nonlaboratory CRF tests with procedures, online calculators, and interpretation explanations that can make the assessment of CRF more time and cost effective: www.exrx.net, www.brianmac. co.uk, and www.topendsports.com.

Once $\dot{V}O_2$max has been estimated, an easy way to predict the percentage of $\dot{V}O_2$max at which a person can work is to use a method first described by Ross and Jackson (1986). In this method, the person is asked to perform a 20- to 30-minute walk/jog evaluation that requires them to measure their heart rate immediately upon completing the assessment. The following equation can be used to calculate the percentage of $\dot{V}O_2$max:

% of $\dot{V}O_2$max = (heart rate in beats/min – k) × 100 / (220 – age – k)

in which $k = 61$ for males and 73 for females and age is expressed in years.

An initial goal for many people would be to maintain ~50% of their initial $\dot{V}O_2$max. A goal of ~75% is reasonable for those who want to improve their CRF. By comparing the relative intensity that people can maintain with the absolute intensity they can achieve, practitioners can help them achieve additional cardiorespiratory health and performance goals.

GENERAL RECOMMENDATIONS FOR CARDIORESPIRATORY HEALTH

As shown in figure 2.3 the relationships between the risks of CHD and stroke, and the volume (frequency, bout intensity, time or duration, and longevity of the program) of physical activity and exercise, are dose dependent. Figure 2.3 shows that CHD risk drops dramatically with moderate amounts of physical activity and exercise volume, whereas the relationship for stroke risk is more of an L shape and drops with a greater (but not too much) physical activity and exercise volume.

As early as 1961, an ad hoc committee of the AHA released a report that recommended that physical activity or exercise be part of the strategies to positively influence plasma lipid and lipoprotein levels that can reduce the risks of heart attacks and stroke (Gotto 1989). The report specifically stated that "overweight persons should decrease their caloric intake and attempt to achieve a desirable body weight, and weight reduction should be facilitated by regular moderate exercise." The work

KEY LEADER PROFILE

I-Min Lee, MBBS, ScD

Why and how did you get into this line of work? Did any one person have an overriding influence on you?

I credit the late Professor Ralph Paffenbarger Jr. for getting me involved in the field of physical activity and health. I met Paff—as everyone calls him—while I was a research assistant at Stanford University, working on a study that was examining occupational exposures in relation to the development of mycosis fungoides (a cutaneous lymphoma). This project had nothing to do with physical activity! At that time, I had just received a master's degree in public health and was interested in pursuing a doctoral degree in epidemiology. Paff was very encouraging of my plans to go back to school and offered me a position as a part-time research assistant on his Harvard Alumni Health Study to help with expenses (being a foreign student, I did not qualify for the NIH training slots available at my school).

Courtesy of I-Min Lee.

So I started working on Paff's study—which was, and is, a prospective cohort study focusing on the relationships between physical activity and health outcomes—and wrote my doctoral dissertation based on these data. At that time, much of the focus in physical activity epidemiology was on cardiovascular disease, and so I decided to conduct research on cancer instead. Cancer is a leading cause of morbidity and mortality worldwide, and little was known then about whether physical activity has any relation to this disease.

What are your current research interests?

I have since continued conducting research in the area of physical activity and chronic disease prevention. I believe physical activity is crucial for health and well-being. As Paff once said, "Everything that gets worse when we grow older gets better with exercise." We now know that physical activity decreases the risk of developing a whole host of chronic diseases and maintains health and function into old age.

What are one or two key issues of importance in our field that must be addressed by 2022?

The challenge facing us today is that despite our knowledge that physical activity is beneficial for health, many individuals worldwide do not get sufficient physical activity. How can we, in public health, get more people to be physically active and to decrease their level of sedentary behavior?

of Mitchell and associates (1966 and 2001) also supports the assertion that physical activity is very important for the maintenance of cardiorespiratory capacity.

Today, almost 50 years later, it is not surprising that many more recommendations to engage in physical activity and exercise to promote public health exist. The following section highlights evidence-based recommendations from the PAGAC (USDHHS, PAGAC 2008) regarding the use of physical activity and exercise to improve cardiorespiratory health.

SCIENTIFIC EVIDENCE

The PAGAC (USDHHS, PAGAC 2008) cited strong scientific evidence that supports an inverse relationship between the volume of physical activity and exercise and incidences of CVD (CHD, stroke, hypertension, and atherogenic dyslipidemia). Regular physical activity and exercise improves cardiorespiratory fitness and lowers the risk for CVD, CHD, and stroke by 20 to 35%. The benefits of physical activity and exercise on cardiorespiratory health apply equally for men and women and people of all ages, and there is no evidence for racial or ethnic differences when adjusted for volume.

The effective dose of physical activity and exercise for cardiorespiratory health was reported to be at least 800 MET-minutes per week (see chapter 2 to review the MET-minute concept) or 12 miles (19.3 km) per week (moderate intensity, vigorous intensity, or a combination). Strong evidence also was found to support positive CRF benefits from walking briskly for at least two hours per week and participation in aerobic activities on top of the usual activities of daily living. The reduction of CVD risk begins to decay once people drop below the dose threshold. Data to support the notion that an accumulation of daily bouts of physical activity and exercise can lower CVD risk are limited, but it makes sense to use this strategy with people who cannot do 20 to 30 minutes of continuous activity because of low fitness levels or with people who can schedule only multiple short bouts daily.

GUIDELINES

This section contains guidelines for physical activity participation to maximize cardiorespiratory health. Special precautions are noted for people with diagnosed or preexisting CVD. For example, for a person with uncontrolled hypertension, beginning an exercise program can be particularly dangerous. In this case, blood pressure should be lowered by pharmaceutical intervention prior to participating in any form of substantial physical activity. Once the blood pressure has been controlled, physical activity can be added as part of the management regimen. The guidelines for cardiorespiratory health can be divided into three parts: children and adolescents (ages 6 to 17), adults (ages 18 to 64), and older adults (>65 years).

Children and adolescents should acquire 60 minutes or more of physical activity and exercise daily for cardiorespiratory health. Most of the 60 minutes should include either moderate- or vigorous-intensity aerobic physical activity or exercise. Youth should include vigorous-intensity physical activity or exercise at least three days per week. Young people should be encouraged to participate in physical activities that are appropriate for their age, are enjoyable, and offer variety. Table 5.1 contains examples of moderate-intensity and vigorous-intensity aerobic activities for children and adolescents, adults, and older adults.

For substantial cardiorespiratory health benefits, adults should do at least 150 minutes (2 hours and 30 minutes) a week of moderate-intensity, or 75 minutes (1 hour and 15 minutes) a week of vigorous-intensity, aerobic physical activity or exercise, or an equivalent combination of moderate- and vigorous-intensity aerobic physical activity or exercise. Aerobic activity should be performed in episodes of at least 10 minutes and preferably it should be spread throughout the week. For additional cardiorespiratory health benefits, adults should increase their aerobic physical activity or exercise to 300 minutes (5 hours) per week of moderate-intensity, or 150 minutes of vigorous-intensity, physical activity or exercise, or an equivalent combination of moderate- and vigorous-intensity aerobic physical activity or exercise. All adults should avoid inactivity. Some physical activity is better than none, and adults who participate in any amount of physical activity or exercise gain some health benefits.

For substantial cardiorespiratory health benefits, older adults should follow the guidelines for adults, but consider the following special situations:

Table 5.1 Examples of Moderate-Intensity and Vigorous-Intensity Aerobic Physical Activity by Age Group

Population	TYPE OF AEROBIC PHYSICAL ACTIVITY	
	Moderate intensity	**Vigorous intensity**
Children and adolescents	• Active recreation such as hiking, skateboarding, in-line skating (or canoeing for adolescents) • Bicycle riding (stationary or road biking for adolescents) • Brisk walking • Housework and yard work, such as sweeping or pushing a lawn mower (adolescents) • Games that require catching and throwing, such as baseball and softball (adolescents)	• Active games involving running and chasing, such as tag (or flag football for adolescents) • Bicycle riding • Jumping rope • Martial arts, such as karate • Running • Sports such as soccer, ice or field hockey, basketball, swimming, and tennis • Cross-country skiing • Vigorous dancing (adolescents)
Adults	• Walking briskly (3 mph, or 4.8 km/h) or faster, but not racewalking • Water aerobics • Bicycling slower than 10 mph (16 km/h) • Tennis (doubles) • Ballroom dancing • General gardening	• Racewalking, jogging, or running • Swimming laps • Tennis (singles) • Aerobic dancing • Bicycling 10 mph (16 km/h) or faster • Jumping rope • Heavy gardening (continuous digging or hoeing, with heart rate increases) • Hiking uphill or with a heavy backpack
Older adults	The intensity of these activities can be either relatively moderate or relatively vigorous, depending on an older adult's level of fitness. • Walking • Dancing • Swimming • Water aerobics • Jogging • Aerobic exercise classes • Bicycle riding (stationary or on a path) • Some gardening activities, such as raking and pushing a lawn mower • Tennis • Golf (without a cart)	

Adapted from USDHHS 2008.

• When older adults cannot do 150 minutes of moderate-intensity aerobic activity a week because of chronic conditions, they should be as physically active as their abilities and conditions allow.

• Older adults should determine their level of effort for physical activity relative to their level of fitness (see the discussion of the OMNI scale in chapter 2).

• Older adults with chronic conditions should understand whether and how their conditions affect their ability to do regular physical activity and exercise safely.

Prevalence and Economic Costs of Metabolic Disease

Various sources estimate that over 1 billion people globally have the cluster of factors as described at the beginning of the chapter associated with metabolic syndrome and 171 million worldwide have diabetes. About one-third of those people are unaware of their condition because the early symptoms are mild. It is estimated that 47 million people in the United States have metabolic syndrome (Ford, Giles, and Dietz 2002). As noted previously in the chapter, metabolic syndrome is associated with a cluster of factors, and researchers have found that these factors are consistent with the dramatic increase in obesity, not only in the United States, but also globally. Cardiovascular disease and type 2 diabetes have been identified as primary clinical outcomes of metabolic syndrome, but those with metabolic syndrome are also at increased risk for other conditions such as hypertension, abnormal lipid levels, asthma, sleep disturbances (sleep apnea), and some forms of cancer.

In 2005, about 7% of the U.S. population was estimated to have type 1 diabetes (5 to 10% of diagnosed cases) or type 2 diabetes (90 to 95% of diagnosed cases). People with diabetes are at high risk for developing heart disease and stroke. Diabetes is the leading cause of adult blindness and kidney failure and causes 60% of nontraumatic lower-limb amputations each year (National Diabetes Education Program 2005).

In the United States, the prevalence of diabetes is higher among Hispanics, African Americans, and Native Americans than among non-Hispanic Caucasians. Figure 5.7 shows the prevalence of diabetes in the United States in 1994, 2000, and 2009. The prevalence of type 2 diabetes in children and adolescents has also increased significantly in the past 10 years, which is alarming because type 2 diabetes was rarely seen previously except in middle-aged adults.

The economic cost of metabolic syndrome is difficult to estimate because of varying methods of establishing a diagnosis (see next section). However, the estimated costs of diabetes in the United States in 2007 (ADA 2011) was estimated to be $174 million: $116 million was related to direct diabetes care, $58 million was related to complications, and $31 million was spent for excess medical costs. Another major economic challenge will be the future cost of diabetes, because experts estimate that 54 million adults age 20 and older have prediabetes, which includes blood glucose levels that are higher than normal but not high enough to indicate a diagnosis of diabetes.

Metabolic Disease Risk Factors

As you learned earlier in the chapter, various risk factors are associated with CVD as well as for metabolic syndrome and for diabetes. Table 5.2 (from the National Cholesterol Education Program) and the highlight box WHO Clinical Criteria for Metabolic Syndrome outline two (there are others) common sets of criteria and the risks associated with developing metabolic syndrome.

Some experts have defined metabolic syndrome as having three of the five ATP III criteria in table

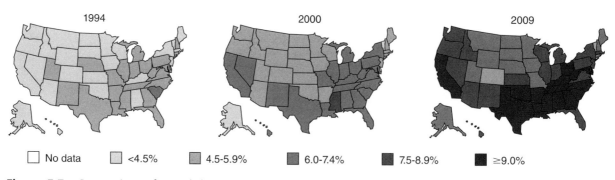

Figure 5.7 Comparison of U.S. diabetes prevalence in adults for 1994, 2000, and 2009.
Reprinted from CDC. Available online at www.cdc.gov/diabetes/statistics.

Table 5.2 ATP III Clinical Identification of the Metabolic Syndrome

Risk factor	Criteria
Abdominal obesity (waist circumference)	
Men	>102 cm (>40 in.)
Women	>88 cm (>35 in.)
Triglycerides	≥150 mg/dl
HDL cholesterol	
Men	<40 mg/dl
Women	<50 mg/dl
Blood pressure	≥130/≥85 mmHg
Fasting glucose	>110 mg/dl

Reprinted from NIH and NHLBI 2004.

5.2. The risk factors for diabetes are essentially the same as those for metabolic syndrome but also include the following:

- Family history of diabetes
- Ethnicity or race: Hispanics, African Americans, Asians, and Native Americans are at higher risk
- Sedentary lifestyle
- History of CVD

The risk factors for metabolic syndrome and diabetes are closely related to those for CVD. However, both physical activity and exercise play a very important role in the prevention and treatment of metabolic syndrome and diabetes.

KINESIOLOGY AND METABOLIC HEALTH

You learned about the relationships among the exercise sciences, and the effects of physical activity and exercise on public health outcomes and cardiorespiratory health, earlier in the chapter. The exercise science–related adaptations that positively influence metabolic health are basically the same as those listed in the Cardiorespiratory Health Benefits From Physical Activity and Exercise Programming highlight box.

You should be able to explain the common metabolic changes that can occur after participation in regular personalized physical activity and exercise programming (see Scientific Evidence and Guidelines sections). Following are specific metabolic adaptations that are expected as a result of physical activity:

- Increased total energy expenditure (helps maintain energy balance)
- Improved protein synthesis rate and amino acid uptake into skeletal muscle
- Reduced low-density lipoprotein levels
- Reduced triglyceride levels
- Increased high-density lipoprotein levels
- Improved glucose tolerance

One way to improve glucose uptake by skeletal muscles is to increase the gene expression of GLUT-4, which is a protein that is metabolically required for insulin to increase glucose uptake. GLUT-4 levels increase with regular physical activity and exercise, and they drop with physical inactivity and weight gain, which promotes insulin resistance.

Scientists have identified or are currently studying many other mechanisms associated with physical inactivity that that can either initiate or compound the effects of insulin resistance. By improving their metabolic health, people also benefit by achieving the movement science–based and behavioral changes associated with participating in regular physical activity and exercise, which can improve their future quality of life.

WHO Clinical Criteria for Metabolic Syndrome

Insulin resistance as defined by one of the following:

- Type 2 diabetes
- Impaired fasting glucose
- Impaired glucose tolerance

Plus any two of the following:

- Antihypertensive medication, high blood pressure (≥140 mmHg systolic or ≥90 mmHg diastolic), or both
- Plasma triglycerides ≥ 150 mg/dl (≥1.7 mmol/L)
- HDL cholesterol < 35 mg/dl (<0.9 mmol/L in men or < 39 mg/dl (<1.0 mmol/L) in women; or both
- BMI > 30 kg/m² or waist:hip ratio > 0.9 in men, > 0.85 in women
- Urinary albumin excretion rate > 20 µg/min or albumin:creatine ratio > 30 mg/g

Derived from AHA (2005).

COMMON TESTS OF METABOLIC FUNCTION

Physicians can evaluate patients for metabolic syndrome using the common clinical measures of blood glucose listed in table 5.2 and the WHO Clinical Criteria for Metabolic Syndrome highlight box. The measurements of height, weight, girth, blood pressure, and metabolic blood values are all part of most regular medical checkups. Although a diagnosis of metabolic syndrome does not necessarily mean that the person has a clinical disorder (e.g., dyslipidemia or diabetes), it does place the person at higher risk for developing preventable chronic diseases.

Fasting blood glucose concentration is a marker of short-term control of glucose, and the glycosylated hemoglobin concentration is a marker of long-term regulation of glucose (glycosylated hemoglobin

Can physical activity and exercise be fun while challenging cardiorespiratory and metabolic systems?

is discussed in more detail a little later). Following are three common tests used to diagnose diabetes or prediabetes (they all require at least two tests conducted on separate days):

- *Fasting plasma glucose (FPG) test.* Blood glucose is measured after an eight-hour fast. If the glucose level is 99 mg/dl or below, the test is normal. A glucose level of 100 to 125 is consistent with prediabetes or impaired fasting glucose; a person with these levels is at higher risk for type 2 diabetes. If the level is 126 or higher, the person has diabetes.

- *Oral glucose tolerance test (OGTT).* Blood glucose is measured after an eight-hour fast and two hours after ingesting 75 grams of glucose dissolved in water. If the glucose level is 139 mg/dl or below, the test is normal. A glucose level 140 to 199 is consistent with prediabetes or impaired fasting glucose. If the level is 200 or higher, the person has diabetes.

- *Random plasma glucose test.* Blood glucose is measured (nonfasting) when a person has diabetic symptoms such as increased urination, increased thirst, unexplained weight loss, fatigue, blurred vision, increased hunger, or sores that do not heal. If the glucose level is above 200 mg/dl, the person probably has diabetes.

Many clinicians prefer the FPG test because it is convenient and inexpensive; however, the OGTT is more sensitive for diagnosing prediabetes. Gestational diabetes is also usually diagnosed with the OGTT.

During the 120-day life span of a red blood cell, glucose molecules bind the hemoglobin contained within it, forming glycosylated hemoglobin. Once a hemoglobin molecule is glycosylated, it remains that way its entire life cycle. The percentage of glycosylated hemoglobin reflects the average level of glucose that the cell is exposed to during its life cycle. The species of glycosylated hemoglobin measured clinically and reported is hemoglobin$_{A1c}$ (Hb$_{A1c}$). In healthy people, Hb$_{A1c}$ is ~5%. Higher levels of Hb$_{A1c}$ are found in people with diabetes depending on their average blood glucose. Most guidelines recommend that Hb$_{A1c}$ be below 7% for most patients, which corresponds to an average blood glucose of ~170 mg/dl.

GENERAL RECOMMENDATIONS FOR METABOLIC HEALTH

Figure 5.8 illustrates the relationship between the risks of having metabolic syndrome and the amount of self-reported physical activity and exercise from several studies evaluating a dose-response relationship (USDHHS, PAGAC 2008). As shown, metabolic syndrome risk drops dramatically with moderate amounts of self-reported physical activity and exercise.

Figure 5.9 illustrates the relationship between developing metabolic syndrome and levels of measured physical fitness from several studies evaluating a dose-response relationship (USDHHS, PAGAC 2008). Metabolic syndrome risk drops dramatically with moderate fitness levels for adults.

In 2004, Dr. John Holloszy (a research exercise physiologist and public health professional) of the Washington University School of Medicine reviewed his career work of studying adaptations of skeletal muscle mitochondria for the American College of Sports Medicine (Holloszy 2004). In his review paper, Dr. Holloszy stated his belief that exercise deficiency is a serious public health problem with regard to the development of chronic diseases and

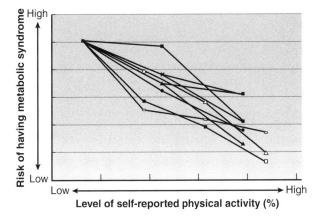

Figure 5.8 Dose-response relationship between self-reported physical activity and the risk of having metabolic syndrome.
Reprinted from USDHHS 2008.

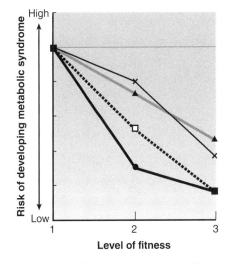

Figure 5.9 Risk of developing metabolic syndrome and long-term physical activity levels, shown in four different research studies.
Reprinted from USDHHS 2008.

the accelerated decline in the function of skeletal muscle, cardiovascular, and metabolic functional capacities with aging. He further asserted that the most important area of future research is to find effective ways to motivate "couch potatoes" to incorporate regular physical activity and exercise into their daily lives.

Evidence from the landmark Diabetes Prevention Program (2002) supports the lifestyle modification observations of Dr. Holloszy. Adults who were overweight and had prediabetes and who lost 7% of their body weight and were physically active 150 minutes per week reduced their risk for developing diabetes by 58%. The section that follows highlights recommendations from the PAGAC (USDHHS, PAGAC 2008) regarding the integration of physical activity and exercise science to promote metabolic health.

Scientific Evidence

The PAGAC reported strong and clear evidence that regular physical activity improves the metabolic health of at least moderately active people by 30 to 40% over that of sedentary people. The benefits of physical activity and exercise on metabolic health apply equally for men and women and people of all ages, and reasonable evidence supports the association for various racial and ethnic groups.

The recommended dose to improve metabolic health is 120 to 150 minutes of moderate- or vigorous-intensity physical activity per week. However, evidence indicates that risk reductions start to be seen at levels below 120 minutes per week in people who engage in leisure time physical activity (LTPA). Evidence that resistance training is effective in treating diabetes is limited, although it can improve glucose control. More studies are needed to determine whether resistance training can prevent type 2 diabetes. Limited data exist regarding the effects of accumulated daily bouts of physical activity and exercise on metabolic risk, as well as on whether physical activity helps control Hb_{A1C} or gestational diabetes.

Guidelines

The guidelines for metabolic health are consistent with those for cardiorespiratory health for youth, adults, and older adults. However, people who may have metabolic syndrome or diabetes should take special precautions before undertaking physical activity or exercise programs.

Insulin concentration is an important determinant of the metabolic response to physical activity and exercise. The maintenance of glucose homeostasis is critical for all people, and a normal response to physical activity or exercise depends on exercise FITT variables, fitness levels, nutritional state, and environmental factors. For those with diabetes, other factors such as the use of insulin or other medications and the temporal relationship to eating are also important to consider during physical activity and exercise.

There are many ways to manipulate the timing and amount of insulin administration and food intake to avoid hypoglycemia or hyperglycemia. It is clear that a reduction in insulin dose in anticipation of exercise decreases the risk of hypoglycemia.

People with type 2 diabetes who are not treated with insulin and who do not have extensive vascular or neurological complications can generally exercise with no more concern than nondiabetic people of equal cardiorespiratory fitness. In addition to a preexercise evaluation, blood glucose monitoring should be performed during and after exercise to minimize the risk of developing hypoglycemia.

Added glucose ingestion prior to, during, or after exercise may be a more practical alternative to lowering insulin dose in the prevention of hypoglycemia. Table 5.3 provides some metabolic control guidelines for people who are active and insulin treated.

Table 5.3 Clinical Blood Glucose Levels for Active Insulin-Treated Clients

Metabolic control	Blood glucose level
Normal blood sugar	80-100 mg/dl
Prediabetic	100-120 mg/dl
High blood sugar (hyperglycemia)	>120 mg/dl
Low blood sugar (hyperglycemia)	<70 mg/dl
Very low blood sugar (unconsciousness)	<40 mg/dl

More specific recommendations on safety are available from the American Diabetes Association (see www.diabetes.org).

CHAPTER WRAP-UP

What You Need to Know

- Cardiovascular diseases (CVDs) are a group of disorders of the heart and blood vessels that include coronary heart disease (CHD, heart attacks), cerebrovascular disease (stroke), raised blood pressure (hypertension), peripheral artery disease, rheumatic heart disease, congenital heart disease, and heart failure. CVD is the number one preventable cause of death globally.

- *Metabolic syndrome* refers to a variety of clinical characteristics that have been defined differently by several organizations (e.g., the National Cholesterol Education Program and the WHO). However, they have similar profiles in adults and adolescents, which include abnormal lipid levels (low high-density lipoprotein [HDL] levels and high triglyceride levels with small, dense low-density lipoprotein [LDL] levels, or atherogenic dyslipidemia), elevated glucose levels, hypertension, and excess abdominal obesity.

- Research clearly demonstrates the importance of avoiding inactivity. The relative risk of dying prematurely is greatly reduced when one is physically active for 1.5 to 2.5 hours per week versus for only 30 minutes per week.

- Personalized plans can be developed to minimize the modifiable and less modifiable risk factors for CVD and metabolic syndrome to ensure future health.

- The physiological, biomechanical, and psychological benefits of participating in regular physical activity and exercise on cardiorespiratory health are numerous.

- Determining $\dot{V}O_2$max (or $\dot{V}O_2$peak) and the percentage of $\dot{V}O_2$max (or $\dot{V}O_2$peak) by direct measures or by various estimations can be helpful for evaluating cardiorespiratory fitness or function (CRF).

- The effective dose of physical activity or exercise for cardiorespiratory health is 800 MET-minutes per week or 12 miles (19.3 km) per week of moderate- or vigorous-intensity activity.

- The *2008 Physical Activity Guidelines for Americans* (USDHHS 2008) state that youth need 60 minutes per day of aerobic physical activity or exercise, and adults and older adults need a minimum of 150 minutes of moderate- and vigorous-intensity activities per week for cardiorespiratory health benefits.

- Metabolic syndrome affects over 1 billion people globally and represents a major future economic challenge related to the development of diabetes in the United States and the world.

- The risk factors for CVD and metabolic syndrome are closely related and are positively affected by participation in regular physical activity and exercise.

- Metabolic function can be evaluated with standard clinical blood test and other measures such as FPG, OGTT, and Hb_{A1C} levels.

- The effective dose of physical activity or exercise for cardiorespiratory health is 120 to 150 minutes per week of moderate- or vigorous-intensity activity.

- The guidelines for metabolic health are consistent with those for cardiorespiratory health, but special precautions may be needed for people who have diabetes and use insulin.

Key Terms

all-cause mortality

cardiovascular disease (CVD)

risk factors

metabolic syndrome

atherogenic dyslipidemia

diabetes mellitus

type 1 diabetes

type 2 diabetes

fasting plasma glucose (FPG) test

oral glucose tolerance test (OGTT)

random plasma glucose test

Study Questions

1. What are the definitions of the terms *cardiovascular disease* and *metabolic syndrome*?

2. What is the difference between type 1 and type 2 diabetes?

3. With regard to all-cause mortality, why are the effects of physical activity and exercise remarkable?

4. What were the 10 leading causes of death in the United States in 2008? (See chapter 1.)

5. What are the differences between modifiable and less modifiable risk factors?

6. What are three physiological, movement science, and behavioral adaptations to physical activity and exercise?

7. How much physical activity or exercise is required for improved cardiorespiratory health?

8. What are two methods used to identify metabolic syndrome in adults?

9. What are three tests commonly used to evaluate metabolic function?

10. With regard to metabolic health, how much physical activity and exercise is needed to see the health benefits?

E-Media

Explore issues related to physical activity, exercise, and public health at the following websites:

Human Kinetics	www.HumanKinetics.com
U.S. Department of Health and Human Services: Physical Activity Guidelines for Americans	www.health.gov/PAGuidelines
International Society for Physical Activity and Health	www.isaph.org
American College of Sports Medicine	www.acsm.org
President's Council on Fitness, Sports & Nutrition	www.fitness.gov
American Heart Association	www.aha.org
World Health Organization	www.who.int
U.S. National Physical Activity Plan: Make the Move	www.physicalactivityplan.org

Bibliography

American College of Sports Medicine. 2010. *ACSM's Guidelines for Exercise Testing and Prescription*, 8th ed. Philadelphia: Lippincott Williams & Wilkins.

American Diabetes Organization. www.diabetes.org. Accessed 9/13/11

American Heart Association. 2005. Leon, A.S. et al. American Heart Association Scientific Statement: Cardiac rehabilitation and secondary prevention of coronary heart disease. *Circulation* 111:369-376.

American Heart Association. 2009. Heart disease and stroke statistics—2009 update. Dallas, TX: American Heart Association.

American Heart Association and the American College of Cardiology. 2006. AHA/ACC Guidelines for Secondary Prevention for Patients with Coronary and Other Atherosclerotic Vascular Disease: 2006 Update, *Circulation* 116:2363-2372.

Ford ES, Giles WH, Dietz WH. 2002. Prevalence of the metabolic syndrome among US adults: findings from the third National Health and Nutrition Examination Survey. *Journal of the American Medical Association* 16; 287 (3): 356-359.

Ford ES, Kohl HW III, Mokdad AH, Ajani UA. 2005. Sedentary behavior, physical activity, and the metabolic syndrome among U.S. adults. *Obesity Research* 13 (3): 608-614.

Gotto AM. 1989. AHA conference report on cholesterol (preface). *Circulation* 80: 716.

Holloszy JO. 2004. Adaptations of skeletal muscle mitochondria to endurance exercise: A personal perspective. *Exercise Sport Science Reviews* 32: 41-43.

McGuire DK, Levine BD, Williamson JW, et al. 2001. A 30-year follow-up of the Dallas bed rest and training study: Effect of age on the cardiovascular response to exercise. *Circulation* 104: 1350-1366.

National Diabetes Information Clearinghouse. 2002. Diabetes Prevention Program. http://diabetes.niddk.nih.gov/dm/pubs/prevention-program/. Accessed on 9/13/11.

Ross RM, Jackson AS. 1986. *Understanding Exercise for Health and Fitness*. Houston: Mac J-R Publishing.

Saltin B, Blomqvist G, Mitchell JH, et al. 1968. Response to exercise after bed rest and after training: A longitudinal study of adaptive changes in oxygen transport and body composition. *Circulation* 37/38 (suppl VII): VII-1–VII-78.

U.S. Department of Health and Human Services. 2008. *2008 Physical Activity Guidelines for Americans*. Washington, DC: U.S. Department of Health and Human Services. www.health.gov/PAGuidelines.

U.S. Department of Health and Human Services, Physical Activity Guidelines Advisory Committee. 2008. *Physical Activity Guidelines Advisory Committee Report, 2008*. Washington, DC: U.S. Department of Health and Human Services. www.health.gov/PAGuidelines.

World Health Organization. Global Status Report on NCDs. 2011. www.who.int/chp/ncd_global_status_report/en/index.html. Accessed 16 June 2011.

PHYSICAL ACTIVITY IN PUBLIC HEALTH SPECIALIST

This chapter covers these competency areas as set forth by the National Society of Physical Activity Practitioners in Public Health:

1.4.1, 2.1.1, 2.3.3, 2.5.2, 3.1.3, 6.1.3, 6.1.4, 6.2.1, 6.2.2, 6.2.3, 6.2.4, 6.3.3, 6.3.5, 6.4.1, 6.4.2

© BananaStock

CHAPTER 6

OVERWEIGHT AND OBESITY

OBJECTIVES

After completing this chapter, you should be able to discuss the following:

» The classification of obesity as a disease or a health risk

» How *overweight* and *obesity* are defined

» Caloric balance and why physical activity and exercise are important in achieving it

» The prevalence, economic costs, and risk factors of overweight and obesity

» The testing methodologies used to evaluate overweight and obesity

» Evidence for a relationship between physical activity and energy balance

» Whether physical activity alone can prevent weight gain, result in weight loss, or keep weight off once it has been lost

» The physical activity and exercise recommendations for achieving energy balance and a healthy weight

Opening Questions

» What criteria are used to determine whether someone is overweight or obese?

» What evidence exists that suggests that physical activity can help with weight maintenance, weight loss, and the prevention of weight regain?

» How much physical activity is consistent with weight maintenance, weight loss, and the prevention of weight regain?

Other than tobacco use, obesity may be the most discussed and debated public health problem in economically advanced countries. The prevalence of obesity has jumped throughout the world; even in countries in which undernutrition has been a recent problem, obesity is starting to take hold. Because it is such a visible condition and because it affects so many people, it has taken a front seat in our health consciousness—perhaps because of its potential health effects, and perhaps for social reasons. Research in this area has exploded; studies related to obesity address such issues as genetics, behavior, and the environment. What makes someone overweight or obese? Why is this such a problem?

According to the U.S. Surgeon General (U.S. Department of Health and Human Services [USDHHS] 2001), overweight and obesity have become major health problems in the United States. The World Health Organization (WHO, 2011) announced that worldwide obesity more than doubled between 1980 and 2011, and that 65% of the world's population lives in countries in which overweight and obesity kill more people than underweight. **Overweight** can be defined as carrying more body fat than is healthy or an amount that increases disease risk. **Obesity** can be defined as having an unhealthy body weight, which is consistent with a variety of disease processes such as CVD, metabolic syndrome, and type 2 diabetes.

Are overweight and obesity diseases themselves? The answer depends on the definition of *disease* and which professional organization, government office, corporation, or foundation you talk to. Overweight is not considered a disease, but has been associated with numerous health consequences, and although obesity has not been officially recognized as a disease in the United States, private and government insurance groups have been reimbursing people

classified as obese by physicians for special foods and weight loss programs and procedures since 2002.

How is *obesity* defined? As it turns out, this is not an easy question because no perfect measure exists. Taller people typically weigh more than shorter people, men typically weigh more than women, and younger people typically weigh more than older people (at a given height). Moreover, statures are different in different cultures across the world. Clearly, body weight is difficult to rely on for a definition of obesity. Later in this chapter, we will review key ways to measure obesity, overweight, and body composition.

Body mass index (BMI) is a frequently used screening measure that takes into account a person's height. Instead of considering only weight, BMI assesses weight for height. The U.S. Centers for Disease Control and Prevention (CDC) advocates the use of BMI as a screening tool to determine the obesity status of adults and children. To calculate BMI, divide weight in kilograms by height in meters squared:

$$BMI = \text{weight (kilograms)} / \text{height (meters}^2)$$

Weight, BMI calculations, and classifications of underweight, healthy weight, overweight, and obese are shown in table 6.1 for adults who are 5 feet 10 inches (175.3 cm) tall. A standard adult classification is that those with a BMI between 18.5 kg/m^2 and 24.9 kg/m^2 are normal weight, and those with a BMI lower than 18.5 kg/m^2 are underweight. Adults with a BMI between 25.0 kg/m^2 and 29.9 kg/m^2 are overweight, and those with a BMI 30 or greater kg/m^2 are obese. Above 30 kg/m^2, there are several additional classifications: class 1 obesity is 30 to 34.9 kg/m^2; class 2 obesity is 35 to 39.9 kg/m^2; and class 3 obesity is greater than 40 kg/m^2. Currently, adults with class 3 obesity are considered

Table 6.1 Weight, BMI, and Status for a Sample Height

Height	Weight range	BMI	Status
70 in. (178 cm)	<129 lb (58.5 kg)	<18.5	Underweight
	129-174 lb (58.5-79 kg)	18.5-24.9	Normal
	175-208 lb (79.3-94.3 kg)	25.0-29.9	Overweight
	>209 lb (94.8 kg)	30 or higher	Obese

at an extremely high health risk and are generally eligible to be referred for surgical intervention.

Although BMI correlates with measures of body composition (e.g., fat percentage, which is discussed later in the chapter), it does not directly measure body composition. BMI measurement does not take into account the specific components of **body composition** such as lean muscle mass versus fat mass. Therefore, a person with high amounts of lean muscle mass (e.g., an intercollegiate athlete) may be misclassified as being overweight or obese using the BMI. This is a key limitation of using BMI, particularly at the individual level. BMI is not meant to be a clinical diagnostic tool. Rather, it is a useful screening mechanism and can be helpful in a complete physiological health assessment.

The situation becomes even more complex when dealing with youth. Children and adolescents mature at different rates; some move through growth spurts early in their teens, whereas others mature at a more gradual pace. For this and other reasons, standard BMI definitions for obesity in children and adolescents are different. Instead of a straight set of criteria (as with adults), the definition of overweight and obesity for youth is based on a relative scale.

Standardized growth charts are used to classify children and adolescents based on BMI. Growth charts consist of a set of percentile curves that illustrate various growth trajectories based on a standardized population. Percentiles are relative to a known distribution or population and are used as a standard against which others can be measured. Knowing sex, weight and height, and age, you can use a growth chart to estimate the percentile of BMI. See figures 6.1.and 6.2 for examples of BMI integrated into growth charts. BMI calculators that can calculate youth BMI based on age, sex, height, and weight are available at the following websites:

- www.cdc.gov/healthyweight/assessing/bmi/index.html
- www.nhlbisupport.com/bmi
- www.mayoclinic.com/health/bmi-calculator/NU00597

Following is a generally accepted classification scheme for BMI and weight status in children and adolescents:

- Underweight: BMI less than the 5th percentile for age and sex
- Healthy weight: BMI between the 5th and 85th percentiles for age and sex
- Overweight: BMI between the 85th and 95th percentiles for age and sex
- Obese: BMI greater than the 95th percentile for age and sex

POPULATION DIFFERENCES

Because of observed health risks (CVD and diabetes) in Asian and Pacific populations with BMI values lower than 25 kg/m², there have been attempts to advance the use of cut points that are different than those developed largely from Western populations. Of basic interest is a division of the "normal" range (18.50 kg/m² to 24.9 kg/m²) into two additional levels. Although the World Health Organization (WHO) does not officially endorse these different cut points (due largely to the variability in the scientific evidence), it does indicate that these lower values may be useful to help people of these ethnicities in an advisory capacity.

2 to 20 years: Boys
Body mass index-for-age percentiles

NAME _____

RECORD # _____

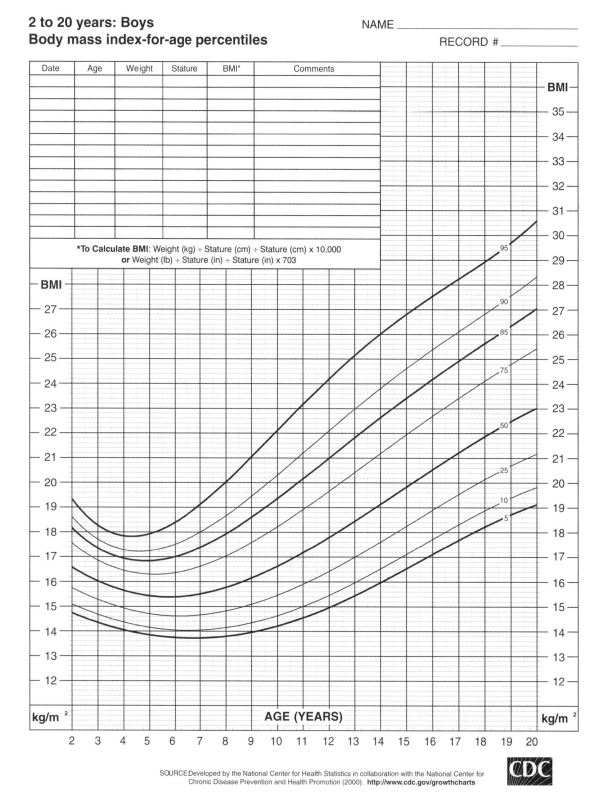

Date	Age	Weight	Stature	BMI*	Comments

*To Calculate BMI: Weight (kg) ÷ Stature (cm) ÷ Stature (cm) x 10,000
or Weight (lb) ÷ Stature (in) ÷ Stature (in) x 703

AGE (YEARS)

SOURCE:Developed by the National Center for Health Statistics in collaboration with the National Center for
Chronic Disease Prevention and Health Promotion (2000). http://www.cdc.gov/growthcharts

Figure 6.1 CDC growth chart and BMI for age percentiles: boys 2-20.
Reprinted from CDC.

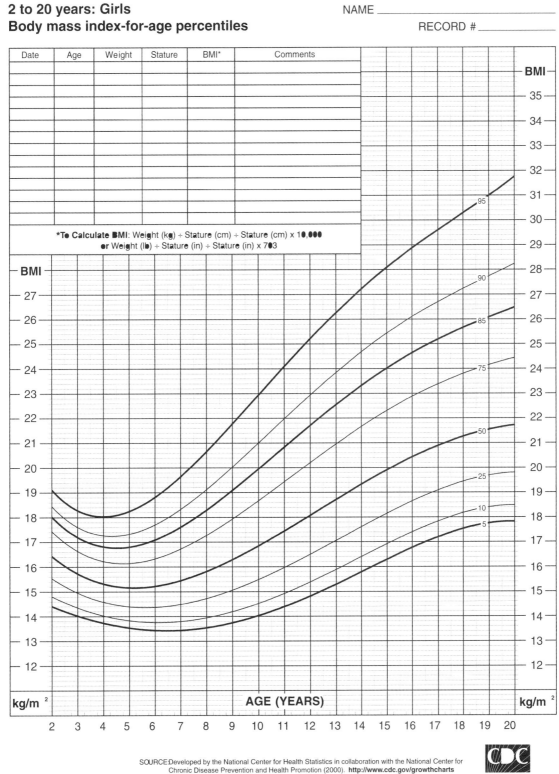

2 to 20 years: Girls
Body mass index-for-age percentiles

NAME _____

RECORD # _____

*To Calculate BMI: Weight (kg) ÷ Stature (cm) ÷ Stature (cm) x 10,000
or Weight (lb) ÷ Stature (in) ÷ Stature (in) x 703

Figure 6.2 CDC growth chart and BMI for age percentiles: girls 2-20
Reprinted from CDC.

CALORIC BALANCE

As discussed in chapters 2 and 4, a basic understanding of energy expenditure is essential for determining how physical activity and exercise can positively affect overweight, obesity, and the achievement of a healthy weight. Although the focus of this text is on energy expenditure to achieve caloric balance, energy intake (i.e., the food you eat) is clearly important as well.

A simple caloric balance equation scale is shown in figure 6.3. For weight loss, all that is required is to create an energy deficit by consuming fewer calories (i.e., eating less), spending more (i.e., being more physically active), or both. Sounds pretty easy. However, on further inspection (and in real life), it is clear that achieving and maintaining energy balance is a complex state that some people never attain—and many cycle in and out of energy balance throughout their lives.

The 2010 *Dietary Guidelines for Americans* (USDHHS, USDA 2010) provides important guidance on energy intake for weight maintenance.

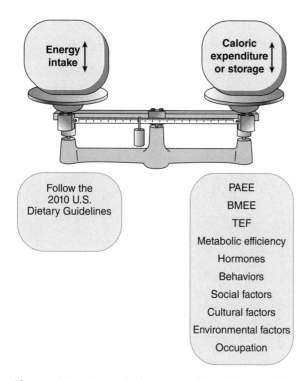

Figure 6.3 Energy balance equation. Please review chapter 4 for discussion of the thermic effect of food (TEF), basal metabolic energy expenditure (BMEE), and physical activity energy expenditure (PAEE).

For energy expenditure related to weight control, a useful reference is the *2008 Physical Activity Guidelines for Americans* (USDHHS 2008). These two references provide information about the variables in figure 6.3 such as behavioral, social, cultural, and environmental factors and how they affect caloric expenditure or increased energy storage.

The increase in the prevalence of overweight and obesity in children and adults worldwide clearly indicates that simple solutions for weight control are not working for a vast majority of people. And, although caloric restriction (i.e., dieting) alone can result in short-term weight loss and management, long-term solutions require physical activity increases and the maintenance of energy expenditure for most to obtain and maintain a healthy weight as well as to achieve other health benefits.

Have you ever thought about how many kilocalories (kcals) you expend per day in physical activity or exercise and how that expenditure affects your body weight and body composition? Although the energy cost of various physical activities (see part I) is known, how can we determine how many calories (kcals) we need to expend, individually or collectively, for effective weight management? Following is an important equation that we can use to answer this question:

1 lb (0.45 kg) of fat = 3,500 kcals

Therefore, to lose 1 pound (0.45 kg) of fat, theoretically, you would need to restrict or maintain your energy intake while expending more kcals in physical activity and exercise to the equivalent of 3,500 kcals. This is not a perfect match because all the while you are breathing, you are burning calories—albeit much more slowly than if you were exercising. A good rule of thumb is that, assuming a caloric intake of 2,500 to 3,000 kcals a day, burning 400 to 500 kcals each day through physical activity, for many people, would be a useful contribution to the maintenance of a healthy weight. For normal weight people, this can be the rough equivalent of walking about 3 miles (4.8 kg) each day at a moderately intense pace.

Body composition is defined as the relative proportion and distribution of fat, lean mass (muscle and bone), and minerals in the body. An important outcome of participating in regular physical activity and exercise is that you can maintain or increase lean body mass while controlling body fat levels (i.e., have a healthy body composition), whereas with dietary-only interventions, lean mass is not maintained.

Estimates of weekly physical activity and exercise caloric expenditure levels, or physical activity energy expenditure (PAEE; see chapter 4), for healthy adults with varying levels of physical activity and exercise behaviors are shown in table 6.2.

The values in table 6.2 are based on data from table 2.4 in chapter 2, which shows that by simply meeting the minimum guideline for physical activity, substantially more kcals are expended (940 versus 190 at rest) by walking at 4 miles per hour (6.4 km/h) for 150 minutes per week than by remaining sedentary. Jogging at 7 miles per hour (11.3 km/h) for 300 minutes per week, to use another example, will burn 3,930 kcals above sedentary levels, which is the energy equivalent of more than 1 pound (0.45 kg) of fat (USDHHS, Physical Activity Guidelines Committee [PAGAC] 2008, p. D-9).

People starting a physical activity or exercise program should try to achieve caloric balance by adjusting their food intake based upon their new energy expenditure levels and weight management goals. Some people may need to set their physical activity and exercise levels above recommended levels to expend enough energy to achieve or maintain a healthy weight. Older adults also need to consider that basal metabolic energy expenditure (BMEE) drops with age, making weight management more challenging and physical activity and exercise even more important for weight management.

PREVALENCE OF OBESITY AND OVERWEIGHT AND ASSOCIATED HEALTH CONSEQUENCES

As noted, the prevalence of obesity and overweight has increased dramatically since the mid-20th century. Although anthropological studies have shown that human populations have been gaining weight (on average) for centuries (due largely to improved nutrition), the tipping point from weight gain to obesity has been only a recent phenomenon.

In the United States, the CDC (Flegal et al. 2010) reported that over 67% of noninstitutionalized people over the age of 20 were overweight or obese. The WHO (2011) estimated that at least 1.5 billion adults over age 20 are overweight, and 200 million men and almost 300 million women throughout the world are obese. Because the human genotype is resistant to short-term changes, it is clear that the global and U.S. trends of increasing obesity are primarily due to external changes (e.g., in the built environment and lifestyle factors). Figure 6.4 shows the obesity trends among adults from the CDC Behavioral Risk Factor Surveillance System for 1990, 1999, and 2010 (USDHHS, CDC 2011a).

The trend of an increasing prevalence of overweight and obesity among children and adolescents over the past 40 years is also disturbing. Obesity trends among children and adolescents from 1963 to 2008 (from the CDC's National Health and Nutrition Examination Survey data) are shown in figure 6.5. In the United States at least 16.3% of youth between the ages of 2 and 19 years are obese. Among children ages 2 to 5 years, the prevalence of obesity has increased from 5 to 12.4%; among

Table 6.2 Expenditure Estimates (PAEE) by Duration and Intensity of Physical Activity

Sedentary – sitting at rest	190 kcals for 150 min/wk (2.5 hr/wk)	380 kcals for 300 min/wk (5.0 hr/wk)
Type of Physical Activity	**DURATION OF PHYSICAL ACTIVITY**	
	150 min/wk (2.5 hr/wk)	**300 min/wk (5.0 hr/wk)**
Walking (4 mph/6.4 km/h)	940 kcals	1,880 kcals
Jogging or running (7 mph/11.3 km/h)	2,155 kcals	4,310 kcals

Based on a 165 lb (75 kg) adult.

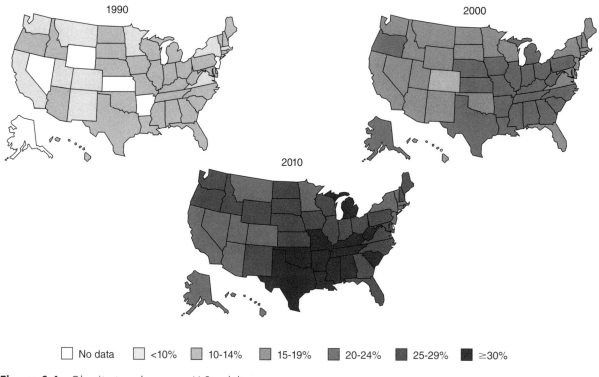

Figure 6.4 Obesity trends among U.S. adults.
Reprinted from CDC.

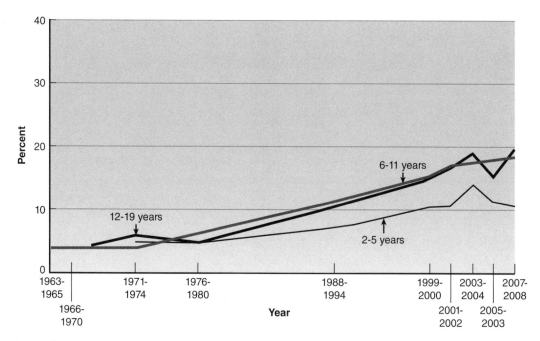

Figure 6.5 Obesity trends among U.S. children and adolescents.
Reprinted from CDC.

children ages 6 to 11 years, it has increased from 6.5 to 17%; and among adolescents ages 12 to 19, it has increased from 5 to 17.6%.

Preventing childhood overweight and obesity is very challenging. Research has demonstrated that obese youth may be more likely to become obese adults than are youth who are not obese. This situation increases the risk of chronic, noncommunicable diseases such as hypertension, high cholesterol, and type 2 diabetes.

The CDC (2011) and others have reported that the prevalence of obesity and overweight varies significantly among U.S. adults, adolescents, and children based on racial and ethnic differences. For example, based on adult surveys from 2006 to 2008, the CDC found that non-Hispanic blacks had a 51% greater prevalence of obesity and Hispanics had a 21% greater prevalence, compared to non-Hispanic whites. The results varied somewhat by state, but were consistent across the United States. These findings support the important goals of the U.S. health promotion initiative Healthy People 2020 (USDHHS 2011), which include reducing the prevalence of obesity in adults in the United States to 15% or less (which no state had achieved by 2007) and eliminating health disparities among racial and ethnic populations that contribute to obesity and overweight.

Interestingly, Wang and colleagues (2008) reported that if the current U.S. overweight and obesity trends continue, by 2030, 86.3% of adults would be overweight or obese, and 51.1% would be obese. They also reported that African American women and Mexican American men would be the subgroups most affected with rates of 96.9% and 91.1%, respectively. The authors projected that by 2048, all American adults would be overweight or obese, and African American women would reach that level by 2030 if trends continue as they are. For children and adolescents, the prevalence of obesity (>95th percentile) will almost double by 2030 to about 30% overall.

The economic costs of overweight and obesity in the United States vary considerably based on how they are categorized and estimated. Obesity and overweight either directly contribute to the development of costly chronic disease processes, or they complicate the medical treatments associated with managing the diseases once diagnosed (or both). The estimated total costs for the United States in 1998 may have been as high as $78.5 billion (Finkelstein, Fiebelkorn, and Wang 2003), which includes medical expenses for both overweight and obesity. The economic costs (estimated at $81.5 billion in 2010) of obesity and overweight have been predicted to double every decade until 2030 and to eventually cost $860 to $956 billion a year, which would account for 15.8 to 17.6% of total U.S. health care costs (Wang et al. 2008).

Data from Chenoweth and Leutzinger (2006) are illustrated in figure 6.6. These authors analyzed the economic costs that physical inactivity and excess weight combined had on the U.S. economy in 2003 and the projected costs in 2004 through 2008 with regard to direct medical care, workers' compensa-

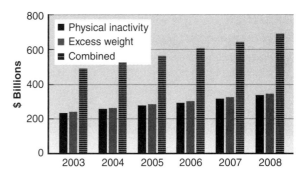

Figure 6.6 Trends in economic costs due to physical inactivity and excess weight.
Reprinted, by permission, from D. Chenoweth and J. Leutzinger, 2006, "The economic cost of physical inactivity and excess weight in American adults," *Journal of Physical Activity and Health* 3 (2): 148-163.

tion, and productivity loss. If medical costs had continued to rise as they had (10.2% per year), along with workers' compensation increases (4.5% per year) and productivity loss increases (4.1%), and if levels of physical inactivity and excess weight prevalence had stayed the same, the projected total costs would have exceeded $708 billion in 2008.

HEALTH CONSEQUENCES OF OBESITY

The health consequences of obesity and overweight include the following physical, psychological, and social challenges for adults, adolescents, and children:

- Coronary heart disease (CHD)
- Type 2 diabetes
- Cancers of the endometrium, breast, and colon
- Hypertension
- Dyslipidemia (high total cholesterol, or high levels of triglycerides)
- Stroke
- Liver and gallbladder disease
- Sleep apnea and respiratory problems
- Osteoarthritis
- Gynecological problems (abnormal menses, infertility)

Source: www.cdc.gov/obesity/causes/index.html, accessed 23 January 2010.

OBESITY AND OVERWEIGHT RISK FACTORS

The risk factors for obesity and overweight are complex and difficult to understand because there is not just one cause. The risk factors for obesity are related to behavior, the environment, and genetic factors, and their interactions. We all know very thin people who don't exercise at all and people who are obese or overweight who report exercising frequently. The cause of obesity in two people is rarely same, and the cause(s) across populations vary in the same way.

A variety of factors that influence the caloric balance equation can cause people to eat too many calories, to not get enough regular physical activity or exercise, or both, resulting in energy imbalance and weight gain. The environment at home, school, work, and in the community can provide numerous barriers and incentives that affect the ability to achieve caloric balance and maintain a healthy weight. Genetics may predispose people to weight gain or obesity, but the lifestyles they adopt can help them achieve caloric balance and a healthy weight based on their body type and body composition (see Common Assessments of Obesity and Overweight later in the chapter for more).

Modifiable Risk Factors for Overweight and Obesity

- Physical inactivity
- Excess caloric intake
- Low socioeconomic status (SES)

Nonmodifiable Risk Factors for Overweight and Obesity

- Age
- Heredity (genetics)
- Ethnicity or race
- Culture
- Metabolism

Following are descriptions of the modifiable and nonmodifiable risk factors for overweight and obesity:

- *Physical inactivity.* Inactivity creates a health cost versus a health benefit, and physical activity helps most people achieve energy balance. Increased levels of physical activity and exercise have been shown to improve long-term weight maintenance and enhance weight loss. Energy expenditure opportunities in modern society have been reduced as a result of technological advances and environmental barriers. Examples include watching TV (more than two hours a day) and excessive media use such as texting and surfing the Internet.

- *Excess caloric intake.* When energy intake exceeds energy expenditure, weight gain occurs.

- *Age.* Obesity and overweight problems increase for many as they age because of the difficulty of achieving caloric balance while dealing with the physical and mental challenges of normal aging. People who are homebound or living alone may be especially susceptible to weight gain.

- *Heredity.* Genes can affect factors such as metabolism, which can predispose a person to obesity or overweight, but research shows that people can adjust their lifestyles and achieve caloric balance and a healthy weight despite hereditary challenges.

- *Race and ethnicity.* Different race and ethnic groups have different prevalences of obesity. However, scientific studies have not revealed whether the differences are due to genetic determinants or to social or cultural disparities.

- *Socioeconomic status (SES).* SES can provide major challenges to acquiring a healthy diet and to finding safe, affordable opportunities for engaging in physical activity or exercise.

- *Culture.* Cultural traditions (e.g., the built environment) and behaviors can provide a variety of challenges to making positive lifestyle changes to achieve caloric balance and a healthy weight.

- *Metabolism.* A person's resting metabolism (i.e., the rate of caloric expenditure at rest) can be affected by a variety of factors including genetics, physical activity, diet, age, and medications.

OBESITY AND OVERWEIGHT CHALLENGES

Physical activity and exercise can affect energy balance related to common obesity and overweight challenges such as weight loss, healthy weight maintenance, the prevention of weight regain, and excessive abdominal fat. Clinically significant **weight loss** has been defined, based on the scientific literature, as at least a 5% loss of body weight (USDHHS, PAGAC 2008). **Weight maintenance** (or *weight stability*) has been defined as a weight change of less than 3%, and prevention of **weight regain** after a substantial loss is consistent with a change in weight of 3% to less than 5%.

Fat is stored throughout the body. It can be found around the organs (visceral fat) or near the skin (subcutaneous fat). Some people seem to store fat preferentially in one location over others. Men, for example, are more likely to store fat in the abdominal region, whereas women are more likely to store fat in the hips, buttocks, and legs. Increased abdominal fat (also called *male pattern fat distribution*) is associated with metabolic disorders including metabolic syndrome (see chapter 5 for more on metabolic syndrome and physical activity) more so than female pattern fat distribution. Research shows that abdominal fat loss is associated with increased levels of physical activity and is proportional to overall fat loss.

Engaging in regular physical activity and exercise, in and of itself, is not a panacea for losing weight, maintaining a healthy weight, preventing weight regain, or decreasing abdominal fat. As described earlier, many other factors influence caloric balance besides physical activity and exercise. Four points to address to help people achieve energy balance are (1) the concept of total energy intake; (2) intervening with physical activity and exercise alone, diet only, or a combination of diet and physical activity and exercise; (3) how nonactive leisure time activities compete with active pursuits; and (4) the benefits of engaging in regular physical activity and exercise other than just preventing obesity and overweight.

Excessive energy intake has become a way of life because of the availability of inexpensive, high-calorie foods that taste good. The trend of overconsumption correlates highly with the prevalence of obesity and overweight; therefore, all weight management strategies should address diet. Numerous excellent resources that focus on healthy eating can be found at www.nutrition.gov.

Physical activity and exercise are associated with caloric expenditure, but these strategies used alone for weight loss, weight maintenance, or the preven-

tion of weight regain are not as effective as when they are combined with diet (energy intake) interventions. Figure 6.7 illustrates weight loss related to a diet intervention (caloric reduction), an exercise intervention, and a diet plus exercise intervention. As you can see, physical activity and exercise are not sufficient for providing clinically significant weight loss. Figure 6.8 illustrates the differences in BMI values between active and less active people; it suggests a dose-response relationship between physical activity and exercise and BMI.

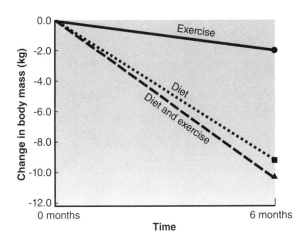

Figure 6.7 Weight loss related to a diet intervention, an exercise intervention, and a diet plus exercise intervention.
Reprinted from USDHHS and PAGAC (2008, p. G4-7); adapted from Wing, RR. 1999. Physical activity in the treatment of the adulthood overweight and obesity: current evidence and research issues. *Medicine and Science in Sports and Exercise* 31 (suppl 11): S547-S552.

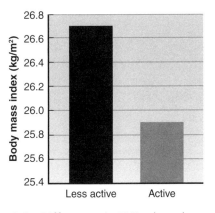

Figure 6.8 Differences in BMI values due to level of physical activity.
Reprinted from USDHHS and PAGAC (2008, p. G4-7); adapted from Kavouras SA, Panagiotakos DB, Pitsavos C, Chrysohoou C, Anastasiou CA, Lentzas Y, Stefanadis C. 2007. Physical activity, obesity status, and glycemic control: The ATTICA study. *Medicine and Science in Sports and Exercise* 39 (4):606-611.

COUNSELING FOR WEIGHT MANAGEMENT

Efforts to increase caloric expenditure with physical activity and exercise can be impeded with competition from all of today's labor-saving attractions (computers, cell phones, and other electronic media) that impede active discretionary time activities. Strategies to help people optimize the leisure time they spend being active can be helpful. Good advice would be: Choose the active choice!

KINESIOLOGY AND BODY WEIGHT

Children, adolescents, and adults can experience the positive exercise-related adaptations associated with regular participation in physical activity and exercise that can help with weight loss, healthy weight maintenance, the prevention of weight regain, and the loss of excessive abdominal fat. The combined benefits of physical activity and exercise listed in the highlight box Adoptations to Fitness Programming Related to Body Composition can help various populations improve their health and quality of life.

Physiologically, participation in regular physical activity can help people lose weight and achieve a healthy weight, or get closer to achieving caloric balance. These basic changes can lead to increases in muscular endurance and $\dot{V}O_2$ max levels, which result in improved functional health (i.e., people can do more work before fatiguing).

Physical activity and exercise can also help people reduce total body fat, waist circumference, and intra-abdominal fat; maintain or increase lean muscle mass; and prevent the regaining of weight lost. Ultimately, participation in physical activity and exercise can lower the risks for such chronic disease processes as type 2 diabetes, metabolic syndrome, and CHD, as well as orthopedic challenges associated with obesity and overweight.

Although the associations are not well understood at this time, overweight and obesity are often linked with multiple factors that negatively affect restful sleep. For example, people who suffer from obstructive sleep apnea (OSA), restless leg syndrome (RLS), and insomnia and who become physically active experience reduced symptoms. Numerous studies have shown that physical activity and exercise are associated with increases in sleep duration and quality of sleep (especially delta sleep). These sleep improvements can help people get a good night's rest and be active and ready to go the next day without the usual chronic fatigue associated with problems such as OSA. The loss of upper abdominal fat, which can be achieved by becoming physically active, is also associated with reduced respiratory problems such as OSA.

Biomechanically, improved economy of movement can be expected with weight loss and the maintenance of a healthy weight. Improved economy of movement is a function of improved efficiency, which is directly related to body weight and body composition, particularly in weight-bearing activities. Losing weight or achieving weight stability can allow people to perform motor skills more efficiently as a result of increases in range of motion, which is limited by carrying excess weight or body fat. Improved biomechanics also inspires confidence to engage in future physical activity and exercise activities. Peripheral proprioception (i.e., sense of position and movement) response, which is associated with balance and fall prevention, often improves as well.

Weight loss and weight stability also have strong psychological effects. Experiencing the unique challenges of the phases of behavioral change (contemplation, preparation, action, maintenance, relapse) gives people coping skills and strategies for achieving and maintaining a healthy weight. In addition, people who have lost weight or succeeded at maintaining a healthy weight experience more self-confidence, greater feelings of self-efficacy, and lower levels of depression and anxiety.

COMMON ASSESSMENTS OF OBESITY AND OVERWEIGHT

Common assessments of obesity and overweight provide estimated body composition values such as the percentage of body fat, lean muscle mass, and

ADAPTATIONS TO FITNESS PROGRAMMING RELATED TO BODY COMPOSITION

Physiological
- Increased muscular endurance
- Increased $\dot{V}O_2$max
- Improved caloric balance
- Improved metabolism
- Lower percentage of body fat
- Smaller waist circumference
- Less intra-abdominal fat
- Maintenance of or increased lean muscle mass
- Maintenance of or loss of weight

Biomechanical
- Improved economy
- Improved balance
- Improved mobility
- Improved proprioception

Behavioral
- Increased self-confidence
- Improved self-efficacy
- Decreased depression and anxiety
- Increased motor skill
- Experience with behavioral change and increased confidence to engage further in physical activity and exercise

total body fat. Everyone needs some fat to maintain normal bodily functions. **Essential fat** is important for stored energy, cushioning and insulation, and vitamin absorption; it is found in and around the nervous system, heart, lungs, kidneys, spleen, intestines, and muscles. The minimal amount of essential fat for men has been estimated to be 3% of body weight; for women the estimate is around 12% of body weight. When essential body fat falls too low, health risks for chronic disease and adverse immune reactions increase.

Ideal body fat and **ideal body weight** are terms that have been used to describe the hypothetical optimal percentage of body fat or body weight. The values for ideal body fat and ideal body weight, however, are highly variable and should be based on factors such as age, sex, personal goals, behaviors, and appropriate educational messaging that does not promote addictive disorders (i.e., eating or exercise). However, most men are considered obese if they are carrying ≥28% body fat, and most women, if they are carrying ≥32% (Jackson and Ross 1997).

The simplest way to assess body composition is visual inspection. If a person's BMI (discussed earlier) is already higher than normal and he or she is gaining weight and not participating in regular physical activity and exercise, he or she is probably carrying too much body fat. Although many of us have noticed visually that we were gaining weight

and were therefore motivated to try to lose weight, we did not have quantitative data (e.g., percentage of body fat, amount of lean muscle mass) to help us set goals and gauge our success. Body composition assessments, described next, provide the quantitative information necessary for setting weight management goals and evaluating program outcomes.

The highlight box Common Methods for Measuring Body Composition lists laboratory-based and field tests that exercise scientists use to measure body composition. See figure 6.9 for photos illustrating skinfold testing and underwater weighing.

Magnetic resonance imaging (MRI) and **computed tomography (CT)** are currently recognized as the gold standard (or best available with minimal error) techniques for measuring body composition. However, because of the expense of these techniques (about $1,000 USD per scan, or $1,000,000 USD for the purchase of a scanning device), they are mainly used in research and as medical diagnostic tools. Both of these techniques rely on X-ray technology to quantify the amount of fat tissue and other tissue in the body or in a region of the body. MRI is thought to be a safer technique largely because it does not rely on ionizing radiation.

Dual-energy X-ray absorptiometry (DXA) full body scans are more affordable than MRIs and CTs, but the device can be expensive. Many exercise physiologists now regularly use DXA in the

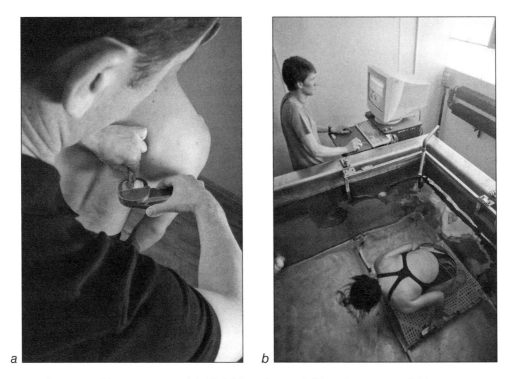

Figure 6.9 Body composition measures: *(a)* skinfold testing and *(b)* underwater weighing.

assessment of body composition. Scanning is done while a person is still and supine on a table; X-ray beams are emitted and data are differentiated into fat mass, fat-free mass, and skeletal (bone) mass in a two-dimensional display. The DXA technique can provide precise data about a person's percentage of body fat, as well as bone mineral density data. The DXA technique usually has an error of about 1% compared to MRI and CT measures for full body scanning, but it can be less accurate if the scan does not include the whole body.

Underwater (or hydrostatic) weighing is based on the principle of water displacement (i.e., when you get in tub of water, the water level rises based on the volume your body displaces). In this technique, a person's weight is measured both in and out of water. A person with more fat (which is less dense than lean body tissue) will be buoyed up more than a leaner person will, and consequently will weigh less underwater. To minimize measurement errors, hydrostatic weighing requires motivated subjects and additional laboratory equipment that can be used to measure residual lung volume (RV). If RV is estimated (as is often done) and not measured, large errors in measurement can result. A hydrostatic tank system costs about $10,000 to $15,000 (USD) and requires regular maintenance.

Hydrostatic weighing used to be considered the gold standard for body composition measurements and has been traditionally used, but the technique has a measurement error of 2 to 3% compared to MRI, CT, and DXA.

Air displacement plethysmography (i.e., the measurement of change in volume) to determine volume and density has become a popular technique used at universities and by many sport teams. A common tool for this technique is the commercially available BOD POD. This technique, based on the same displacement principles as hydrostatic weighing, uses multiple sensors in the measurement unit to measure air displacement in a known period of time. Body fat is then calculated based on those data. Air displacement plethysmography is relatively easy to use, but the devices are expensive (about $35,000 USD) and not readily available. Compared to the MRI, CT, and DXA techniques, this technique has about a 3% measurement error.

Skinfold measurement has been used by exercise scientists and clinicians for many years to estimate body composition based on population-specific and generalized equations. Measuring skinfold thickness at various sites on the body (e.g., the tricep, abdomen, and thigh) provides an estimate of subcutaneous fat (about 50% of total body fat) and therefore an estimate of body density and the percentage of body fat. Using the equations developed by Pollock and Jackson (1978, 1980), the percentage of body fat can be estimated from the sum of skinfold thickness, age, and sex. Skinfold calculators can be found at websites such as www. exrx.net. Skinfold measurements correlate well with hydrostatic weighing measures, but have at least a 3% error rate.

Bioelectrical impedance analysis (BIA) sends a low-amperage electrical current through surface electrodes on the body (e.g., the wrist and ankle). Measurements of the resistance to the current permit the estimation of body composition using prediction equations. Tissues of different densities conduct electricity at different rates, and thus body composition can be estimated. The BIA technique can be highly variable based on the following factors:

- Quality of the BIA instrument used
- Fluid balance (normal hydration or dehydration) of the subject
- Recent food consumption
- Effects of recent bouts of physical activity and exercise

The cost of BIA instruments ranges from $100 to several thousand dollars (USD); the more expensive models provide better measures when subjects' hydration levels are controlled (about 3% error rate). Some simple BIA instruments (e.g., bathroom scales) predict the percentage of body fat based on BMI, but they have about a 6% error rate.

Waist circumference measures have become common ways to determine when people are carrying too much abdominal fat for good health. For waist circumference, it is generally recommended to measure girth at the level of the lowest rib or umbilicus level using a cloth measuring tape with a spring-loaded handle that costs about $15 to $20 USD. Waist circumferences are measured in inches or centimeters. Adults should have waist measures of ≤40 inches (102 cm) for men, and ≤35 inches (89 cm) for women (Kenney, Wimore, and Costill 2012).

Waist circumference values for children and adolescents (ages 2 to 18 years) have been published (Fernandez et al. 2004) and are expressed as percentile ratings by age, sex, and ethnicity (African American, European American, and Mexican

COMMON METHODS FOR MEASURING BODY COMPOSITION

- Visual inspection
- BMI
- Magnetic resonance imaging (MRI); computed tomography (CT); dual-energy X-ray absorptiometry
- Hydrostatic weighing
- Air plethysmography
- Skinfold measurement
- Bioelectrical impedance analysis (BIA)
- Circumferences (waist and hip)

American). Youth can be classified by waist girth as being in the 10th through 90th percentile. Although there is no professional consensus on an ideal waist circumference measure for youth because of growth and development issues, it is reasonable to encourage youth to maintain a girth close to the 50th percentile.

Waist circumferences can be used in combination with youth BMI measures to clarify higher-than-expected BMI measures (e.g., in athletic youth with high levels of muscle mass). For example, if a youngster has a high BMI and a high waist measurement (>90th percentile), she should not be encouraged to gain more weight. A youngster with a low BMI (<5th percentile for age and sex) and low waist measurement should not be encouraged to lose more weight.

The **waist-to-hip ratio (WHR)** is another simple way to evaluate the distribution of body fat in adults. Health risk increases as WHR increases, and standards vary by age and sex. According to the guidelines of the American College of Sports Medicine (ACSM; 2010), young adult men and women should have ratios of lower than 0.95 and 0.86, respectively, to be categorized as having a low health risk. Older adult men and women (ages 60 to 69 years) should have ratios of lower than 1.03 and 0.90, respectively, to be classified as having a low health risk.

PHYSICAL ACTIVITY GUIDELINES FOR A HEALTHY WEIGHT

For many adults, obesity is associated with significant increases in abdominal fat that increase the risk for metabolic syndrome and type 2 diabetes (see chapter 5 for more information). Regular participation in aerobic physical activity and exercise can decrease total body fat and abdominal fat, and these changes are consistent with improved metabolic function. Generally, the greater the volume of physical activity or exercise acquired by individuals or populations, the greater the reductions in body and abdominal fat.

For children and adolescents, the prevention of excessive weight gain during maturation is critical to prevent obesity and overweight in adulthood. Unfortunately, recent public health surveys note that the parents of these at-risk children are not aware of the problem or do not recognize that their children are overweight. Recent U.S. national policy

statements targeting pediatricians and other health care providers assert that children and adolescents who are overweight or obese in their early teens should be identified as early as possible and given some sort of weight management plan through the intervention of parents, teachers, coaches, and family physicians.

One of the most obvious influences on weight management is how people perceive their ideal weight or physique. Research findings indicate a great disparity between reasonable weight loss or weight gain goals and people's "dream weight." Even though some people may not reach their goal weight, they will likely report positive physical, social, and psychological benefits from weight loss. The following scientific evidence and guidelines can help individuals and populations establish realistic goals for weight management (e.g., achieving a healthy weight) and meet their specific needs (e.g., weight loss, weight stability, prevention of weight regain, or loss of excessive abdominal fat) through regular participation in physical activity and exercise.

SCIENTIFIC EVIDENCE

The PAGAC (USDHHS, PAGAC 2008) noted strong scientific evidence that supports a consistent effect of aerobic physical activity and exercise on weight maintenance (otherwise known as *weight stability* and defined as a <3% change in weight) for adults. Evidence that resistance training helps with weight maintenance is not as strong, given that resistance training can increase lean muscle mass, and the volume of resistance training regimes is usually less than that for aerobic training.

Although scientific evidence supports the utility of physical activity in weight maintenance, the evidence for the specific amount necessary or for a dose-response relationship between physical activity and exercise and weight maintenance is lacking. However, the evidence that is currently available suggests that a volume of aerobic activity between 13 and 26 MET-hours (see chapter 2 for more information on METs) is associated with weight stability. Thirteen MET-hours is approximately equivalent to walking at 4 miles per hour (6.4 km/h) for 150 minutes per week or jogging at 6 miles per hour (9.7 km/h) for 75 minutes per week.

Can a walking program be effective for weight management without also controlling diet?

Strong evidence suggests that the accumulation of energy expenditure (total volume of physical activity) is a key factor in achieving energy balance. Details such as how that total volume of energy expenditure is attained (all at once or in multiple, shorter bouts) are less clear.

A person intending to lose weight solely through physical activity or exercise (without dietary changes) would have to expend a large volume of energy to achieve even a 5% weight loss. However, if that person maintains an isocaloric diet from baseline (i.e., calorie intake matches calorie expenditure) and maintains a physical activity or exercise intervention, the evidence is strong that weight loss can exceed 5%.

Somewhat surprisingly, the relation between weight maintenance and physical activity appears to be somewhat different than the situation with weight loss and physical activity. Assuming appropriate caloric restriction, the evidence for a consistent dose-response relationship between aerobic physical activity and exercise and weight loss is strong. Physical activity and exercise volumes of 26 MET-hours per week are associated with weight loss (moderate evidence). This relatively large volume of physical activity and exercise is the equivalent of walking 45 minutes daily at 4 miles per hour (6.4 km/h), or about 70 minutes daily at 3 miles per hour (4.8 km/h), or jogging 22 minutes daily at 6 miles per hour (9.7 km/h). The accumulation of physical activity and exercise can have positive effects on weight loss, but accumulating large volumes of physical activity without concentrated bouts can be difficult.

Yet another consideration is the relation between physical activity and prevention of weight gain after a substantial amount of weight has been lost. The evidence of an effect of aerobic physical activity and exercise on weight maintenance after weight loss (or the prevention of weight regain) is moderate at best. Moderate evidence also suggests a dose-response effect of larger volumes of physical activity and exercise on weight regain after weight loss.

Physical activity and exercise volumes of 30 MET-hours or more per week are associated with the prevention of weight regain (moderate evidence); this is the equivalent of walking about 50 minutes daily at 4 miles per hour (6.4 km/h), or about 80 minutes daily at 3 miles per hour (4.8 km/h), or jogging 25 minutes daily at 6 miles per hour (9.7

km/h). The accumulation of physical activity and exercise can have positive effects on preventing weight regain after weight loss, but again, accumulating large volumes of physical activity without concentrated bouts is difficult.

Decreases in abdominal fat and intra-abdominal fat are consistently associated with aerobic physical activity and exercise, but such decreases are less well documented for resistance training. Although many people claim that it is possible to spot-reduce in certain parts of the body, only moderate scientific evidence supports this idea. Physical activity may be able to reduce abdominal obesity.

Physical activity and exercise volumes of 13 to 26 MET-hours per week are associated with decreases in abdominal fat and can improve metabolic function. The equivalent of 13 MET-hours is approximately equivalent to walking at 4 miles per hour (6.4 km/h) for 150 minutes per week or jogging at 6 miles per hour (9.7 km/h) for 75 minutes per week. Larger volumes of physical activity or exercise, such as 42 MET-hours per week, can result in three to four times more abdominal fat loss than seen at 13 to 26 MET-hours per week. The equivalent of 42 MET-hours is approximately equivalent to engaging in well over 300 minutes of physical activity and exercise per week. For example, a person who walks at 4 miles per hour (6.4 km/h) would require about 375 minutes per week to achieve this threshold. The effects of an accumulation of physical activity and exercise on abdominal obesity have not been tested.

Few studies exist on the effects of physical activity and exercise on people of different sexes, ages, or races or ethnicities. Thus, the evidence is not currently sufficient to recommend different regimens for physical activity or exercise based on sex, age, race or ethnicity, or socioeconomic status.

As mentioned earlier in the chapter, regular physical activity and exercise are also important to prevent and control obesity and overweight in children and adolescents. The specific scientific evidence of the positive effects of physical activity and exercise on youth body composition is strong and is consistent with acquiring at least 30 minutes of regular moderate- or vigorous-intensity physical activity per day. See chapter 12 for more physical activity and exercise recommendations and interventions for youth.

KEY LEADER PROFILE

Loretta DiPietro, PhD, MPH

Courtesy of Loretta DiPietro.

Why and how did you get into this field?

In 1983, I was convinced that regular exercise was the key to longevity (at least it was in my family tree). I was interested in whether this theory applied to all families or just to mine. Exercise epidemiology was not considered an area of public health research at the time—at least that is what I was told on my first day at Yale. Moreover, the belief was that any possible relation between exercise and health was attributable to heredity. So, those negative attitudes toward my interests made me even more determined to succeed in this line of work. Just after I graduated in 1988, a paper by Blair and colleagues was published, which detailed the inverse relation between cardiorespiratory fitness and all-cause mortality. I felt so vindicated and excited to be involved in this new area of research. I spent three postdoctoral years doing epidemiology, physiology, and then surveillance work in order to refine my skills and interests.

Did any one person have an overriding influence on you?

It took a village to train me. The three people with a major influence on me were Adrian Ostfeld, Albert Stunkard, and Ethan R. Nadel. They all colored outside the lines and refused to engage in safe science. It was a privilege and a thrill to have worked with these men. Then there were countless senior scientists who opened doors for me in ways that are invaluable to the career of a young investigator: Steve Blair, Ralph Paffenbarger, Ken Powell, Bill Haskell, and Carl Caspersen. These are the people who invited me to speak at meetings, write review papers, and included me on papers or grant applications. The relation of these single and collective opportunities to a young person's trajectory of success cannot be matched, and I am so grateful to these gentlemen for their interest in my growth. Finally, I continue to rely on a battalion of contemporary colleagues for their greater intelligence and ability to argue with me. You know who you are.

What are your current research interests?

My current research interests focus on two lines of work: the role of exercise and physical activity on metabolic flexibility and resiliency in aging, and physical activity and the trajectory of weight gain in middle age or of weight loss in older age.

What drives you as a researcher and activist?

The fact that we have lost a second generation of children to the preventable health risks of obesity is a key driving force. Public health research must be action oriented, and the results of research in this area are immediately applicable for improving health. Second, easily enacted and evidence-based environmental strategies to promote physical activity are underused and underappreciated for their effectiveness. These are not intrusive on personal choice; rather, they are helping to make the active choice the easy choice.

What are one or two key issues to be addressed by 2022?

Here are the two key research issues that must be addressed in the upcoming decade: How effective for health promotion and disease prevention among adults are multiple short activity breaks throughout the day? What strategies are effective individually and collectively in reducing the incidence and prevalence of insulin resistance in children?

GUIDELINES

The physical activity guidelines for youth and adults regarding weight control are the same as for other health outcomes with one notable exception. The guidelines recommend that youth participate in 60 minutes of moderate- or vigorous-intensity physical activity daily to maintain a healthy weight. Adults (and older adults) who are trying to control their weight should include dietary considerations, and they should follow the 2010 U.S. *Dietary Guidelines for Americans* for more about weight management and how to determine a healthy weight. Many adults who meet the minimal physical activity and exercise guidelines of acquiring 150 minutes of moderate-intensity physical activity, 75 minutes of vigorous-intensity physical activity, or a combination of both, and two days of muscle-strengthening activities will achieve caloric balance and lose weight.

Given the effect of body weight on both energy expenditure and energy intake, the minimal physical activity guidelines may not be enough in some cases to create an energy deficit or balance. Thus, for weight control, 150 minutes per week of moderate-intensity physical activity, 75 minutes per week of vigorous-intensity physical activity, or a combination of both, is a good start. If this does not help with weight loss or prevent weight gain, an increase in physical activity may be necessary.

Older adults may have physical disabilities that limit the intensity at which they can engage in physical activity and exercise, making weight loss more of a challenge than it is for younger adults. The physical activity guidelines stress that older adults (as well as youth and adults 18 to 65) should remember that caloric expenditure results from engaging in all types of physical activities, and not just exercise. For example, taking short walks throughout the day and making active choices such as taking the stairs instead of the elevator expend calories and can be helpful in weight control versus remaining sedentary. Older adults who can engage in vigorous physical activity or exercise may find that this strategy is more effective (time efficient) for weight control than working at lower-intensity activities for more time. In any case, older adults should try to participate in regular physical activity and exercise that is sustainable and safe.

CHAPTER WRAP-UP

What You Need to Know

- A basic understanding of energy expenditure and caloric balance is essential for a more detailed understanding of the relationships between exercise and physical activity, and overweight and obesity.
- BMI is a common measure to classify adults and youth as obese, overweight, normal weight, or underweight.
- The prevalence of obesity and overweight is at pandemic levels globally and in the United States for youth and adults.
- By 2030, if obesity prevalence levels in the United States do not change, the economic costs will reach 15.8 to 17.6% of total U.S. health care costs.
- The health consequences of obesity include CHD, type 2 diabetes, cancer, hypertension, dyslipidemia, stroke, liver and gallbladder disease, sleep apnea and respiratory problems, osteoarthritis, and gynecological problems.
- The environment at home, school, work, and in the community provides numerous barriers and incentives to achieve caloric balance and maintain a healthy weight. Genetics may increase the predisposition for weight gain or obesity, but a healthy lifestyle can help people achieve caloric balance.
- Physical activity and exercise are associated with caloric expenditure, but using these strategies alone for weight loss, healthy weight maintenance, or the prevention of weight regain is not as effective as combining them with diet (i.e., energy intake) interventions.

- The physiological, biomechanical, and psychological benefits of participating in physical activity and exercise on obesity and overweight are numerous and achievable for most people.
- Common laboratory-based and field tests used to measure body composition are visual inspection, BMI, MRI, CT, DXA, hydrostatic weighing, air plethysmography, skinfold measurement, BIA, waist circumference, and waist-to-hip ratios.
- Engaging in regular moderate- to vigorous-intensity physical activity for 150 minutes per week can help many adults achieve energy balance, weight loss, weight maintenance, and the prevention of weight regain after weight loss, as well as reduce abdominal fat. However, some people may require more activity (≥300 minutes).
- Positive changes in youth body composition have been associated with at least 30 minutes of regular moderate- to vigorous-intensity physical activity per day.
- Everyone should try to follow the recommendations in the *2008 Physical Activity Guidelines for Americans* to achieve and maintain a healthy weight.

Key Terms

overweight

obesity

body mass index (BMI)

body composition

weight loss

weight maintenance (weight stability)

weight regain

essential fat

ideal body fat

ideal body weight

magnetic resonance imaging (MRI)

computed tomography (CT)

dual-energy X-ray absorptiometry (DXA)

underwater (hydrostatic) weighing

air displacement plethysmography

skinfold measurement

bioelectrical impedance analysis (BIA)

waist circumference

waist-to-hip ratio (WHR)

Study Questions

1. How can we define the terms *obesity* and *overweight* for adults using BMI?

2. How can we define the terms *obesity* and *overweight* for youth using BMI?

3. What is the current U.S. prevalence of obesity or overweight in adults?

4. What is the current U.S. prevalence of obesity or overweight for children and adolescents?

5. What are five common consequences of obesity?

6. What are six common tests of body composition to help people with weight control?

7. What are five common body composition related adaptations to fitness programming?

8. How can obesity and overweight be controlled by regular participation in physical activity and exercise?

9. How much physical activity and exercise is required for weight loss or weight maintenance?

10. How much physical activity and exercise is needed to prevent weight regain after weight loss?

E-Media

Explore issues related to physical activity, exercise, and public health at the following websites:

Human Kinetics	www.HumanKinetics.com
U.S. Department of Health and Human Services: Physical Activity Guidelines for Americans	www.health.gov/PAGuidelines
International Society for Physical Activity and Health	www.ispah.org
American College of Sports Medicine	www.acsm.org
National Strength and Conditioning Association	www.nsca-lift.org
President's Council on Fitness, Sports & Nutrition	www.fitness.gov
U.S. Centers for Disease Control and Prevention Behavioral Risk Factor Surveillance System (BRFSS)	www.cdc.gov/brfss
U.S. Centers for Disease Control and Prevention National Health and Examination Survey	www.cdc.gov/nchs/nhanes.htm
U.S. Department of Agriculture: Choose My Plate	www.choosemyplate.gov
U.S. National Agricultural Library	www.nutrition.gov

Bibliography

American College of Sports Medicine. 2010. *ACSM's Guidelines for Exercise Testing and Prescription*, 8th ed. Philadelphia: Lippincott Williams & Wilkins.

Centers for Disease Control and Prevention. 2011. Data from the National Health and Nutrition Examination Survey, 2005–2008. Available at www.cdc.gov/obesity/data/adult.html. Accessed on 9/13/11.

Chenoweth D, Leutzinger J. 2006. The economic costs of physical inactivity and excess weight in American adults. *Journal of Physical Activity and Health* 3: 148-163.

Fernandez JR, Redden DT, Pietrobelli A, Alliso, DB. 2004. Waist circumference percentiles in nationally representative sample of African-American, European-American, and Mexican-American children and adolescents. *Journal of Pediatrics* 145: 439-444.

Finkelstein E, Fiebelkorn I, Wang G. 2003. National medical expenditures attributable to overweight and obesity: How much and who's paying? *Health Affairs* 3: 219-226.

Flegal KM, Carroll, MD Ogden, CL Curtin LR. 2010. Prevalence and trends in obesity among US adults, 1999-2008. *Journal of the American Medical Association* 303(3):235-241.

Fleck, SJ, Kraemer, WJ 2004. *Designing Resistance Training Programs,* 3rd ed., Champaign, IL: Human Kinetics Publishing.

Jackson AS, Pollock, ML. 1978. Generalized equations for predicting body density of men. *British Journal of Nutrition* 40:497-504.

Jackson AS, Pollock, ML, Ward A. 1980 Generalized equations for predicting body density of women. *Medicine and Science in Sports and Exercise* 12:175-182.

Kavouras SA, Panagiotakos DB, Pitsavous C, Chrysohoou C, Anastasiou CA, Lentzas Y, Steanadis C. 2007. Physical activity, obesity status, and glycemic control: The ATTICA study. *Medicine & Science in Sports & Exercise* 39: 606-611.

Kenney, WL., Wilmore JH, Costill DL, 2012. *Physiology of Sport and Exercise* 5th ed. Champaign, IL: Human Kinetics.

Morbidity and Mortality Weekly Report (MMWR). 2009. Differences in prevalence of obesity among Black, White, and Hispanic adults— United States, 2006-2008. *Morbidity and Mortality Weekly Report (MMWR)* 58: 740-744.

U.S. Department of Health and Human Services. 2001. *The Surgeon General's call to action to prevent and decrease overweight and obesity.* Rockville, MD: U.S. Department of Health and Human Services, Public Health Service, Office of the Surgeon General.

U.S. Department of Health and Human Services. 2008. *2008 Physical Activity Guidelines for Americans.* Washington, DC: U.S. Department of Health and Human Services. www.health. gov/PAGuidelines.

U.S. Department of Health and Human Services, Centers for Disease Control and Prevention. 2011. *Behavioral Risk Factor Surveillance System (BRFSS)*. http://www.cdc.gov/brfss/. Accessed 12 August 2011.

U.S. Department of Health and Human Services, Centers for Disease Control and Prevention. 2011. *National Health and Nutrition Examina-tion Survey (NHANES)*. http://www.cdc.gov/ nchs/nhanes.htm. Accessed 12 August 2011.

U.S. Department of Health and Human Service, Office of Disease Prevention and Health Promotion. 2011. *Healthy People 2020.* www. healthypeople.gov/2020/. Accessed 12 August 2011.

U.S. Department of Health and Human Services, Physical Activity Guidelines Advisory Committee. 2008. *Physical Activity Guidelines Advisory Committee Report, 2008.* Washington, DC: U.S. Department of Health and Human Services. www.health.gov/PAGuidelines.

U.S. Department of Health and Human Services, U.S. Department of Agriculture. 2010. *Dietary Guidelines for Americans, 2010,* 7th ed. Washington, DC: U.S. Government Printing Office.

Wang Y, Beydoun MA, Liang L, Caballero B, Kumanyika SK. 2008. Will all Americans become overweight or obese? Estimating the progression and cost of the U.S. obesity epidemic. *Obesity* 16: 2323-2330.

World Health Organization. 2011. www.who.int/ en. Accessed 23 June 2011

PHYSICAL ACTIVITY IN PUBLIC HEALTH SPECIALIST

This chapter covers these competency areas as set forth by the National Society of Physical Activity Practitioners in Public Health:

2.2.3, 2.3.2, 2.5.1, 6.2.1, 6.2.2, 6.2.3, 6.2.4, 6.4, 6.4.1, 6.4.2, 6.4.3, 6.4.4, 6.5.2

CHAPTER 7

MUSCULOSKELETAL AND FUNCTIONAL HEALTH

OBJECTIVES

After completing this chapter, you should be able to discuss the following:

» The prevalence of musculoskeletal disorders and related health challenges

» The physical challenges associated with low levels of musculoskeletal fitness, as well as common testing methodologies to assess musculoskeletal fitness

» How physical activity affects bone, joint, muscle mass, and muscle function in relationship to musculoskeletal health

» How physical activity can influence functional and role ability

» How exercise adaptations positively influence functional health

» The physical activity guidelines for promoting physical activity related to functional health, and the evidence behind them

OPENING QUESTIONS

» Can exercise improve the strength and quality of bone tissue?

» Why is muscle quantity (muscle mass) and muscle quality (muscle function) important for good health?

» How much physical activity is needed for good musculoskeletal health?

» What is functional health? How can functional health be improved to reduce the risks for physical disabilities?

Have you ever watched a grandparent, aunt, or other older relative struggle to rise from a sitting to a standing position because of a lack of upper body strength? Have you ever watched (or been) a child who was not selected for an athletic team or pickup game for lack of strength? Do you want to be able to enjoy life, be as independent as possible, and do all the things you want to without physical limitations into old age? Each of these questions directly relates to the human musculoskeletal system (i.e., bones, muscles, joints, and connective tissue) and its ability to help us do the physical things we want to do.

Although much of the scientific and popular attention on physical activity and health has (appropriately) focused on how exercise helps the cardiovascular and metabolic systems (see chapter 5), we are learning that the benefits to the musculoskeletal system are just as important. Scientific evidence clearly supports the importance of maintaining physical activity as we age to optimize physiological capacity, minimize physiological limitations, maintain functional and role ability, and reduce the risks of falls (U.S. Department of Health and Human Services [USDHHS], Physical Activity Guidelines Advisory Committee [PAGAC] 2008). This chapter introduces these topics and presents the scientific rationale for engaging in physical activity for musculoskeletal health.

PREVALENCE OF MUSCULOSKELETAL DISORDERS AND RELATED HEALTH CHALLENGES

The most common musculoskeletal disorders that result from low physical activity levels are **osteoporosis** (bone health), **osteoarthritis** (joint health), and low levels of the quantity and quality of **muscle mass**. Osteoporosis (low bone mass and structural deterioration of bone tissue) is estimated to be a public health threat to 44 million Americans, or 55% of those over 50 years of age (National Osteoporosis Foundation 2010). Clearly established diagnostic criteria for osteoporosis exist: people with a bone mineral density (BMD) below the normal range have the disease. Women are much more likely than men to have osteoporosis. Of the 10 million people estimated to have osteoporosis in the United States in 2010, approximately 80% were women and 20% were men.

Why is osteoporosis a problem? Most important, osteoporosis and low **bone mineral density (BMD)** contribute to increased bone fracture risk including fractures of the hip, vertebral column, and wrist. The risk of falling is increasingly a problem as we age, and low BMD can accelerate the risk of fractures that are either spontaneous or secondary to a fall. In 2005, the National Osteoporosis Foundation reported that osteoporosis was related to more than 2 million fractures in the United States in the following categories:

• Hip: 297,000

• Vertebrae: 547,000

• Wrist: 397,000

• Pelvis: 135,000

• Other sites: 675,000

Due primarily to the aging of the U.S. population, the incidence of bone fractures resulting from osteoporosis and low BMD is expected to increase by a million more cases by 2025, which will contribute to lower quality of life, higher disease management costs, and increased mortality for those affected. The economic cost of osteoporosis and low BMD associated with osteoporosis-related fractures was $19 billion (USD) in 2005, and has been predicted to affect 3 million people at a cost of $25 billion by 2025.

Although there are several types of arthritis, osteoarthritis (OA) is the most common form of joint disease and is associated with joint pain and dysfunction along with irreversible loss of articular cartilage. Osteoarthritis is often thought of as a mechanical joint disease affecting primarily the weight-bearing joints (knees and hips). Clinically, OA symptoms include joint pain, swelling, stiffness, and weakness. The development of OA is associated with increased disability and lifestyle challenges.

Approximately 27 million U.S. adults (or ~12% of the U.S. population) are affected by this most common form of arthritis, and it is also the fifth leading cause of disability. To put this into perspective, California is the only state in the United States (according to the 2010 U.S. Census) with more than 27 million people! Women are affected with OA more than men are, especially related to the knee. By the year 2030, it has been predicted that 67 million people, or 25% of the U.S. population, will suffer from OA.

The economic costs of OA are difficult to estimate because it is often reported medically as a musculoskeletal condition, and not differentiated from other disease processes. Moreover, falls and fractures can occur in the absence of OA. Many of the economic costs of OA are work related (i.e., absenteeism and presenteeism, a term used in the workplace to define decreased on-the-job performance as a result of health problems). In 2007, U.S. costs were estimated at more than $185 billion a year including physician office visits, hospital and outpatient treatments, medication use, diagnostic tests, and other related medical treatments (Kotlarz et al. 2009). Obviously, for those who have OA and require arthroscopic surgery, joint replacement surgery, or reconstruction, the costs can accumulate rapidly.

RISK FACTORS ASSOCIATED WITH MUSCULOSKELETAL DISORDERS AND ASSOCIATED HEALTH CHALLENGES

Chapter 5 highlighted the major risk factors for cardiovascular disease (CVD) and diabetes. Although a complete review of all the risk factors associated with low levels of musculoskeletal health and poor

functional health is beyond the scope of this text, we will review here the risk factors associated with the musculoskeletal health challenges of osteoporosis, osteoarthritis, and low levels of the quantity and quality of muscle mass. As you might have guessed already, many, but not all, musculoskeletal health challenges are associated with sedentary lifestyles or low levels of physical activity.

OSTEOPOROSIS

According to the U.S. National Osteoporosis Foundation (2010), there is an extensive list of risk factors associated with the development of osteoporosis. As with CVD and diabetes mellitus, these factors can be grouped into modifiable and nonmodifiable factors.

Modifiable Risk Factors for Osteoporosis
- Physical inactivity
- Tobacco use
- Being thin or underweight
- Alcohol abuse
- Low sex hormone (estrogen or testosterone) levels
- Low calcium or vitamin D intake or absorption
- Excessive caffeine intake

Nonmodifiable Risk Factors for Osteoporosis
- Age
- Sex
- Heredity (genetics)
- Ethnicity or race
- History of fractures

Following are descriptions of the modifiable and nonmodifiable risk factors for osteoporosis:

• *Physical inactivity.* People who are physically inactive are less likely to have optimal bone mineral density (BMD) than their more active peers.

• *Tobacco use.* Smokers reduce their ability to absorb calcium, which is important for good bone health, and smoking also tends to interfere with estrogen and bone protection.

• *Being thin or underweight.* Being underweight is associated with small bone structure and less weight and force on bones, and may be associated with poor nutrient intake or eating disorders.

- *Age.* Advancing age (men >70 years, women >50 years and postmenopause) is associated with increased osteoporosis risk because of physiological changes associated with aging and bone loss.

- *Sex.* Women are at a higher risk for osteoporosis than men at an earlier age; however, men with lower testosterone or estrogen levels, or both, are also at increased risk.

- *Nutrition.* Low calcium and vitamin D intake and availability are risk factors for osteoporosis. Excessive alcohol and caffeine intake have also been implicated.

- *Heredity (genetics).* A family history of osteoporosis can increase the risk for developing the condition.

- *Ethnicity or race.* Evidence suggests that some groups have lower rates of osteoporosis than others (e.g., African Americans have lower rates than American Caucasians, Asians, or Latinos).

- *Low estrogen or testosterone levels.* Normal levels of sex hormones, by sex, (i.e., estrogen and testosterone) help protect bone; therefore, women who are postmenopausal, women with amenorrhea (i.e., are not menstruating regularly), and some men with lower testosterone levels may be at higher risk for osteoporosis.

- *History of fractures.* A history of fractures in adulthood may indicate that the person already has osteoporosis.

Osteoarthritis

Osteoarthritis (OA) is one of many forms of arthritis characterized by excessive (and lasting) stiffness and swelling in the joints. Unlike other forms of arthritis, OA is thought to result from abnormal stresses on the joints. According to the U.S. National Arthritis Foundation (2011), the primary risk factors associated with the development of OA are age, obesity, injury or overuse, genetics or heredity, and muscle weakness. Not surprisingly, some risk factors for OA can be modified, whereas others cannot.

Modifiable Risk Factors for OA

- Physical inactivity
- Excessive physical activity, or overuse
- Excess body mass

Nonmodifiable Risk Factors for OA

- Age
- Sex
- Heredity (genetics)
- History of joint injury

Following are descriptions of the modifiable and nonmodifiable risk factors for osteoarthritis:

- *Physical inactivity.* Physical inactivity is a two-edged sword when it comes to OA. Some physical activity is associated with a *lower* risk of OA, whereas too much provides excessive joint stress and may *increase* the risk. Physical activity helps increase muscular strength, which is important for supporting joints (e.g., strong quadriceps for the knee joint).

- *Excessive physical activity.* Too much physical activity, or a history of athletic overtraining (i.e., excessive exercise that produces negative effects) is associated with an increased risk of OA. Some sports and recreational activities have been found to be associated with the development of OA.

- *Excess body mass.* Overweight (i.e., high BMI) is associated with higher rates of OA. This is thought to also be related to joint overload.

- *Age.* Joints degenerate over time. This natural degeneration may be compounded by previous joint injury history, excessive physical activity or exercise, or both.

- *Sex.* Women are at a higher risk than men for most types of OA, which is probably due in part to lower levels of quadriceps strength on average as compared with males, which can influence OA development in the hips or knees.

- *Heredity (genetics).* A family history of OA may predispose a person to developing OA.

- *History of joint injury.* Previous joint injuries associated with weight-bearing movement (especially at the knee) increase the risk for developing OA.

- *Occupational load.* Some occupations require workers to carry heavy loads and this form of overuse may contribute to OA risk.

LOW MUSCLE MASS

Musculoskeletal health is not necessarily linked to a specific disease process. Low muscle mass (also known as *sarcopenia*) is thought to be an important determinant of functional health, particularly among older adults. Therefore, this section reviews selected factors that are associated with low levels of functional health, specifically falls, which correlate with low levels of muscle mass quantity and quality. According to the Physical Activity Guidelines Advisory Committee (PAGAC) (USDHHS, PAGAC 2008), the three primary risk factors associated with low levels of functional health are aerobic capacity, muscular strength, and balance. Following are the major modifiable and nonmodifiable risk factors for low muscle mass, as well as other potential risks.

Modifiable Risk Factors for Low Muscle Mass

- Physical inactivity
- Tobacco use

Nonmodifiable Risk Factors for Low Muscle Mass

- Age
- Sex
- Heredity (genetics)

Following are descriptions of the modifiable and nonmodifiable risk factors for low muscle mass:

- *Physical inactivity.* Physical inactivity is associated with lower muscle mass, particularly in aging, frail people. The "use it or lose it" principle is highly appropriate here.

- *Age.* Although aging is inevitable, the quantity and quality of muscle mass and aerobic capacity in most people can be maintained at high levels well past the age of 65, if they maintain a normal BMI (18.5 to 25), remain physical active, and do not smoke.

- *Sex.* Women, on average, have a lower muscle mass than men do, so it is even more important that they become physically active (in their youth) and remain so throughout their lives.

- *Genetics (heredity).* Functional health depends on genetics; however, practicing positive behaviors such as being physically active, maintaining a healthy weight, and minimizing risky activities can increase the quantity and quality of muscle mass.

KINESIOLOGY AND MUSCULOSKELETAL HEALTH

Musculoskeletal function has been repeatedly demonstrated to improve from engaging in activities such as strength and resistance training. Even with moderate-intensity physical activity, most people see improvements in strength (muscle mass quantity) and muscular endurance (muscle mass quality related to function). One does not necessarily need to lift weights to see musculoskeletal benefits; other activities such as working with resistance bands, doing calisthenics (push-ups, pull-ups, sit-ups), carrying heavy loads, climbing stairs, and heavy gardening (digging and hoeing) are also associated

STRENGTH TRAINING AND FRAIL OLDER ADULTS

Strength training was once thought to be too dangerous and unproductive for frail older adults. A landmark study by Dr. Maria Fiatarone and her colleagues at Tufts University (Fiatarone et al. 1994) was quick to disprove that notion. In this study, 100 nursing home residents who averaged 87.3 years of age were randomized to weight training or a nonexercise control for a 10-week period. Muscular strength and muscle mass were measured before and after the study. Not only did the training significantly improve muscular strength in those who did the strength training, but muscle mass increased (2.7%) as well. The outcomes were much better in the strength training group than in the group that received dietary supplements alone (nonexercise). This study helped change many preconceptions about strength training and older adults.

with positive changes in strength and muscular endurance. Older sedentary adults can see improvements of 50 to 100% in certain measures of strength and muscular endurance (e.g., hand grip and leg extension and flexion) in 8 to 12 weeks.

The highlight box Musculoskeletal Adaptations to Physical Activity and Exercise contains some of the long-term exercise-related adaptations acquired by engaging in muscle-strengthening activities, which can positively influence musculoskeletal health. The amount of physiological adaptation related to each of the benefits is dose dependent, meaning that low doses yield lower results and fewer changes than do higher doses. Chapter 2 discusses physical training principles in more detail.

Although cardiorespiratory adaptations (see chapter 5) to musculoskeletal strengthening activities are less than those seen with aerobic (cardiorespiratory) physical activities, sedentary people can increase their $\dot{V}O_2$max (or $\dot{V}O_2$peak) from 5 to 8% with resistance training. Those who engage in circuit training (a form of interval training using resistance training and aerobic exercise; see chapter 2 for more) often increase their $\dot{V}O_2$max (or $\dot{V}O_2$peak, ~5% change) (Gettman and Pollock 1981).

Increases in the rate of force (strength) development and power (time rate of work) are observed over time during physical activity and resistance training. Musculoskeletal strengthening results in the ability to recruit more motor units (the nerve and the fibers it controls), increased individual muscle fiber size (fast and slow twitch), increased numbers of anaerobic enzymes, and higher amounts of anaerobic energy stores. All of these factors are important for performing high-intensity, short-duration musculoskeletal activity.

Connective tissue changes associated with the specific muscles used during musculoskeletal strengthening activities may include increased ligament strength, tendon strength, and collagen content. Resistance training can also cause positive hormonal changes (e.g., up-regulation of anabolic hormone receptors) that allow for the remodeling of bone. Further, resistance activities can increase bone mass (or BMD) in some people, or may at least delay bone mass loss in people at risk for osteoporosis.

Exercise physiology research has clearly shown that regular participation in physical activity and

musculoskeletal strengthening increases lean muscle mass (quantity and quality) and reduces body fat, thereby helping with weight management. Figures 7.1 through 7.3 illustrate common changes in musculoskeletal fitness or function after physical activity or exercise training.

Biomechanically, improved economy or efficiency (i.e., reduced energy cost at a given workload) can be expected after musculoskeletal strengthening. For elderly people, this may mean simple yet important improvements in day-to-day activities. Balance and stability are often improved by engaging in resistance training activities. Improved mobility (flexibility and range of motion) and motor skills obtained through strengthening activities can give people the confidence to engage in future physical activity and exercise activities. Their peripheral proprioception (i.e., sense of position and movement) response often improves as well.

Behaviorally, people often feel better and have more self-confidence after several weeks of participation in physical activity that includes resistance activities. Some evidence suggests that people who participate in muscle-strengthening and resistance training activities experience

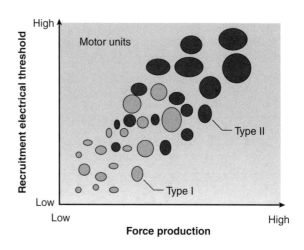

Figure 7.1 Motor units (i.e., the nerve plus the muscle fibers it controls) that include slow-twitch fibers (aerobic) and fast-twitch fibers (anaerobic) and the relationship between force production and motor unit recruitment. Physical activity and exercise improve the ability to increase the recruitment of motor units and therefore improve force production.

Reprinted, by permission, from National Strength and Conditioning Association, 2008. Adaptations to anaerobic training programs, by N.A. Ratamess. In *Essentials of strength training and conditioning*, 3rd ed. (Champaign, IL: Human Kinetics), 97.

MUSCULOSKELETAL ADAPTATIONS TO PHYSICAL ACTIVITY AND EXERCISE

Physiological

Increased muscular strength: Ability to move more weight

Increased muscular endurance: Ability to sustain muscular contractions over time

Increased $\dot{V}O_2$max: Increased aerobic capacity

Increased muscle force and power: Increased amount of force generated by the muscle when contracting, remaining static, or elongating

Increased muscle fiber size: Skeletal muscle **hypertrophy** can occur with or without measurable gains in strength.

Improved neural recruitment: Recruitment of more nerves to stimulate muscle contraction

Increases in anaerobic enzymes: Increases in compounds that can use fuel without the need for oxygen

Increased anaerobic energy stores: Increased ability to store fuel when needed for high power output

Improved hormone-mediated bone remodeling: Increases in bone matrix and mineral turnover

Improved connective tissue function: Improved ability of tissues to connect, bind, support, and anchor the body

No change, or increase, in BMD: Increase in bone quality

Maintenance of, or increase in, lean muscle mass: Increase in total muscle

Maintenance or loss of weight: Weight loss even with an increase in muscle tissue

Biomechanical

Improved economy: Ability to move more efficiently

Improved balance: Maintenance of even weight distribution to prevent falling

Improved mobility: Increased range of motion of muscles and connective tissue around joints

Increased motor skill function: Improved communication between the brain and muscles for smoother, more efficient operation

Improved proprioception: Improved ability of the body to sense movement and space; helps in movement and balance

Behavioral

Increased self-esteem and self-confidence: Improved satisfaction with or confidence in oneself, particularly related to physical activity and exercise

Improved self-efficacy: Improved attitudes, abilities, and cognitive skills, particularly related to physical activity and exercise

Decreased depression and anxiety: Less moodiness, loss of interest, despondency, nervousness, unease, and worry

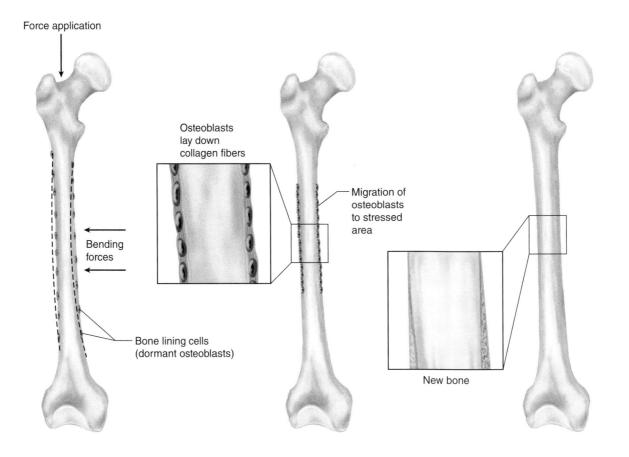

Figure 7.2 Bone remodeling, progressing from left to right, in response to mechanical loading such as that from participating in regular physical activity and exercise.
Reprinted, by permission, from National Strength and Conditioning Association, 2008. Adaptations to anaerobic training programs, by N.A. Ratamess. In *Essentials of strength training and conditioning,* 3rd ed. (Champaign, IL: Human Kinetics), 103.

lower levels of depression and anxiety, and they also probably experience higher self-efficacy (i.e., a sense of personal accomplishment and well-being) levels.

COMMON TESTS OF MUSCULOSKELETAL FITNESS OR FUNCTION

Many tests have been developed over the years to evaluate musculoskeletal fitness and function. Some are fairly easy to use, whereas others require a laboratory and medical personnel. Field tests, which can be done fairly easily and with large numbers of people, may not be as accurate as clinical ones. A familiarity with some of the common tests of musculoskeletal fitness or function (and the interpretation of their results) will help with the following:

- Establishing baseline levels of strength and muscular endurance
- Understanding the strengths and weaknesses of the scientific literature in this area
- Determining the extent of disabilities or limitations that can influence the ability to engage in physical activity and exercise
- Developing a plan for including regular musculoskeletal strengthening activities in an overall physical activity program
- Understanding the need to seek musculoskeletal rehabilitation or medical advice

The references and websites listed at the end of this chapter provide more comprehensive information on musculoskeletal fitness and function tests, and how to successfully administer and interpret them. This section provides a brief overview of the most common muscle and bone assessments.

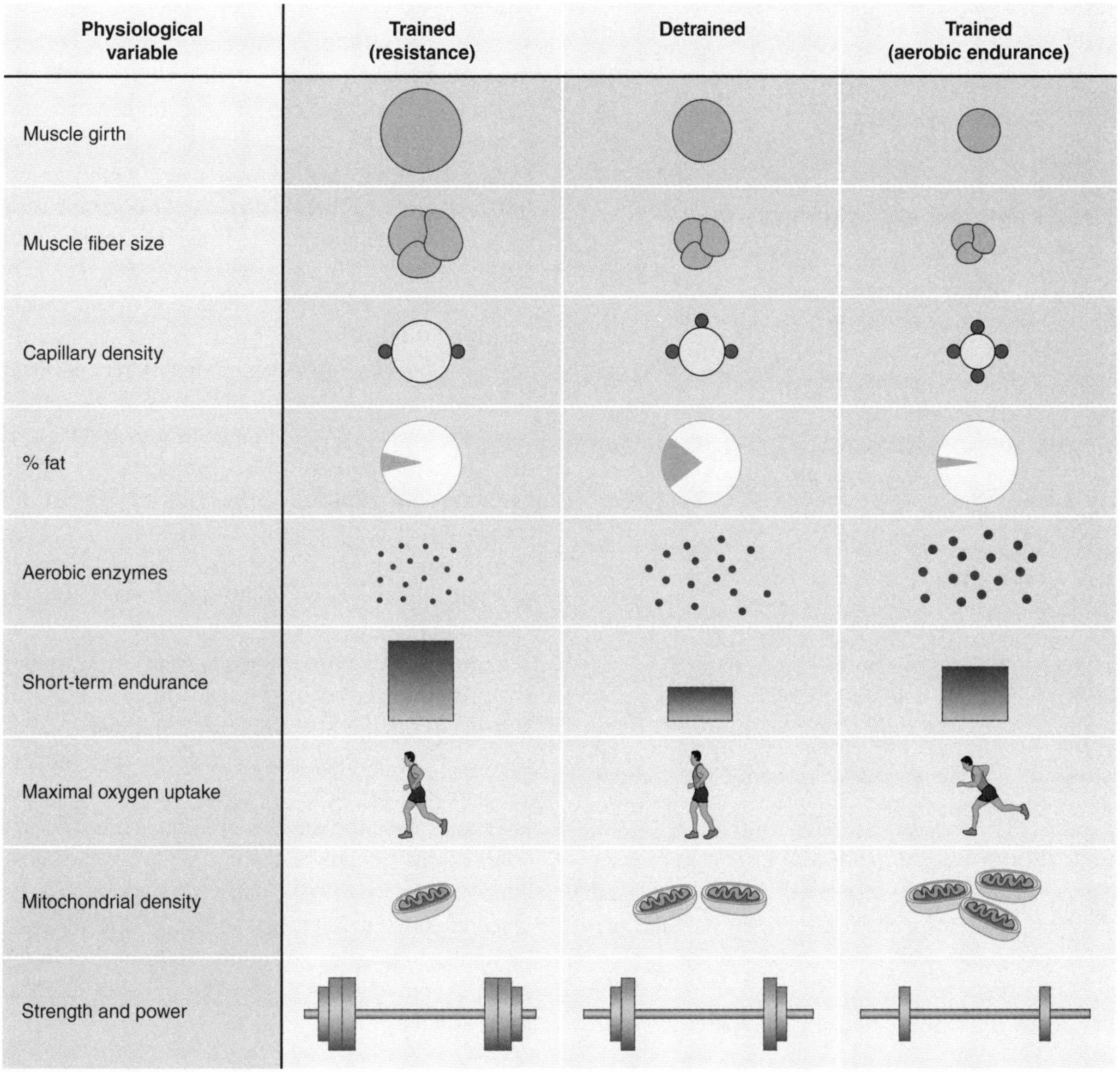

Physiological variable	Trained (resistance)	Detrained	Trained (aerobic endurance)
Muscle girth			
Muscle fiber size			
Capillary density			
% fat			
Aerobic enzymes			
Short-term endurance			
Maximal oxygen uptake			
Mitochondrial density			
Strength and power			

Figure 7.3 Relative responses of physiological variables to training and detraining (see chapter 2).

Reprinted, by permission, from S.J. Fleck and W.J. Kraemer, 2003, *Designing resistance training programs,* 3rd ed. (Champaign, IL: Human Kinetics), 115.

A list of assessments for strength, muscular endurance, power, balance, gait, mobility, and bone strength can be found in the highlight box Assessments of Musculoskeletal Fitness or Function. The discussion that follows briefly describes the assessments and how they might be used clinically (see www.acsm.org and www.nsca-lift.org for more details on musculoskeletal testing). Several commercial websites offer free performance evaluation calculators (see the E-Media section at the end of this chapter).

Muscular strength is essentially the ability of a muscle or set of muscles to generate adequate force to move a predetermined weight. People with weightlifting experience often perform muscular strength testing with free weights or machine weights. The very common 1-repetition maximum (1RM) test indicates how much a person can lift one time; 1RM can be determined for several muscle groups (e.g., arms, legs, chest) and is a useful indicator of overall muscular strength. The 1RM value is easy to use when developing a muscle-strengthening plan. People without

ASSESSMENTS OF MUSCULOSKELETAL FITNESS OR FUNCTION

- *Strength*: 1RM, 5RM, 10RM, handgrip dynamometry, isokinetic dynamometry
- *Muscular endurance*: Number of lifts using 70% of 1RM, push-ups, pull-ups, sit-ups, sit-to-stand tests, sport-specific tests
- *Power*: Wingate anaerobic power test, standing long jump, vertical jump
- *Balance*: One-leg stand, gait speed, Berg balance scale
- *Gait*: Get-up-and-go, curve course walk
- *Mobility*: Goniometer test
- *Bone strength*: DEXA, ultrasound

weightlifting experience or who have safety issues can perform other lifts such as a 5RM or a 10RM, and their 1RM can be estimated from their performance on these lifts (see Baechle and Earle 2008).

Handgrip dynamometry is a simple static method of assessing grip strength using commercially avail-

able devices; it is correlated to other measures of static strength. Isokinetic dynamometry is a test of dynamic strength in which the subject performs an exercise through a range of motion at a constant speed. Because isokinetic testing requires expensive equipment, it is most often performed in clinical and rehabilitation settings. Photos of some common tests of musculoskeletal fitness or function are shown in figure 7.4.

Contrary to muscular strength, *muscular endurance* refers to the ability of a muscle or set of muscles to repeatedly generate a submaximal force (perform repeated contractions) or to sustain a contraction for a period of time. Muscular endurance testing can also be performed using free weights or machine weights (once 1RM has been established) by lifting a weight that is a percentage of 1RM (e.g., 70%) as many times as possible. The percentage of 1RM used to determine muscular endurance varies based on the person being tested.

Simple and common weight-bearing muscular endurance tests such as push-up, pull-up, and sit-up (or curl-up) tests are traditional assessments of muscular endurance, and specific protocols exist for testing various populations. Sit-to-stand tests (e.g., the number of repetitions someone can complete in 30 seconds) for evaluating the muscular endurance of the elderly can be found in the research and clinical literature. Sport-specific tests for higher-fit people that assess muscular endurance are also readily available (see the websites listed in the E-Media section).

a　　　　　　　　　　　*b*　　　　　　　　　　　*c*

Figure 7.4 Musculoskeletal fitness and function may be assessed in many ways, including *(a)* handgrip dynamometry, *(b)* 1RM, and *(c)* the vertical jump.

Measuring power in younger people is usually easier than for older people because of the need for explosive effort (>95%). The Wingate anaerobic power test can be used in laboratory settings to determine maximal and average muscular power. Traditional power tests such as the standing long jump and vertical jump can provide simple measures of power that are easily interpreted.

Balance assessments are primarily for older people who may be at higher risk for falls (i.e., have low functional health), people who are in physical rehabilitation programs (e.g., physical therapy), and athletes who are striving for high levels of performance (e.g., dancers, gymnasts, martial artists). The one-leg stand is a balance test that can be performed with the eyes open or closed; the time the person can stand in the proper position is recorded. Gait speed on a straight or curved course can be used to determine the balance abilities of elderly people. The Berg balance scale is a functional test that provides a composite balance score of 14 items for evaluation of the elderly.

Gait assessments are also used primarily for older people to evaluate their functional health. Common tests related to gait analysis are the get-up-and-go assessment and various curved walking course assessments.

Numerous mobility (flexibility and range of motion) tests have been developed to evaluate people of all ages. Perhaps the most common assessment of mobility is goniometry, which involves having a professional (e.g., a physical therapist or exercise physiologist) measure joint angles, movement limitations, or both.

The only sure way to assess bone mineral content is to surgically remove a piece of bone (by making an incision or inserting a needle into a bone close to the skin's surface) and analyzing it in a laboratory. This is obviously a painful procedure; it is most frequently done to diagnose bone diseases such as cancer. Technological advancements have resulted in the development of BMD screening tools that are quite accurate, painless, and fairly low cost. Bone strength screening and assessment are common clinical and medical assessments and best accomplished with the use of dual-energy X-ray absorptiometry (DXA). DXA is the most valid and reliable screening measure for BMD. Other techniques such as ultrasound (i.e., sound waves) are available, but currently are less accurate than DXA.

Figure 7.5 illustrates the relationship between BMD and aging, as well as a strategy (i.e., increased physical activity including musculoskeletal activities) to prevent or delay osteoporosis by increasing peak bone mass in youth. The figure shows normal BMD loss with aging in a healthy, inactive individual and the increases and maintenance of higher levels of BMD for a healthy, physically active individual with aging. Peak bone mass (or BMD) occurs on average for men and women by age 30 and begins to drop after menopause in women, and by age 70 in most men. Teenage girls, as a population, are an excellent target group to encourage regular participation in musculoskeletal (as well as other physical activity and exercise) activities throughout life, because they should be able to increase their peak bone mass. This, at least theoretically, might delay the point at which they would develop osteoporosis or experience negative musculoskeletal symptoms or disabilities.

PHYSICAL ACTIVITY AND MUSCULOSKELETAL HEALTH

Although extensive research documents the benefits of resistance training, particularly for athletic competition, only since the early 1980s has this evidence been developed enough to make specific

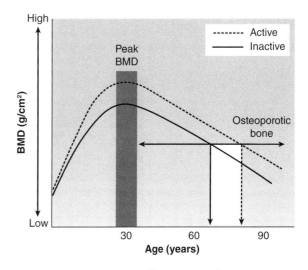

Figure 7.5 Schematic illustration of a strategy to prevent or delay the onset of osteoporosis by increasing peak bone mass during youth.
Reprinted, by permission, from C.J.R. Blimkie and O. Bar-Or, eds. 1995. *New horizons in pediatric exercise science* (Champaign, IL: Human Kinetics), 78.

recommendations for engaging in musculoskeletal activities. Historically, it was thought that the only purpose of the skeletal system was to provide structure for muscles; its role in physical activity and exercise was considered minimal. We obviously now know that the skeletal system is a living organ that can change substantially based on external stimuli, such as physical activity. Similarly, the connective tissues of muscles, tendons, and ligaments, as well as weight-bearing bones, have been found to become thicker and develop greater tensile strength in response to dynamic exercise training. In addition, bones' response to resistance exercise has been to become thicker and stronger regardless of their weight-bearing function. That is, the benefits are not all due to the constant bearing of

body weight: Bone health can be improved through physical activity.

SCIENTIFIC EVIDENCE

The PAGAC (USDHHS, PAGAC 2008) noted moderate scientific evidence of an inverse relationship between physical activity and exercise and the risk of hip fractures in adults. The evidence of a relationship between physical activity and the risk of vertebral fractures exists, but it is not as strong as that for hip fractures. There is no direct evidence that regular moderate-intensity physical activity promotes the development of OA. Participation in low or moderate levels of physical activity may protect against the development of OA, whereas

SPACE, BED REST, AND BONE HEALTH

Much of what we know about the effects of physical activity and exercise on bone health includes information not on activity but on *inactivity.* Studies have looked at BMD and bone mineral content (BMC) in astronauts who have spent multiple weeks or months in space with zero gravity and thus no force on their skeletal systems. These studies consistently show that people subjected to these environments (1) show measurable bone loss that is not uniformly distributed across the skeleton, (2) absorb vitamin D and calcium (two micronutrients important for bone strength) less efficiently, and (3) need much more time to build the lost bone back up than it took to lose it. Other studies have specifically looked at the effects of several days or weeks of complete bed rest on physiological factors like $\dot{V}O_2$max. In the study results shown in figure 7.6, fitter (trained) individuals actually had greater decrements with inactivity (20 days of bed rest) than less fit (sedentary) individuals, and the fitter individuals required 55 days to regain their baseline fitness levels. Clearly, these conditions can accelerate the risk of osteoporosis, osteoporotic fractures, or both.

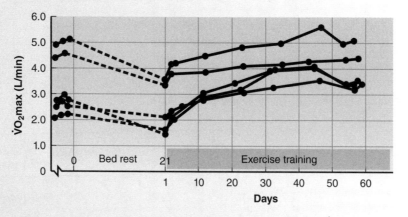

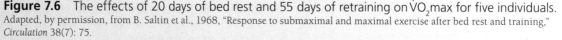

Figure 7.6 The effects of 20 days of bed rest and 55 days of retraining on $\dot{V}O_2$max for five individuals.
Adapted, by permission, from B. Saltin et al., 1968, "Response to submaximal and maximal exercise after bed rest and training," *Circulation* 38(7): 75.

participation in moderate-intensity, low-impact physical activity has been shown to decrease pain and increase function, quality of life, and mental health in people with OA, rheumatoid arthritis, and fibromyalgia (i.e., overall muscular pain and aching). The current evidence that physical activity delays the onset of disability from OA is weak.

Although the scientific evidence is never as strong as we want it to be, existing studies show fairly consistently that participation in regular physical activity and exercise can reduce the risk of hip fractures anywhere from 36 to 68%, and regular physical activity can increase BMD consistently by 1 to 2%. These data are based on short-term exercise training studies, most of which have been less than one year in duration. For this reason, the exact benefits of a lifetime of physical activity participation are unknown. Further, benefits of physical activity on BMD have been found for premenopausal women, menopausal women, and adult men.

What about physical activity for people who already have OA? Does it help? There is fairly strong evidence that people of any age with preexisting OA seem to benefit from aerobic and resistance exercise. Muscular strength improves, as does BMD. Women with OA may benefit more than men from resistance training because they likely have lower baseline strength levels. General muscular strength benefits of physical activity are similar for men and women across the life span, although they diminish with older age. Information on the benefits of physical activity on the bone and general muscular health of people of various races or ethnicities is lacking.

How much physical activity is necessary? The effective dose of physical activity and exercise for musculoskeletal health varies depending on the desired outcome (bone, joint, or muscle health). For bone health and depending on the study, four hours of walking per week, 2 to 4 hours of moderate- or vigorous-intensity physical activity each week, and one hour of weekly physical activity have each been associated with a 36 to 41% reduction in the risk of hip fracture. Weight-bearing endurance and resistance physical activity of moderate intensity three to five days per week for 30 to 60 minutes per session increases BMD. Walking-only protocols may improve spinal BMD (moderate evidence of bone changes).

Evidence strongly suggests an association between physical activity and relief from pain in people with arthritis. The suggested dose of physical activity and exercise for adults with arthritis (to reduce pain and disability and increase function) is 130 to 150 minutes per week of moderate-intensity, low-impact activity; experts suggest 30 to 60 minutes per session, three to five days per week. Both aerobic and muscle-strengthening activities improve joint function and reduce pain. Progressive, high-intensity (60 to 80% of 1RM) muscle-strengthening activities can preserve or increase skeletal muscle mass, power, and intrinsic neuromuscular activation.

The effects of an accumulation of physical activity and exercise throughout the day on musculoskeletal health have not been tested, or research is limited. The scientific evidence from randomized controlled trials and laboratory animal studies has shown the intensity of loading forces to be the key determinant for skeletal response. Joint injuries and carrying excess body mass are more important risk factors for OA than sport participation. Finally, endurance types of physical activity do not increase muscle mass, but they may slow the rate of muscle mass loss with aging, while preserving function.

GUIDELINES

Resistance training and the options available related to muscle-strengthening activities were discussed earlier in the chapter. This section contains the *2008 Physical Activity Guidelines for Americans* recommendations in lay terms. The guidelines for musculoskeletal health have been divided into three parts: children and adolescents (ages 6 to 17), adults (ages 18 to 64), and older adults (>65 years).

Like aerobic activities (see chapter 2), muscle-strengthening activities should be based on the dose-response concept—or in this case, intensity, frequency, and repetitions. Following are definitions related to resistance training:

- *Intensity:* How much weight or force is lifted or used relative to how much a person is able to lift or exert
- *Frequency:* How often (expressed usually per week) a person does muscle-strengthening activities

- *Repetitions:* How many times a person lifts a weight in a given set and how many sets the person performs (related to rest between reps and sets, or groups of repetitions)

The benefits of muscle-strengthening activities are limited to the muscle groups that are worked. Therefore, people need to work the major muscle groups, which include the legs, hips, back, abdomen, chest, shoulders, and arms.

Bone-strengthening activities can include both aerobic and muscle-strengthening activities, because both promote growth and strength, particularly compared to sedentary living. Bone-strengthening activities include jumping jacks, running, brisk walking, and weightlifting. Any weight-bearing activity can count as a bone-strengthening activity—even stair climbing.

Children and adolescents should participate in muscle-strengthening activities at least three days per week, as part of the overall recommendation

of 60 minutes or more of physical activity and exercise per day. They should also participate in bone-strengthening activities at least three days per week, as part of the same overall recommendation. Table 7.1 contains specific examples of muscle-strengthening and bone-strengthening activities for children and adolescents, as well as for adults and older adults. As shown, children and adolescents do not need to engage in formal resistance training to acquire musculoskeletal benefits. Adults and older adults should engage in muscle-strengthening activities involving all major muscle groups at moderately to vigorously intense levels at least twice per week to see improvements in muscular strength and endurance.

No specific number of repetitions has been recommended for resistance training, but people should perform to the point at which performing another repetition without help would be difficult. Resistance training involving one set of 8 to 12 repetitions using several muscle groups has increased

Table 7.1 Examples of Muscle- and Bone-Strengthening Activities by Age Group

Population	TYPE OF ACTIVITY	
	Muscle strengthening	**Bone strengthening**
Children	<ul><li>Games such as tug-of-war</li><li>Modified push-ups (with knees on the floor)</li><li>Resistance exercises using body weight or resistance bands</li><li>Rope or tree climbing</li><li>Sit-ups (curl-ups or crunches)</li><li>Swinging on playground equipment or bars</li></ul>	<ul><li>Games such as hopscotch</li><li>Hopping, skipping, jumping</li><li>Jumping rope</li><li>Running</li><li>Sports such as gymnastics, basketball, volleyball, tennis</li></ul>
Adolescents	<ul><li>Games such as tug-of-war</li><li>Push-ups and pull-ups</li><li>Resistance exercises with exercise bands, weight machines, handheld weights</li><li>Climbing wall</li><li>Sit-ups (curl-ups or crunches)</li></ul>	<ul><li>Hopping, skipping, jumping</li><li>Jumping rope</li><li>Running</li><li>Sports such as gymnastics, basketball, volleyball, tennis</li></ul>
Adults and older adults	<ul><li>Exercises using exercise bands, weight machines, handheld weights</li><li>Calisthenic exercises (body weight provides resistance to movement)</li><li>Digging, lifting, and carrying as part of gardening</li><li>Carrying groceries</li><li>Some yoga exercises</li><li>Some tai chi exercises</li></ul>	

Adapted from USDHHS 2008.

strength, although performing two or three sets, with the appropriate amount of rest between sets, may be more effective. Muscular strength and endurance changes occur progressively over time, and increases in the amount of weight, frequency, or both, can result in greater changes.

Adults and older adults may benefit from engaging in warm-up (prior to physical activity bout) and cool-down (after physical activity bout) activities that include muscle-strengthening activities. These activities slowly increase the blood flow to the working muscles, thereby helping to deliver fuel and take away metabolic waste. Adults and older adults may also benefit from performing flexibility activities (e.g., stretching), although no scientific evidence documents health benefits related to stretching, and stretching does not seem to reduce the risk of injury associated with physical activity.

Older adults who are at risk for falls should include balance activities such as backward walking, sideways walking, heel walking, toe walking, and standing from a sitting position in their physical activity and exercise plans at least three times per week. Combining balance and muscle-strengthening activities for 90 minutes per week, along with moderate-intensity walking for 60 minutes per week, can maintain functional health and may reduce the incidence of falls. Emerging evidence indicates that higher-velocity resistance movements at lower intensities may improve power more in the elderly than lifting at higher intensities and lower velocities.

The specific combinations of type, amount, and frequency of activities that might reduce falls are, unfortunately, unknown. However, some evidence shows that tai chi exercises may help prevent falls. Older adults can start balance and muscle-strengthening activities by holding on to stable supports (e.g., furniture) and wean themselves away from the supports over time.

FUNCTIONAL HEALTH

Musculoskeletal health is not necessarily linked to a specific disease, but low levels of musculoskeletal health (low muscle mass and poor muscle function) can contribute to poor functional health, functional ability, and role ability. **Functional health** (sometimes called *health-related quality of life*) is a concept that suggests that an otherwise

At what age do many older people you know lose their functional health? Do you think regular physical activity helps maintain functional health longer? Why or why not?

healthy person may live with some type of functional disability (USDHHS, PAGAC 2008) and includes two key subitems: functional ability and role ability. In a practical sense, functional health includes being able to physically do the things one wants to do without pain or limitation. The concept of functional health bridges the gulf between the performance-related orientation of physical activity and the health-related orientation of public health. It is the key reason physical activity is important.

Functional ability refers to the capacity to perform a task, activity, or behavior independently. For example, an elderly woman who is unable to go grocery shopping, lift and store her groceries at home (e.g., a heavy milk jug), or move around a two-story house the way she wants to because of musculoskeletal limitations has a loss of functional ability. **Role ability,** on the other hand, refers to the ability to perform **activities of daily living (ADLs)** and **instrumental activities of daily living (IADLs;** USDHHS, PAGAC 2008). Examples include being able to play with her grandchildren or perform self-care tasks such as bathing and laundry. The loss of functional ability and role ability would obviously impair functional health, although factors other than physical changes (e.g., the mental health challenges of aging) could also negatively affect functional health.

The prevalence of poor functional health is difficult to determine in any population because it is multifactorial and a variety of definitions exist. However, it has been reported that the economic cost of fall injuries of those over 65 in the United States, which are often due to poor functional health, exceeded $19 billion in 2000. As the elderly population continues to grow, it is expected that the costs will reach $54.9 billion by the year 2020.

Figure 7.7 illustrates a model that links the relationships among lifestyle behaviors, health status, exercise science principles, and outcomes such as

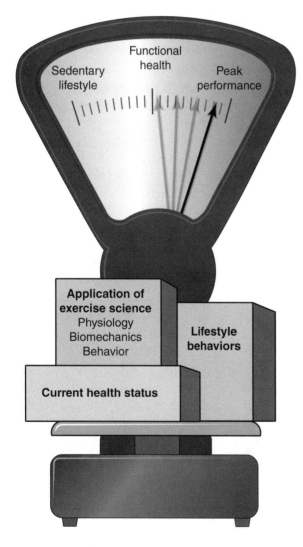

Figure 7.7 Relationships among lifestyle behaviors, health status, exercise science principles, and outcomes such as functional health and peak performance.

functional health and peak performance. Basic functional health should be a primary goal for everyone; beyond that, further participation in physical activity and exercise may lead to higher levels of performance. An understanding of basic exercise science principles is necessary for understanding how to maintain functional fitness, how to exercise, and how to face the challenges to maintaining physical activity and exercise (at least for health benefits), such as mobility challenges, medical management issues, and exercise rehabilitation challenges. For example, managing disease processes (e.g., type 2 diabetes) or maximizing the benefits of rehabilitation (after athletic injury or relapse) requires that people first maintain or regain their basic functional health; only then can they optimize performance. A primary message to share regarding developing and maintaining functional health and optimizing performance is that research has clearly demonstrated the importance of avoiding inactivity for maintaining and improving functional health (USDHHS, PAGAC 2008).

RISK FACTORS FOR POOR FUNCTIONAL HEALTH

Many of the risk factors for poor functional health overlap with those for osteoporosis, osteoarthritis, and low muscle mass as discussed earlier. This is because poor functional health is often a key by-product of these (and other) chronic conditions. Functional health is not necessarily linked to a specific disease process; however, being physically inactive for long periods of time reduces functional health. The basic risk factor for poor functional health is physical inactivity; existing or developing mobility challenges, medical management issues, and rehabilitation challenges further add to the risk.

Low levels of functional health, which can lead to falls and other disabling conditions, correlate highly with low levels of muscle quantity and quality. According to PAGAC (USDHHS, PAGAC 2008), the three primary risk factors associated with low levels of functional health are low aerobic capacity, a lack of muscular strength, and poor balance. In addition, cultural and social factors can affect functional health, including lack of social support (the absence of a human support network or assistance).

COMMON TESTS OF FUNCTIONAL HEALTH

The tests highlighted earlier in the chapter for osteoporosis, osteoarthritis, and muscle mass focus largely on physiological parameters. Functional health status is a complex concept that encompasses quality of life issues such as role functioning, physical functioning, social functioning, and emotional status. Because these issues go beyond physiological concepts (although they are very much related to them), they are best measured using tests that focus on self-assessment or the assessment of a third party (e.g., a physician or researcher). These tests do not isolate one aspect of functional health (e.g., a muscle or joint), but rather, attempt to integrate observations into broad indexes of function. The highlight box Assessments of Functional Health Status lists some of the more common measures used to assess functional health status.

FITNESS RECOMMENDATIONS FOR FUNCTIONAL HEALTH

Figure 7.8 shows the results from several research studies concerning the relationship between the amount of physical activity or exercise performed regularly and mobility limitations. The individuals who were in the lowest physical activity or exercise category were also those who needed medical evaluation due to their mobility limitations. Figure 7.9 also shows that more active people were at

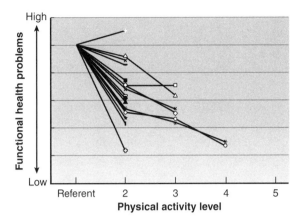

Figure 7.9 The relationship between increased physical activity levels and the risk of functional health problems as shown from several studies.
Adapted from USDHHS 2008.

lower odds ratios, or less risk, for functional health problems as measured by various functional health, ADL, or IADL outcomes.

Figures 7.8 and 7.9 show clearly that higher levels of participation in physical activity and exercise are significantly associated with measures of better functional health. However, the diversity of methods used in the many research studies summarized does not allow us to speculate on a specific dose-response relationship between the amount of physical activity or exercise and functional health. However, the findings in figures 7.8 and 7.9 do illustrate that modest levels of physical activity and exercise are associated with lower risk for functional and role limitations.

SCIENTIFIC EVIDENCE

The PAGAC (USDHHS, PAGAC 2008) noted moderate scientific evidence that physical activity and exercise, at midlife and beyond, reduce the risk of moderate to severe functional limitations. There is also evidence that regular physical activity or exercise is safe and improves functional ability. There is not currently enough evidence to show that physical activity or exercise improves or maintains role ability or prevents disability in older adults who already have functional limitations. Strong evidence suggests that regular physical activity or exercise is safe and reduces the risk of falls in older adults by 30%.

Participation in regular physical activity and exercise can prevent or delay function or role limitations (or both) by 30% (moderate to strong

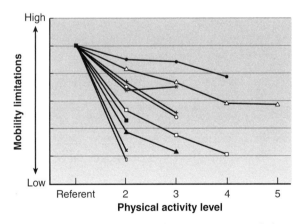

Figure 7.8 The relationship between increased physical activity levels and the risk of mobility limitations.
Adapted from USDHHS 2008.

ASSESSMENTS OF FUNCTIONAL HEALTH STATUS

SF-36 (long) and SF-12 (short)

Measures multiple subdomains of functional health status including physical and role function. Originated in the Medical Outcomes Study (SF-36.org 2011).

Functional Independence Measure (FIM)

Measures physical and cognitive disability (Center for Outcome Measurement in Brain Injury 2011a).

Functional Assessment Measure (FAM)

Meant to supplement FIM by integrating additional concepts such as community integration, orientation, and emotional status (Center for Outcome Measurement in Brain Injury 2011b).

Quality of Well-Being Scale (QWB)

One of the earliest functional health status tools; measures comprehensive quality of life and focuses to a large extent on disease symptoms (University of California at San Diego Health Services Research Center 2011).

evidence). Risk reduction appears to be similar for men and women; evidence regarding relationships between physical activity and exercise and functional health for racial and ethnic groups is limited.

Moderate evidence suggests a dose-response relationship between physical activity and exercise and the prevention or delay of function and role limitations. The dose-response relationship between physical activity and exercise and the prevention of falls in older adults has not been tested. The dose of physical activity or exercise needed for prevention of falls is not known, but participation in walking activities can improve functional health. The maintenance or improvement of functional health requires at least 30 minutes of moderate- or vigorous-

intensity physical activity three to five days per week, emphasizing aerobic and muscle-strengthening activities (strong to moderate evidence).

To lower the risk of falls, exercise programs should include balance training and muscle-strengthening activities three times per week for 30 minutes each session (strong evidence); adding moderate-intensity walking activities two or more other times per week is also recommended. Some evidence suggests that participation in tai chi one to three or more times per week may prevent falls.

The evidence for the effects of the accumulation bouts of physical activity and exercise throughout the day on functional health is not currently available. It is important to remember that most midlife and older adults have very low fitness levels. For this reason, any program of physical activity or exercise focused on improving or maintaining functional health should include slow progressions in the volume of activities to reduce the risk of adverse events.

GUIDELINES

People who have lost some ability to perform a task of everyday life, such as climbing stairs, have a functional limitation. In older adults with existing functional limitations, scientific evidence indicates that regular physical activity is safe and improves functional ability. The *2008 Physical Activity Guidelines for Americans* offers the following guidelines for achieving or maintaining functional health in adults with disabilities (USDHHS 2008):

- Aerobic activity is recommended to be done throughout the week in sessions lasting at least 10 minutes, at one of the following levels and durations:
  - Moderate-intensity aerobic exercise totaling 150 minutes per week
  - Vigorous-intensity aerobic exercise totaling 75 minutes per week
  - A combination of the preceding
- Muscle-strengthening activity for all major muscle groups is recommended at moderate or high intensity on two or more days per week.

Adults with a disability who are not able to meet these recommendations should be encouraged to avoid inactivity by performing as much physical activity as they are able. They should discuss with their health care provider the type and duration of physical activity that is appropriate for their ability.

KEY LEADER PROFILE

Sandra Marcela Mahecha Matsudo, MD, PhD

Why and how did you get into this field? Did any one person have an overriding influence on you?

In my medical school in Bogotá (Colombia), as in most medical schools in our region, no information was given about chronic disease prevention, and we were totally unaware of the relationship of physical activity to health.

Shortly after I had graduated and worked in the compulsory social medical service to the government of Colombia, an opportunity arose to work as a doctor in the Ministry of Sports. I learned from my colleagues about sport traumatology and physical education. That's when I decided to specialize in this area.

In my country, at that time, we did not have this option. The only place to study sports medicine was in Porto Alegre, Brazil, and so I decided to take a plane out of my country for the first time. However, before going to Brazil, I traveled by bus to Ecuador to participate in the Pan American Congress of Sports Medicine, where I met James Skinner and Victor Matsudo. I had the opportunity to attend a lecture by Victor, in which he showed how in São Caetano do Sul, Brazil, in a particular center (CELAFISCS) without economic resources, they were able to do research and to make a difference in health just using two tools: the neuron (knowledge) and the heart (passion).

The last slide of Victor's presentation changed my life forever. In that moment I decided not to go to Porto Alegre but to São Caetano do Sul, which turned my dreams into reality. Thus began my contact with this area, where I did my first scientific paper, with Victor's own proposal, on the validity of self-assessments of sexual maturation. This study was later published in the *Annals of Human Biology* with the support of Bob Malina. It was my first international publication, and I did not speak English at that time!

What are your current research interests?

After many years of researching physical activity, biological maturation, and growth in children and adolescents, I decided to explore issues related to aging. Thus was born the Longitudinal Project on Physical Fitness and Aging in São Caetano do Sul, the first of its kind in a developing country. That has been my main area of research in recent years.

What drives you as a researcher and activist?

After six years of work at the CELAFISCS Research Center, our coordinator was invited by his former professor of public health and then the secretary of health of São Paulo (a state with 645 municipalities and 40 million inhabitants) to implement a program to promote, maintain, and restore health in the community through physical activity. Thus arose the Agita São Paulo Program, which soon became a model intervention recognized by the World Health Organization (WHO) and used to inspire other Latin American countries to implement similar strategies. This led to the consolidation of a network of organizations devoted to the promotion of physical activity for the Americas: RAFA/PANA (www.rafapana.org), as well as the international network Agita Mundo (www.agitamundo.org).

What are one or two key issues to be addressed by 2022?

The great challenge of our area is transforming all of this knowledge into a real change of behavior of the population, which undoubtedly requires effective strategies that can be applied in various settings (e.g., schools, workplaces, home health care centers) and an understanding of what strategies really work. I strongly believe that our role is to put physical activity into the daily routine of people along with eating, sleeping, and brushing teeth, and the best way to do so is to support each person to find his or her way to be physically active with pleasure!

Adults with chronic conditions should engage in regular physical activity because doing so can improve their quality of life and reduce the risk of developing new conditions. The type and amount of physical activity should be determined by their abilities and the severity of the chronic condition. The *2008 Physical Activity Guidelines for Americans* emphasize the following three messages for people with chronic medical conditions (USDHHS 2008):

- Regular physical activity can confer important health benefits.

- Physical activity is safe if done according to abilities.

- People with chronic conditions should always consult with their health providers about which types of physical activity are appropriate.

For many chronic conditions, physical activity provides therapeutic benefits and is part of the recommended treatment for the condition. The *2008 Physical Activity Guidelines for Americans* do not specifically address therapeutic exercise or rehabilitation.

CHAPTER WRAP-UP

What You Need to Know

- The prevalence of common musculoskeletal challenges such as osteoporosis (bone health), osteoarthritis (joint health), and low levels of the quantity and quality of muscle are high in the U.S. population.

- Osteoporosis (i.e., low bone mass and structural deterioration of bone tissue) is estimated to be a public health threat to 44 million Americans, or 55% of those over 50 years of age.

- Osteoarthritis (OA) is the most common form of joint disease and is associated with joint pain and dysfunction along with an irreversible loss of articular cartilage.

- Functional health is a concept that suggests that an otherwise healthy person may have some form of functional disability and includes the maintenance of functional ability and role ability. *Functional ability* refers to the capacity to perform a task, activity, or behavior independently. *Role ability* refers to the ability to perform activities of daily living (ADLs) and instrumental activities of daily living (IADLs).

- Common risk factors for most musculoskeletal disorders are physical inactivity, overweight, sex, heredity, and age.

- The physiological, biomechanical, and psychological benefits of participating in physical activity and exercise on musculoskeletal health are numerous.

- Common musculoskeletal tests can be used to determine levels of strength, muscular endurance, power, balance, gait, and mobility.

- Youth should engage in muscle-strengthening activities that include the major muscle groups at least three days per week, and adults should participate at least two days per week.

- Common risk factors for low levels of functional health are physical inactivity, low physical fitness, tobacco use, age, sex, and heredity or genetics.

- Simple tests of functional health address muscular strength, muscular endurance, balance, gait, and mobility.

- Moderate levels of physical activity or exercise are associated with lower risk for functional and role limitations.

- Participation in regular physical activity and exercise can prevent or delay function and role limitations and reduce the risk of falls in older adults by 30%.

Key Terms

osteoporosis

osteoarthritis (OA)

muscle mass

bone mineral density (BMD)

hypertrophy

functional health

functional ability

role ability

activities of daily living (ADLs)

instrumental activities of daily living (IADLs)

Study Questions

1. What are the definitions of *osteoporosis* and *osteoarthritis*?

2. How do functional health, functional ability, and role ability differ?

3. What are common risk factors for musculoskeletal health challenges?

4. What five long-term adaptations to physical activity and exercise are associated with increased musculoskeletal health?

5. What are six common tests of musculoskeletal fitness or function?

6. What are the dose-response relationships between physical activity or exercise and the various indicators of musculoskeletal health?

7. What are three common negative physiological changes associated with the environments of space or bed rest?

8. What are the annual economic costs associated with falls and low functional mobility?

9. How much physical activity or exercise is recommended to prevent loss of functional ability?

10. How much physical activity or exercise is recommended to prevent falls in older adults?

E-Media

Explore issues related to physical activity, exercise, and public health at the following websites:

Human Kinetics	www.HumanKinetics.com
U.S. Department of Health and Human Services: Physical Activity Guidelines for Americans	www.health.gov/PAGuidelines
International Society for Physical Activity and Health	www.ispah.org
American College of Sports Medicine	www.acsm.org
National Strength and Conditioning Association	www.nsca-lift.org
President's Council on Fitness, Sports & Nutrition	www.fitness.gov
National Osteoporosis Foundation	www.nof.org
National Arthritis Foundation	www.arthritis.org
Commercial website for personal trainers	www.exrx.net
Commercial websites for sports training	www.brianmac.co.uk
	www.topendsports.com

Bibliography

Baechle TR, Earle RW, eds. National Strength and Conditioning Association. 2008. *Essentials of Strength Training and Conditioning*, 3rd ed. Champaign, IL: Human Kinetics.

Center for Outcome Measurement in Brain Injury. 2011a. Introduction to the FIM. www.tbims.org/combi/FIM. Accessed 21 June 2011.

Center for Outcome Measurement in Brain Injury. 2011b. Introduction to the Functional Assessment Measure. www.tbims.org/combi/FAM. Accessed 21 June 2011.

Fiatarone MA, O'Neill EF, Ryan ND, et al. 1994. Exercise training and nutritional supplementation for physical frailty in very elderly people. *New England Journal of Medicine* 330:1769-1765.

Gettman LR, Pollock ML. 1981. Circuit training: A critical review of its physiological benefits, *Physician and Sportsmedicine* 9: 44-60.

Kotlarz H, Gunnarsson CL, Fang H, Rizzo JA. 2009. Insurer and out-of-pocket costs of osteoarthritis in the U.S. *Arthritis and Rheumatism* 60 (12): 3546-3553.

National Arthritis Foundation. 2011. www.arthritis.org. Accessed 24 September 2011.

National Osteoporosis Foundation. 2010. www.nof.org. Accessed 16 December 2011.

SF-36.org. 2011. www.sf-36.org. Accessed 21 June 2011.

Tosteson AN, Melton LJ, Dawson-Hughes B, Balm S, Favus MJ, Khosla S, Lindsey RL. 2008. Cost-effective osteoporosis treatment thresholds: The United States perspective. *Osteoporosis International* 19: 437- 447.

University of California at San Diego Health Services Research Center. 2011. Quality of Well-Being Scale–Self Administered (QWB-SA). https://hoap.ucsd.edu/qwb-info. Accessed 21 June 2011.

U.S. Department of Health and Human Services. 2008. *2008 Physical Activity Guidelines for Americans*. Washington, DC: U.S. Department of Health and Human Services. Available online at www.health.gov/PAGuidelines.

U.S. Department of Health and Human Services, Physical Activity Guidelines Advisory Committee. 2008. *Physical Activity Guidelines Advisory Committee Report, 2008*. Washington, DC: U.S. Department of Health and Human Services. Available online at www.health.gov/PAGuidelines.

PHYSICAL ACTIVITY IN PUBLIC HEALTH SPECIALIST

This chapter covers these competency areas as set forth by the National Society of Physical Activity Practitioners in Public Health:

2.2.3, 2.3.2, 2.5.1, 6.2.1, 6.2.2, 6.2.3, 6.2.4, 6.4, 6.4.1, 6.4.2, 6.4.3, 6.4.4, 6.5.2

CHAPTER 8

CANCERS

OBJECTIVES

After completing this chapter, you should be able to discuss the following:

» What cancer is and which types of cancer are affected by physical activity and exercise

» The prevalence of these cancers

» Possible mechanisms by which physical activity or exercise can lower cancer risk

» Physical activity guidelines for cancer prevention

» How physical activity can be a part of cancer survivorship

OPENING QUESTIONS

» What is cancer, and how can physical activity play a role in its prevention?
» What evidence exists that suggests that physical activity can reduce cancer risk?
» How much physical activity is sufficient to reduce cancer risk?
» How can physical activity help people who have survived a bout with cancer?

Cancer. Just the mention of the word evokes many images and emotions for many people. Most people have known a relative, friend, or acquaintance who has had, is currently diagnosed with, or has died from some type of cancer. Many cancers can recur after successful treatment or can spread to organs and systems other than their origin (also known as **metastases**). Many types of cancer are rapidly fatal, whereas others offer a very positive prognosis, particularly when diagnosed early. Although referring to *cancer* as a disease in the singular is convenient, many types exist, and they are unique. Cancer is actually a group of diseases, each with its own risks, etiologies, and pathologies.

Although much of the work in the area of physical activity and health has focused on the cardiovascular system and traditional physiological responses to exercise training (see chapters 2 and 5), growing evidence from emerging literature suggests that physical activity may help prevent certain cancers, particularly colon cancer and breast cancer. The proposed mechanisms of these associations between physical activity and cancers are intriguing because they involve systems in the body that exercise physiology research has not focused on. In addition to questions of primary prevention, a growing body of data supports a role for physical activity in improving cancer prognoses and the quality of life of people who have cancer (U.S. Department of Health and Human Services [USDHHS], Physical Activity Guidelines Advisory Committee [PAGAC] 2008).

PREVALENCE OF CANCERS

Cancers are diseases with processes associated with uncontrolled abnormal cell growth and proliferation. Different cancers have different causes in different people. Many cancers are idiopathic—meaning that the cause in a given person may never be known. The concepts of risk factors and probability have helped our understanding of cancers. For example, although cigarette smoking is one of the most insidious risk factors for lung cancer and the probability of cancer is much higher among cigarette smokers, people who have never smoked a cigarette in their lives are diagnosed with lung cancer. Cancers are caused by both internal factors (e.g., heredity, immune dysfunction, and abnormal metabolism) and external factors (e.g., behaviors such as smoking and a sedentary lifestyle, pollution, and radiation exposure). These factors can act alone or in synergy over time to produce uncontrolled cell growth and proliferation.

Evidence is very clear that the interaction between the environment and genetics is a possible trigger for cancerous cell changes. For example, a person may have a family history (heredity) of a certain type of cancer, thus possibly carrying a genetic code that may increase cancer risk. That genetic disposition toward cancer may never be expressed if it does not come into contact with an environmental exposure (e.g., cigarette smoke); such contact, though, may trigger the expression of the gene, causing the cells to become cancerous.

Cancers were the second leading cause of death in the United States in 2009, affecting 45% of men and 38% of women during their lifetimes. Only about 10 to 15% of cancers are linked to heredity; the remaining cases are thought to be related to external factors such as lifestyle or environmental factors. In the United States (2009) 1,500 people per day were predicted to die from cancers. Cancer accounts for 25% of all deaths in the United States (American Cancer Society 2009a). The estimated deaths from cancers by site in 2009 for the United States are shown in table 8.1. Globally, cancer killed an estimated 7.6 million people in 2007, and the rate is expected to increase to 17.5 million as a result of the growth and aging of the world's population.

Table 8.1 Estimated Deaths From Cancers by Site

MALE			FEMALE		
Site	Male deaths	Percentage	Site	Female deaths	Percentage
Lung and bronchus	88,900	30%	Lung and bronchus	70,490	26%
Prostate	27,360	9%	Breast	40,170	15%
Colon and rectum	25,240	9%	Colon and rectum	24, 680	9%
Pancreas	18,030	6%	Pancreas	17, 210	6%
Leukemia	12,590	4%	Ovary	14,600	5%
Liver and intrahepatic bile duct (LIBD)	12,090	4%	NHL	9,670	4%
Urinary bladder	10,180	3%	Uterine corpus	7,780	3%
Non-Hodgkin's lymphoma (NHL)	9,830	3%	LIBD	6,070	2%
Kidney and renal pelvis	8,160	3%	Brain and nervous system	5,590	2%

Data from American Cancer Society 2009a.

Importantly, research has demonstrated that many cancers may be due to health behaviors and, thus, are theoretically preventable. The American Cancer Society (2009a) estimated that one-third of the projected 562,340 U.S. deaths from cancers in 2009 would be related to lifestyle factors such as overweight and obesity, physical inactivity, and unhealthy eating. Vainio and Bianchini (2002) reported that 25% of the cancer cases globally were due to excess weight and a sedentary lifestyle.

The economic burden of cancer is, not surprisingly, a complicated topic. Different cancers have different health effects—some are rapidly fatal, others are treatable but carry a long burden of illness, and still others are curable if diagnosed early. Thus, direct costs (i.e., how much money is needed for treating the disease in the hospital, pay for physicians and nursing home services, and pay for pharmaceuticals) as well as indirect costs (e.g., changes in health-related quality of life, lost productivity due to disability and death) differ by cancer type. Age is also an important issue. Cancers that occur later in life, on a population level, cost less than those in younger people. Cancers occurring in countries with medical care delivery systems that are different from that of the United States have

different economic costs than similar cancers in the United States.

The American Cancer Society (2009b) estimated that, in 2009, the economic costs of all cancers was $228.1 billion (USD). This included an estimated $93.2 billion for direct medical care, $18.8 billion due to lost productivity among people with cancer, and $116.1 billion due to lost productivity among people who died from cancer. Although these numbers are staggering, it may help to put them in context. A total of $228 billion (USD) in 2009 was roughly equivalent to the total gross domestic product (i.e., the market value of all goods produced) of the entire country of Portugal. Clearly, the economic burden, in addition to health and social burdens, is important to address as a public health problem.

Because cancers are such an important health burden, and because disease surveillance is so central to public health, many useful sources of cancer statistics are available. These statistics are important in many ways, but one key way is to help us distinguish new cases of cancer (**incidence**) from existing cases (**prevalence**). Because more people are living longer with cancers than at any other time in history, which is due to impressive advances in

Many risk factors for cancers are modifiable. What are the barriers to reducing these risks?

therapy, the number of existing cases of cancers is on the rise (prevalence). If we could not distinguish these existing cases from the new cases (incidence) we would incorrectly assume that cancer is an increasing problem simply because more people are living longer with cancer.

One excellent resource for cancer information (in the United States) is the National Cancer Institute, one of the institutes at the National Institutes of Health (NIH). The Cancer Surveillance Epidemiology and End Results (SEER) program has been operating since the early 1970s and is the definitive web resource for cancer mortality, incidence, and prevalence data by sites affected and overall (http://seer.cancer.gov).

CANCER RISK FACTORS

Before considering the risk factors for cancers, we should have an understanding of how cancer starts and progresses in general. Despite multiple types of cancers with various etiologies, the process by which normal cells change and become cancerous is thought to be common to all cancers. In this **multistage model of carcinogenesis**, outlined in figure 8.1, a subset of normal cells becomes initiated for a cascade toward uncontrolled proliferation in stage 1. This **initiation** can be the result of a genetic mutation, a spontaneous change, or an external cause. The genetic material is altered, which makes the affected cells more likely to grow more rapidly than unaffected cells.

In the second stage of the multistage process of carcinogenesis, **promotion,** some of these initiated (converted) cells become precancerous through additional genetic changes as a result of the altered state they entered in the initiation stage. The promotion stage is characterized by the rapid proliferation of these altered cells.

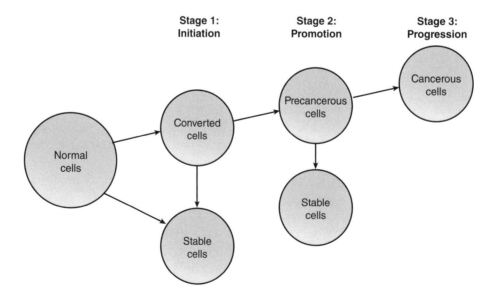

Figure 8.1 Multistage model of carcinogenesis.
Adapted from Rogers et al. 2008.

Some of the cells that have proliferated during the promotion stage will progress to the third stage of **progression**. In this stage, proliferating precancerous cells become full, invasive tumors, and cancer is subsequently diagnosed.

Why is this model important? The multistage model of carcinogenesis, which includes initiation, promotion, and progression, allows us to hypothesize about and study the possible mechanisms by which physical activity and other factors may interrupt this process and prevent cancers. Although they are still working largely with animal models, researchers are aggressively studying how physical activity may affect (and prevent) key genetic changes and their cellular expressions at each of these three stages (Rogers et al. 2008).

A list of general risk factors associated with developing cancers follows. Specific cancers may have additional risk factors that can cause cellular damage internally (e.g., mutations and immune conditions), such as hormonal regulation or nutrient metabolism, or externally, such as tobacco use, chemicals, and sun exposure. As discussed earlier, even though heredity is listed here as a nonmodifiable risk factor, how inherited genes are expressed depends on interactions with environmental exposures.

Modifiable Risk Factors for Cancer

- Physical inactivity
- Obesity
- Tobacco use
- Poor nutrient intake
- Excessive sun exposure
- Toxic environmental exposure

Nonmodifiable Risk Factors for Cancer

- Age
- Heredity (genetics)
- Sex

Following are descriptions of the modifiable and nonmodifiable risk factors for cancer:

- *Physical inactivity.* Physical inactivity is emerging as an important risk factor for several prevalent cancers; meeting physical activity guidelines may lower the risk of developing these cancers.

- *Obesity.* As with CVD risk, having a BMI greater than or equal to 25 (overweight) or 30 (obesity) significantly increases the risk of most cancers.

- *Tobacco use.* Chronic smoking is associated with lower levels of aerobic capacity and functional health. Tobacco use was associated with 169,000 cancer deaths in 2009 and is the main risk factor associated with cancers of the lung and bronchus.

- *Poor nutrient intake.* The consumption of a diet that is low in essential nutrients (e.g., those found in fresh fruits and vegetables), practicing risky behaviors (e.g., excessive alcohol intake), or both, can increase the risk of cancers.

- *Sun exposure.* Excessive exposure to the sun's rays and indoor tanning (i.e., nonionizing radiation) are the primary risk factors associated with skin cancers.

- *Toxic environmental exposure.* Long-term exposure to environmental toxins such as chemicals, ionizing radiation, and infectious diseases can increase the risk of most cancers.

- *Age.* The risk of developing cancer increases the longer one lives, because most cancers develop over time from damaged genes. The majority of cancers (>70%) occur in adults who are over 55 years of age.

- *Heredity (genetics).* Although cancer risk depends somewhat on genetics, engaging in positive behaviors such as being physically active, eating healthfully, maintaining a healthy weight, and minimizing risky activities can lessen overall risk by improving functions such as circulation, ventilation, bowel transit time, energy balance and immune function, and DNA repair (Thune and Furberg 2001).

- *Sex.* Some cancers are sex specific. Prostate cancer in men and reproductive system cancers in women (ovarian, endometrial) are examples of sex-specific cancers.

KINESIOLOGY AND CANCERS

As noted by the PAGAC (USDHHS, PAGAC 2008), the relationship between physical activity and reduced cancer risk has been found most consistently for colon and breast cancers. Emerging evidence suggests an association between being

The positive effects of physical activity for cancer survivors are becoming clearer each year with new research.

physically active (versus inactive) and a lowered risk for lung and endometrial cancers. However, little evidence supports relationships between risks for other cancers and physical activity.

The earliest and strongest evidence of a link between physical activity and cancer showed an inverse relationship between cancer of the colon and physical activity. Many studies using various designs and types of participants have been conducted over the years. In a recently published quantitative summary of those studies, Wolin and colleagues (2009) looked at 52 of the best studies that focused only on colon cancer. Taking all these results together, and controlling for the differences in study design and other factors, the authors estimated that people who are most physically active had a 24% lower risk of developing colon cancer than did people who were inactive in these studies. Men who were active had a slightly lower risk (24%) compared to similar men who were inactive than did active women compared to inactive women (21%).

What's more, Wolin and colleagues found that, among the studies with a satisfactory design, the association with physical activity was best described as dose-response. That is, the most active people had the lowest risk for colon cancer, the inactive people had the highest risk, and those in between had a risk that was also in between. This is a very powerful finding that strongly suggests that physical activity may be causally related to colon cancer.

The observation that higher levels of physical activity are associated with lower risks of breast cancer is an exciting one, particularly given the prevalence of breast cancer among women. The U.S. National Cancer Institute (2011) estimated that in 2011 the average likelihood that a woman living in the United States would be diagnosed with breast cancer was 12.2%. Stated another way, about one in eight women living today will be diagnosed with breast cancer during her lifetime. With recent time trends for the disease being relatively stable, breast cancer clearly is an important health problem to address.

A review of 87 studies of physical activity and female breast cancer (Friedenreich and Cust 2008) found the average risk of breast cancer to be 25 to 30% lower among the most physically active women compared to similar inactive women. The amount of physical activity needed to reduce the risk of breast cancer has been estimated to be the equivalent of brisk walking 45 or 60 minutes or more per day, five or six days per week (McTiernan 2008). Because of the wide range of studies and designs examined, Friedenreich and Cust were also able to examine the risk of breast cancer by the specific types of physical activity evaluated in the studies. Some studies focused on exposure to occupational physical activity, whereas others quantified physical activity as that done in leisure time (i.e., recreational physical activity and exercise), during active transportation (e.g., walking and bicycling), or during household-related activity.

Friedenreich and Cust showed that all types of physical activity were associated with a lower risk of breast cancer—a remarkably consistent finding. Women who were more physically active in their jobs or who had a measureable amount of physical activity while traveling had a lower risk of breast cancer. Women participating in recreational physical activity seemed to have a higher reduction in risk. Why might this be?

PHYSICAL ACTIVITY EXPOSURE IN CANCER STUDIES

Cancer is an extremely difficult disease to study in free-living populations. This is true in studies of physical activity exposure and cancer risk. The multistage model of carcinogenesis can occur fairly rapidly (e.g., over the course of a few years) or over a lifetime. This means that knowing *when* the exposure to a risk factor such as physical inactivity occurred is very important. For example, several researchers have hypothesized that physical activity may be more important for cancer prevention in certain periods of life than in others (e.g., it may be more important during early adulthood than during later adulthood, which is closer to when the disease develops). Others take the contrary view that a constant and sustained exposure to physical activity is most important—that lifetime exposure to physical activity confers the lower risk. Either way, long-term studies that follow people over their lifetimes are critical for answering these questions. What do you think? Would you be willing to be followed throughout your life to participate in such a study?

Interestingly, in the analysis by Friedenreich and Cust, the time of life in which women engaged in physical activity did not seem to matter. The extent of the lower risk of breast cancer was roughly equivalent in studies of young women, middle-aged women, and older (postmenopausal) women. The risk reductions were somewhat larger in studies of postmenopausal women. This observation supports the idea that there is no particular life phase in which physical activity is more likely to reduce the risk of breast cancer.

Although strong scientific evidence supports a dose-response relationship between physical activity and lower risk of breast cancer and colon cancer, the specific physiological mechanisms by which this lower risk might be operating are unknown. In the last 10 years, researchers have started to study the biological mechanisms of, and linkages between, the specific dose-response relationships between physical activity and cancer risks. Some of the chronic exercise-related adaptations that have been reported to lower the risks for cancers are shown in the highlight box Potential Mechanisms Through Which Physical Activity Lowers Cancer Risk.

Most of the contributions of kinesiology to our understanding of cancer mechanisms come from exercise physiology. Research laboratories around the world have focused attention on how the exposure to physical activity may block key steps in the multistage carcinogenesis process. Biomechanical and behavioral factors, although important, have not been identified as central to the mechanisms of physical activity that affect the development of cancer.

The scientific literature seems to support at least six likely explanations—through two different pathways—of how physical activity reduces the risk of cancers. One pathway is an indirect one: physical activity affects body composition and adiposity (weight loss). These changes in body composition lower cancer risk indirectly by changing factors associated with higher body fatness, such as

POTENTIAL MECHANISMS THROUGH WHICH PHYSICAL ACTIVITY LOWERS CANCER RISK

- Avoidance of weight gain or weight loss
- Reduced insulin resistance
- Lower systemic low-grade inflammation
- Lower colon transit time
- Lower production of sex hormones
- Improved immune function

positively affecting biomarkers of systemic inflammation (a marker for several types of chronic disease including cancers), improving **insulin resistance** (i.e., improving the ability of the hormone insulin to clear glucose from the body), increasing blood **insulin** levels, and lowering the production of sex hormones (estrogens and androgens).

Physical activity can also reduce the risk of cancer directly (i.e., not through reducing adiposity). The direct physiological effects of physical activity on skeletal muscle improve insulin resistance. Physical activity has also been shown (independent of adiposity) to improve levels of sex steroid hormones (androgens and estrogens), improve biomarkers of low-grade inflammation, and improve immune function. Further, physical activity can work mechanically to reduce colon transit time, thus reducing the time potential carcinogenic compounds are in contact with the digestive system. This is one of the key hypothesized mechanisms for physical activity preventing colon cancer.

It is fairly clear that sex hormones are important in cancer risk. Higher levels of circulating estrogens and androgens place women at higher risk of breast cancer. This appears to be especially true for postmenopausal women. Exercise training has been shown to decrease circulating sex hormones directly and through fat loss. The positive effects of physical activity on sex hormones can be expected as a result of moderate-intensity as well as vigorous-intensity physical activity.

Higher insulin resistance has been associated with several types of cancer, and insulin has been shown to increase cell proliferation coincident with the multistage model of carcinogenesis (see figure 8.1). Physical activity and exercise very clearly lower insulin resistance (allowing for glucose to be cleared from the body more efficiently) via improved cellular metabolism both acutely (i.e., immediately after a single exercise bout) and chronically (i.e., among those who are habitually physically active). Interestingly, this effect has been shown for aerobic physical activity as well as resistance training.

The interrelationship among immunity, cancer, and physical activity is a complicated topic that researchers are only just now starting to investigate. Because of the uncontrolled differentiation and multiplication of cells that characterize most cancers, the body's immune system has long been a target for understanding cancer prevention and treatment. This is because we normally depend on the immune system to isolate and eliminate foreign or abnormal cells.

Small, short-term exercise training studies have been shown to increase individual markers of immune function, and this response has been maintained for several hours after the cessation of exercise. Further, a positive dose-response association between physical activity intensity and immune function has been shown; more improvement has been observed with higher-intensity activity. The longer-term association with chronic exercise is unclear. However, bouts of vigorous-intensity exercise (such as those performed by elite athletes who may overtrain) over time may actually lower immune function and increase the risk of acute infections such as those of the upper respiratory tract. Clearly, much research is needed on this promising topic.

PHYSICAL ACTIVITY AMONG CANCER SURVIVORS

Is there a benefit to being physically active after being diagnosed with cancer, perhaps during and after treatment? With improved cancer treatments and earlier, more effective diagnostic techniques, more and more people are living with cancer. In 2009, an estimated 11 to 12 million people in the United States were living with cancer or had survived a bout of cancer.

A cancer diagnosis is a difficult situation filled with many psychological and physiological changes—as a result of both the disease and the therapeutic regimens used to control the disease. Chemotherapy, radiation, and surgery are standard cancer treatments, and all can negatively affect the body in the quest to reduce and or eliminate a cancerous tumor. Hormone therapy and steroid treatments that accompany some cancer treatments can also have deleterious effects. Issues such as fatigue, lymphedema (i.e., localized fluid retention secondary to radiation therapy), cardiorespiratory fitness, muscular strength and endurance, quality of life, self-esteem, and safety (i.e., risk of adverse events) are most important to cancer survivors.

Speck and colleagues (2010) published a comprehensive review of studies of cancer survivors ($n = 6,838$). These researchers reviewed 82 studies that examined some effect of exercise training or physical activity during and after cancer treatments. Overwhelmingly, breast cancer has been the cancer most frequently examined for benefits of physical activity (>80% of studies). Because cancers are unique diseases, however, the effects of physical activity on breast cancer survivors may not be attributable to survivors of colon cancer, endometrial cancer, prostate cancer, or other types of cancer.

A listing of the health benefits of physical activity for cancer survivors is shown in table 8.2. Factors are categorized according to when their effects were studied: during or after cancer treatment. Although dozens of health and physiological outcomes have been investigated, the factors listed in the table are those that have been consistently shown to result from increased exercise or physical activity across several studies.

Table 8.2 suggests that physical activity has some consistent positive effects on several physical and psychological parameters among cancer survivors. Clear and substantial gains in upper and lower body strength can be expected among cancer survivors after treatment, and small to moderate gains can be expected even during treatment. Although observed

SCREENING FOR EARLY DIAGNOSIS OF CANCER

Screening is an extremely effective way to catch several types of cancers early. A variety of regular screening tests are used to detect precancerous growths and stage 1 cancers. (Cancer occurs in four stages: stage 1 is early-stage cancer, and stage 4 is late-stage, or advanced, disease.) As noted in "Cancer Facts and Figures" (American Cancer Society 2009a), cancers of the cervix, colon, and rectum can be prevented by removing precancerous tissue. Cancers of the breast, colon, rectum, cervix, prostate, oral cavity, and skin can also be diagnosed through screening. Self-screenings for many cancers (e.g., breast and skin) are valuable for personal awareness and may result in the early detection of disease.

effects are not large for other physiologic outcomes, this emerging evidence gives substantial credibility to the belief that physical activity programming should be part of cancer survivors' therapy.

Table 8.2 Summary of Physical Activity Effects Among Cancer Survivors

Parameter	Improvement after cancer treatment	Improvement during cancer treatment
Upper body strength	Large	Small to moderate
Lower body strength	Large	Small to moderate
Breast cancer–specific concerns	Large	——
Cardiorespiratory fitness	Small to moderate	Small to moderate
Fatigue	Moderate	Small to moderate
Overall quality of life	Small to moderate	Small to moderate
Anxiety	——	Small to moderate
Self-esteem	——	Small to moderate
Physical activity participation	Small to moderate	Small to moderate
Symptoms and side effects*	Small to moderate	——

*Symptoms and side effects include nausea, lymphedema, and pain.

Data from Speck et al. 2010.

KEY LEADER PROFILE

Barbara E. Ainsworth, PhD, FACSM, FNAK

Courtesy of Barbara E. Ainsworth.

Why and how did you get into this line of work?

My love of sports led me to major in physical education as an undergraduate. There I realized I did not want to teach in the public schools, so I went to graduate school for my master's degree. I taught physical education in a small college and realized I needed a doctoral degree to advance my career in higher education. I earned a PhD in exercise physiology at the University of Minnesota. In one of my classes, I learned that physical inactivity was a public health problem. Because I wanted to learn more about public health strategies to address physical inactivity, I secured a postdoctoral position at the University of Minnesota's School of Public Health, where I earned a master of public health (MPH) degree specializing in epidemiology. There I learned about the need to validate physical activity questionnaires used in public health surveillance settings and in observational studies to assess the prevalence of physical inactivity and to understand how regular physical activity can prevent the development of chronic diseases and premature death. These educational experiences set the path for my career in university teaching and research.

Did any one person have an overriding influence on you?

Several people have been mentors during my career. Undergraduate professors at Fresno State University taught me about the importance of being an excellent teacher and how to be a professional in my field. Dr. Rose Lyon and Dr. Carol Gulyas encouraged me to go to graduate school. My uncle, who was a university professor, told me to always do my best and that I could make a difference. At the University of Minnesota, Dr. Robert Serfass guided me through my PhD, and Dr. Arthur Leon opened my eyes to physical activity and public health. I regard Dr. Leon as my first mentor in physical activity and public health because he provided my postdoctoral experience and opened doors for me in the public health field. I regard Dr. Carol Macera as my other mentor in physical activity and public health. She guided me toward success in higher education and included me as a collaborator on the CDC's physical activity surveillance activities. Both of my mentors continue to provide sage advice on many topics.

What are your current research interests?

The assessment of physical activity, the validity of physical activity questionnaires, and physical activity patterns in women.

What drives you to be a researcher and activist in the field of physical activity and health?

I have a passion to increase the accuracy of physical activity measurements to improve the quality of physical activity assessment and to help other researchers find the best ways to measure physical activity for their research and practice needs.

What are one or two key issues of importance in our field that must be addressed by 2022?

First, we need to identify the most accurate ways to measure physical activity for use in public health surveillance and intervention settings, and second, we need to understand effective behavioral approaches to get people to be less sedentary and more physically active in their daily lives.

PHYSICAL ACTIVITY GUIDELINES FOR CANCER PREVENTION

Although scientists first suggested that physical activity plays a role in cancer prevention at least 300 years ago, specific mechanisms of how this may happen and recommendations about the levels required for prevention have been articulated only in the past 10 years or so.

SCIENTIFIC EVIDENCE

The Physical Activity Guidelines Advisory Committee (USDHHS, PAGAC 2008) noted strong scientific evidence of an inverse relationship between participation in physical activity and exercise and the risk of breast and colon cancer. Participation in regular physical activity and exercise can lower the risk of colon cancer by 30% and the risk for breast cancer by 20%. The association is strong regardless of sex or age and reasonable when race and ethnicity are taken into consideration.

A reasonable amount of evidence indicates that a dose-response association between physical activity and exercise and a lower risk of developing colon or breast cancer, which is consistent with the recommendation of 30 to 60 minutes of moderate- to vigorous-intensity physical activity or exercise every day. Data for this effect are strongest in relation to aerobic leisure time (recreational) physical activity. Data supporting the effect of an accumulation of short bouts of physical activity and exercise on cancers risk are currently limited. A small body of strong evidence suggests that breast cancer survival (i.e., quality of life and fitness) is associated with participation in regular physical activity and exercise. Finally, growing evidence suggests that increased physical activity and exercise is associated with a reduced risk of cancers of the endometrium and the lung.

GUIDELINES

The levels of physical activity and exercise associated with cancer risk reduction can be attained by following the *2008 Physical Activity Guidelines for Americans* as described for cardiorespiratory health (see chapter 5) and musculoskeletal health (see chapter 7). Adults should engage in 150 minutes of moderate-intensity or 75 minutes of vigorous-intensity aerobic activity per week, or an equivalent amount of mixed moderate-intensity and vigorous-intensity aerobic activity. They should also try to participate in muscle-strengthening activities involving the major muscle groups on two days or more per week.

Cancer survivors who engage in physical activity and exercise can reduce their risks for new chronic diseases, and participation may reduce the adverse effects of cancer treatment. Cancer survivors should consult with their health care providers to verify that their physical activity or exercise plans are consistent with their current physical abilities and health status.

CHAPTER WRAP-UP

What You Need to Know

- Cancers are disease processes associated with uncontrolled abnormal cell growth and proliferation. Cancers are caused by both internal factors (e.g., heredity, immune dysfunction, abnormal metabolism) and external factors (e.g., behaviors such as smoking and a sedentary lifestyle, radiation exposure). These factors can act alone or in synergy over time to produce carcinogenesis.
- Common risk factors for most cancers include age, physical inactivity, obesity, heredity, sex, tobacco use, sun exposure, and poor nutrient intake.
- The risk of colon and breast cancer can be reduced by participating in at least 150 minutes of moderate-intensity physical activity per week.
- Cancer survivors who engage in physical activity and exercise can reduce their risks for new chronic diseases, and participation may reduce the adverse effects of cancer treatment.

Key Terms

cancer
metastases
incidence
prevalence

multistage model of
  carcinogenesis
initiation
promotion

progression
insulin resistance
insulin

Study Questions

1. What is the definition of *cancer*?
2. What is the prevalence of cancers in the United States?
3. What are five common risk factors associated with the development of most cancers? Describe them.
4. Which cancers can be prevented by regular participation in physical activity and exercise?
5. What is the multistage model of carcinogenesis?
6. What physiological adaptations may be mechanisms by which physical activity may affect cancer risk?
7. Why is self-screening important in cancer prevention?
8. How much physical activity and exercise is needed to reduce cancer risks and improve the quality of life for those who survive a bout of cancer?
9. What are the scientifically observed health benefits of physical activity among cancer survivors?
10. What physical activity and exercise guidelines can reduce cancer risk?

E-Media

Explore issues related to physical activity, exercise, and cancer at the following websites:

American Cancer Society	www.cancer.org
U.S. National Cancer Institute's Surveillance Epidemiology and End Results (SEER) Program	http://seer.cancer.gov
World Health Organization: On Physical Activity	www.who.int/topics/physical_activity/en
U.S. National Cancer Institute	www.cancer.gov

Bibliography

American Cancer Society. 2009a. Cancer Facts and Figures 2009. www.cancer.org/Research/CancerFactsFigures/cancer-facts-figures-2009. Accessed 3 July 2011.

American Cancer Society. 2009b. Economic Impact of Cancer. www.cancer.org/cancer/cancerbasics/economic-impact-of-cancer. Accessed 5 July 2011.

Friedenreich CM, Cust AE. 2008. Physical activity and breast cancer risk: Impact of timing, type and dose of activity and population subgroup effects. *British Journal of Sports Medicine* 42: 636-647.

McTiernan A. 2008. Mechanisms linking physical activity with cancer. *Nature Reviews Cancer* 8: 205-211.

National Cancer Institute. 2011. Probability of Breast Cancer in American Women. www.cancer.gov/cancertopics/factsheet/detection/probability-breast-cancer. Accessed 2 July 2011.

Rogers CJ, Colbert, LH, Greiner JW, Perkins, SN, Hursting SD. 2008. Physical activity and cancer prevention: Pathways and targets for intervention. *Sports Medicine* 38: 271-296.

Speck RM, Courneya, KS, Masse, LC, Duval S, Schmitz KH. 2010. An update of controlled physical activity trials in cancer survivors: A systematic review and meta analysis. *Journal of Cancer Survivorship* 4: 87-100.

Thune I, Furberg AS. 2001. Physical activity and cancer risk: Dose–response and cancer, all sites and site-specific. *Medicine & Science in Sports and Exercise* 33: S530–S550.

U.S. Department of Health and Human Services, Physical Activity Guidelines Advisory Committee. 2008. *Physical Activity Guidelines Advisory Committee Report, 2008.* Washington, DC: U.S. Department of Health and Human Services. www.health.gov/PAGuidelines.

U.S. Department of Health and Human Services, 2008. *2008 Physical Activity Guidelines for Americans.* Washington, DC: U.S. Department of Health and Human Services. www.health.gov/PAGuidelines.

Vainio H, Bianchini F. 2002. *International Agency for Research on Cancer: Weight Control and Physical Activity.* Lyon: IARC Press.

Wolin KY, Yan Y, Colditz GA, Lee I-M. 2009. Physical activity and colon cancer prevention: A meta-analysis. *British Journal of Cancer* 100: 611-616.

PHYSICAL ACTIVITY IN PUBLIC HEALTH SPECIALIST

This chapter covers these competency areas as set forth by the National Society of Physical Activity Practitioners in Public Health:

1.4.2, 2.1.1, 2.2.1, 2.2.3, 2.3.3, 3.2.1, 3.2.2, 6.1.4, 6.1.5, 6.2.1, 6.4.1, 6.4.3

MENTAL HEALTH

OBJECTIVES

After completing this chapter, you should be able to discuss the following:

» The prevalence and economic costs of mental health disorders and related health challenges

» The factors associated with common mental disorders, and how they are generally assessed

» How physical activity affects depression, anxiety, psychological distress, cognitive function, dementia, and sleep disorders in relationship to mental health

» The relationship of physiological adaptations to mental health

» The physical activity guidelines related to mental health, and the evidence behind them

» Physical activity and exercise recommendations for mental health

OPENING QUESTIONS

» Can regular physical activity have a positive effect on mental health? If so, how?
» What are the associations among exercise, depression, and anxiety?
» Can regular participation in exercise prevent age-related declines in cognitive function and dementia?
» Can exercise prevent Alzheimer's disease?
» Does participation in regular physical activity improve sleep?
» How much physical activity is consistent with good mental health?

Have you ever noticed that exercising helps you forget about a problem that was causing you a great deal of mental stress, at least for a little while? Do you feel calmer or happier after a visit to the gym? What factors in your life cause you stress? Does regular physical activity and exercise help you reduce or control your stress levels? Have you ever known someone who was addicted to exercise and noticed the adverse psychological effects of that addiction? All of these questions are related to the relationship between being physically active and good mental health.

Because so much attention has been focused on the physical health benefits of physical activity, we know much less about the mental health benefits. The emerging clues we do have are most enticing, however. Although most people may assume that physical activity enhances mental well-being, evidence of the biological mechanisms that explain the effects of physical activity or exercise on mental health is very limited. The scientific evidence does support the importance of maintaining physical activity for some mental health conditions such as depression and anxiety, but only limited data are available on other indicators of mental health (U.S. Department of Health and Human Services [USDHHS], Physical Activity Guidelines Advisory Committee [PAGAC] 2008). This chapter addresses the concepts related to physical activity and mental health, and gives a background on the scientific evidence concerning the relationships between physical activity and mental health.

PREVALENCE AND ECONOMIC COSTS OF MENTAL HEALTH DISORDERS

What is mental health? We may each have an idea of what "good" mental health and "bad" mental health is. The World Health Organization (WHO) has defined *mental health* as a state of well-being in which the individual realizes his or her own abilities, can cope with the normal stresses of life, can work productively and fruitfully, and is able to make a contribution to his or her community (WHO 2001a, p. 1).

Mental health is an essential part of overall well-being, and mental disorders or problems can limit people's ability to obtain or maintain functional health. The number and types of mental disorders are numerous and their prevalence increases as people age. Mental disorders include emotional and behavioral symptoms as described in the *Diagnostic and Statistical Manual of Mental Disorders, Fourth Edition* (American Psychiatric Association 2000) and the International Classification of Diseases (ICD) (WHO 2011). Common mental health disorders include schizophrenia, dementia, depression, anxiety disorders, substance dependence, and substance abuse. The material in this chapter is limited to common mental health disorders and problems that have been addressed by a strong or emerging body of research in the relevant literature.

Understanding the prevalence of common mental health disorders is difficult largely because of the cultural stigma associated with being diagnosed with a mental health problem, difficulties people may have in seeking treatment, and differences in assessing the severity of a disorder. Other issues, such as a lack of health insurance, further compound the problem of getting stable national and international estimates of the scope of the problem. The best information available in the United States suggests that common mental health conditions affect 26.2% of American adults in any one year (National Institute of Mental Health [NIMH], USDHHS 2011a). This percentage was the equivalent of 57.7 million Americans in 2004.

Mental disorders were one of the five most costly medical conditions in the United States from 1996 to 2006 (Soni 2009). The number of people associated with these costs for mental disorders almost doubled from 19.2 million to 36.2 million in 10 years. The relative increase in medical expenditures for mental disorders rose from $35.2 billion in 1996 (in 2006 U.S. dollars) to $57.5 billion in 2006. The total economic costs of depression alone have been estimated to be in the tens of billions of dollars each year in the United States.

The largest component of this economic burden derives from lost work productivity due to depression (Wang et al. 2003). In 1990, the U.S. estimated economic cost of anxiety disorders was $46.6 billion, which accounted for 31.5% of the total expenditures for mental illness (Dupont et al. 1996). Obviously, those costs have most likely increased dramatically, although the actual costs in today's dollars have not been reported.

COMMON MENTAL HEALTH DISORDERS

Following are common mental disorders or problems that have been studied in relation to physical activity and exercise (USDHHS, PAGAC 2008):

- Mood disorders
- Anxiety disorders
- Psychological distress
- Age-related decline in cognitive function
- Low self-esteem
- Eating or exercise-related disorders

Mood disorders include depression, bipolar or manic–depressive disorders, medical conditions related to mood changes, and substance-induced mood disorders (American Psychiatric Association 2000). Depression can be classified as mild (also known as dysthymia) or as major depressive disorder (MDD). **Dysthymia**, defined as having depression symptoms for the past two years, affects 1.5% of the U.S. adult population, and 16% of adults experience MDD yearly with symptoms lasting two weeks or more. Adolescents (ages 13 to 18) in the United States have a lifetime combined prevalence of 11.2% for dysthymia and MDD (NIMH, USDHHS 2011a).

The symptoms of depression are diverse and can include difficulty concentrating and making decisions, loss of interest in hobbies and activities, feelings of hopelessness and helplessness, insomnia, and even thoughts of suicide. The worries that are accompanied by depressive symptoms can also lead to physical symptoms such as fatigue, headaches, muscle tension and aches, difficulty swallowing, trembling, twitching, irritability, sweating, and hot flashes. Depression can leave a person emotionally numb or suicidal, and may be related to other factors such as abuse of alcohol or drugs, phobias, obsessions, and preoccupation with physical challenges. Periodic feelings of mild depression are normal for us all, and can be caused by grief due to a loss or a medical condition. However, depression or mood disorders that persist beyond two months may indicate major mood change problems.

Anxiety can be broadly defined as a condition of nervousness, uneasiness, or apprehension about a future event or events. Anxiety, although a predictable part of everyday life, can become a mental disorder over time that can hinder daily functional abilities. Anxiety is usually classified as either state anxiety or trait anxiety. **State anxiety** refers to a person's existing or current emotional state. **Trait anxiety** is specific to a person's personality and has been described in general as type A (aggressive, high-stress personality) or type B (low-key, low-stress personality). Chronic anxiety disorders can lead to specific phobias, social phobias, panic disorders, obsessive-compulsive disorder, or post-traumatic stress disorders.

Psychological distress refers to mental stressors that are not congruous with good health. Feelings you may have when you are sick or facing medical conditions such as surgery are examples of distress. Psychological distress is often measured as subjective feelings of a lack of well-being in subjects in exercise studies. Higher reported levels of well-being are usually associated with a higher quality of life.

Age-related decline in cognitive function refers to negative changes that occur over time in the ability to process, select, manipulate, or store information; it affects both behavior and functional ability. Central nervous system (CNS) disorders associated with genetics and aging that have been linked to mental disorders include dementia (i.e., a loss of brain function that affects memory, thinking, language, judgment, and behavior), multiple sclerosis, Parkinson's disease, and Alzheimer's disease.

Self-esteem refers to feelings of self-worth and value that can influence mental health positively. For example, studies show that people who begin an exercise program may experience higher self-esteem than nonexercisers. Further, more experienced exercisers may maintain higher levels of self-esteem over time if they continue to exercise compared to people who stop exercising. If adverse events such as injury or the adoption of addictive behaviors (e.g., compulsive running, exercise addiction, disordered eating) occur, self-esteem levels may drop or become inconsistent with good mental health. Other addictive behaviors that are associated with lower levels of mental health are **anorexia nervosa** (limiting food intake and becoming excessively lean), **bulimia** (bingeing and purging), and **muscle dysmorphia** (a preoccupation with muscularity).

RISK FACTORS ASSOCIATED WITH MENTAL HEALTH DISORDERS

A full review of the risks associated with mental health disorders is beyond the scope of this text; however, descriptions of some of the risk factors associated with common mental health are provided here. Many mental health disorders are associated with sedentary lifestyles or low levels of physical activity.

Numerous risk factors are associated with mental disorders. Although each disorder has its own unique risk factors, several consistent themes appear across major mental health diagnoses.

Modifiable Risk Factors for Mental Disorders

- Physical inactivity
- Substance abuse (including alcohol)
- Low self-esteem
- Distress
- Negative lifestyle behaviors

Nonmodifiable Risk Factors for Mental Disorders

- Age
- Sex
- Heredity (genetics)
- Undergoing traumatic experiences
- Chronic medical conditions

Following are descriptions of the modifiable and nonmodifiable risk factors for mental disorders:

- *Physical inactivity.* For some mental health conditions, some physical activity appears to be very helpful, although too much exercise may aggravate existing mental health problems.

- *Substance abuse.* The abuse of any legal or illegal substance, particularly over a period of years, can lead to mental health disorders.

- *Low self-esteem.* Negative feelings of one's capabilities, goals, accomplishments, place in the world, and relationships with others can have a major effect on mental health.

- *Distress.* Perceptions of and the ability to cope with the various stressors of life can positively or negatively affect the function of the central nervous system (CNS) and the adoption of positive or negative health behaviors.

- *Negative lifestyle behaviors.* Overcoming negative lifestyle behaviors by adopting positive ones can improve mental health.

- *Age.* Many mental health disorders (e.g., major depressive disorder) are more commonly seen in younger adults; older adults appear to be less affected, with the exception of those who become challenged to maintain their functional health.

- *Sex.* Women are at a higher risk for some mental disorders (e.g., major depressive disorder) than men.

- *Heredity (genetics).* A family history of mental health or sleep disorders may predispose a person to these conditions.

- *Traumatic experiences and medical conditions.* Traumatic experiences and poor health conditions are known risk factors for mental disorders, most of which are nonmodifiable by the individual. Previous suicidal thoughts are also a nonmodifiable risk factor for future episodes. Age, self-esteem, genetics, and current health status all affect how people cope with life stresses, the risk of suicidal thoughts, and how they respond to medical treatments such as surgery and mental health therapy.

PHYSICAL ACTIVITY, EXERCISE, AND MENTAL HEALTH

Prior to a discussion of the physiological mechanisms that may explain the observed relationship between physical activity and some mental health disorders, it is helpful to conceptualize the breadth of work in the area. Perhaps more than any other health outcome in this text, the bulk of the scientific work related to physical activity and mental health has addressed physical activity as a possible *treatment* for the disorder. For example, does a physical activity or exercise program improve the sense of well-being among people diagnosed with trait anxiety? How much improvement might be expected? How long will the improvement last? Does it depend on the dose (i.e., the amount) of physical activity? Can physical activity have an additive effect in an existing treatment regimen—that is, can physical activity or exercise improve the effects of a standard treatment for a mental health disorder?

These questions do not address the issue of whether physical activity can *prevent* some mental health disorders from occurring at all. Clearly, preventing a disease is preferable to treating it once it has been diagnosed (this is a central tenet of public health). However, much less research has been conducted on the preventive role of physical activity in mental health disorders. What we know (or don't know) from studies of people with mental health disorders (e.g., anxiety or depression) is sometimes the best current evidence for efforts to promote physical activity for prevention of mental health disorders.

Dunn and Jewell (2010) created a useful framework for conceptualizing existing (and future) studies of physical activity as a treatment modality for mental health disorders. This framework takes into account the three ways exercise and physical activity may be used with people with mental health disorders: as a monotherapy (i.e., the sole treatment under investigation), as an augmentation therapy (i.e., to add to existing treatments such as prescription drugs), or as an adjunct therapy (i.e., having health benefits other than helping to treat the disease). These three distinctive types are then placed into a 3-by-3 table with the following lengths of effects: acute (short-term) effects (the kind one might see with a standard laboratory-based exercise training study), continuation effects (moderate-term effects that might be expected when patients begin to exercise on their own away from a laboratory), and maintenance effects (longer-term effects that might signify the effectiveness of the physical activity behavior in controlling the condition under study).

The framework in figure 9.1 is particularly useful because it helps us explore and categorize the physiological and behavioral effects that may be at work as mechanisms for any associations between physical activity and mental health outcomes. Acute exercise-related adaptations are most likely to be initially apparent in short-term training studies. The extent of physiological adaptations, as with other health outcomes, are likely dose dependent—that is, higher doses and intensities of physical activity result in greater physiological changes (see chapter 2). These adaptations should remain with a continued dose of exercise into the continuation and maintenance periods, but behavioral changes should also be apparent as the exercise training theoretically evolves into a physically active lifestyle.

What physiological adaptations resulting from physical activity may explain its association with some mental health disorders? As noted in previous chapters, even moderate-intensity physical activity results in improvements in strength and muscular endurance, $\dot{V}O_2$max, force, and power in most previously sedentary people. The extent of these adaptations can be expected to correlate closely with the dose of exercise: the higher the dose and the more intense the physical activity, the greater the

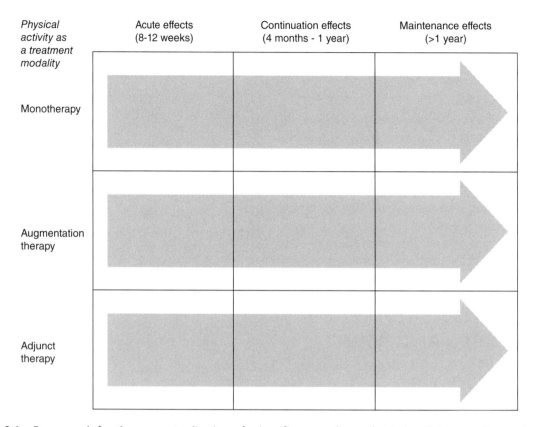

Figure 9.1 Framework for the conceptualization of scientific research on physical activity, exercise, and mental health.
Based on Dunn and Jewell 2010.

physiological response. To understand the impact of these changes, we must also understand their effects on the brain and nervous system—the center of most mental health disorders.

The physiological adaptations that result from physical activity have been shown to improve cerebral capillary growth and development (also called *angiogenesis*), brain blood flow, and oxygenation. A popular hypothesis is that this increase in cerebral blood flow increases cerebral metabolism, and that this increased cerebral activity (particularly in older adults) may be partially responsible for a protective effect of physical activity against mental health disorders (Deslandes et al. 2009). Other, related hypotheses include the notion that exercise improves the regulation of neurotransmitters (i.e., chemical substances that assist in the transfer of nerve impulses across synapses), the growth and maintenance of brain nerve cells, and the ability of nerves to conduct impulses across synapses.

Somewhat distally related to the physiologic adaptations are the biomechanical improvements

that result from physical activity. Biomechanically, people can expect to see improved economy or efficiency (i.e., reduced energy cost at a given workload) after aerobic and musculoskeletal strengthening activities. Improvements in balance, stability, mobility (flexibility and range of motion), and peripheral proprioception (i.e., sense of position and movement) can also help people develop or maintain positive levels of self-esteem and well-being. The central hypothesis is that the development of motor skills allows people to participate in a greater variety of physical activity and exercise activities with more confidence, and that these increases in self-efficacy can result in positive changes in mental health for some people.

How intense does physical activity need to be to have an effect on these physiological markers of mental health? Clearly, the answer varies with the marker, but the best evidence from neurobiology studies is that light physical activity (strolling, performing activities of daily living) is not enough to elicit the necessary physiological responses. The

EXERCISE, PHYSICAL ACTIVITY, AND BRAIN FUNCTION

Exercise and physical activity may also affect brain function in people without mental health disorders. Many have wondered: Can physical activity make us smarter or help us remember more things for longer? Do people who are more physically active have better cognitive function than similar, but inactive people? Do physically active children do better in school than inactive children? Clearly, these questions can be related to mental health in that compromised brain function may be a subclinical precursor to more serious mental disorders.

Cognitive function declines accelerate rapidly in older adults. With populations in developed countries aging rapidly, questions of how physical activity and exercise affect brain function become most important as we seek to keep our parents and grandparents functionally independent for as long as possible.

Brain function outcomes that researchers have investigated for an association with physical activity and exercise are shown in figure 9.2. This is not an exhaustive list, and studies have varied from single-bout exercise studies to short-term training studies. Although definitive conclusions regarding the role physical activity plays in brain function remain to be put forward, the topics highlighted in this figure are some of the most promising areas for future work.

Regular physical activity has measurable and substantial positive effects on mental health.

physical activity must be at 3.0 METs (see chapter 2) or higher (i.e., vigorous or moderate intensity) to generate the physiological stimulus necessary to promote mental health.

TESTING FOR MENTAL HEALTH DISORDERS

Tests to evaluate, diagnose, and treat mental health disorders are typically conducted by physicians or researchers with expertise in the area. Screening tests (to identify cases to refer for more extensive diagnostic workups) are often used as the first step in evaluating mental health disorders. These tests range from self-administered questionnaires to in-person evaluations and observation studies. An example of a simple mood disorder screening assessment is the Profile of Mood States (POMS), which is a questionnaire-type instrument that assesses fluctuating mood states on key markers such as tension-anxiety, anger-hostility, fatigue, inertia, vigor-activity, and confusion-bewilderment. The POMS has been used in a variety of clinical mental health settings including those involving the relationships between physical activity and exercise interventions or participation levels (see www.mhs.com/product.aspx?gr=cli&prod=poms&id=overview for more about the POMS).

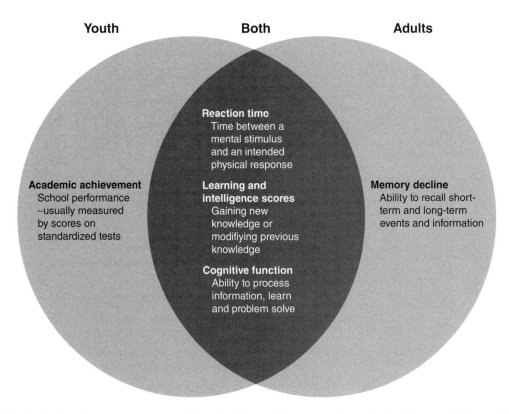

Figure 9.2 Brain function outcomes in youth and adults that have been investigated for associations with physical activity and exercise.

Physical Activity Guidelines for Mental Health

Scientific evidence supports the assertion that physical activity lowers the risk of (1) anxiety symptoms, (2) anxiety disorders, (3) depressive symptoms, (4) major depressive disorder, and (5) age-related decline in cognitive function. There is also adequate evidence that physical activity is associated with enhanced psychological well-being and a delay in the psychological effects of dementia (USDHHS, PAGAC 2008). Despite substantial research in the area, however, physical activity has not been shown to be effective in the treatment of mental health disorders. Currently, physical activity is recommended only as an adjunct therapy to other treatment modalities (USDHHS, National Guideline Clearinghouse 2006).

Figure 9.3 is a summary of recent research on the dose-response relationship between physical activity and exercise and feelings of distress. As shown, modest levels of physical activity are associated with significantly lower odds of distress, or higher

odds of well-being, than low levels of physical activity are. Further, it appears (at least based on these study results) that higher doses and intensities of physical activity do not confer additional benefits. The data in this figure provide more evidence of the public health benefits of physical activity; the

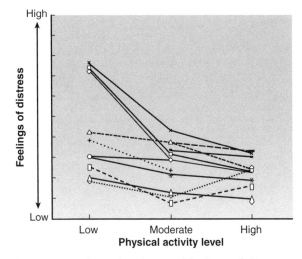

Figure 9.3 Physical activity and feelings of distress—prospective cohort studies 1995-2007.
Adapted from USDHHS, PAGAC 2008.

mental health benefits are not limited only to athletes and people who exercise at the highest levels.

Although extensive research on and scientific evidence of the health benefits of physical activity and exercise on many physical health outcomes exist, only since the 1980s have positive relationships between physical activity and exercise and mental health begun to emerge in the literature. The development of new exercise science technologies and techniques along with the greater public health emphasis on the management of mental health disorders should help us optimize physical activity and exercise interventions for those with mental health disorders.

SCIENTIFIC EVIDENCE

The PAGAC (USDHHS, PAGAC 2008) noted strong scientific evidence that physical activity reduces the risk of depression and cognitive decline in adults and older adults. Limited evidence suggests that physical activity and exercise reduce feelings of distress or increase feelings of well-being. There is moderate evidence that physical activity and exercise improve sleep.

Participation in regular physical activity and exercise can lower the risk for depression, distress and lack of well-being, and dementia by 20 to 30%. Risk reduction for men and women appears to be similar, and there is limited evidence that blacks, Hispanics, and white Caucasians benefit alike.

The dose of physical activity or exercise needed to ameliorate mental health disorders (e.g., depression and distress) is three to five days per week of 30- to 60-minute sessions of moderate to vigorous activities (moderate evidence). Aerobic or muscle-strengthening activities, or both, appear to have similar effects, although the minimal or optimal type or amount of physical activity or exercise for mental health benefits is not yet known. Scientific evidence of a dose-response relationship between physical activity and exercise and lessened anxiety, better cognitive health, and improved sleep patterns is currently insufficient.

The evidence for positive effects on mental health of shorter bouts of physical activity and exercise accumulated throughout the day is limited or insufficient. There is some scientific evidence that physical activity or exercise may reduce the onset, progression, or adverse impact of central nervous system (CNS) disorders such as multiple sclerosis and Parkinson's disease.

© Getty Images

How does participation in physical activity help maintain or improve self-esteem?

GUIDELINES

The following recommendations are from the *2008 Physical Activity Guidelines for Americans* (USDHHS 2008, p. 14):

> Physically active adults have lower risk of depression and cognitive decline (declines with aging in thinking, learning, and judgment skills). Physical activity also may improve the quality of sleep. Whether physical activity reduces distress or anxiety is currently unclear. Mental health benefits have been found in people who do aerobic or a combination of aerobic and muscle-strengthening activities 3 to 5 days a week for 30 to 60 minutes at a time. Some research has shown that even lower levels of physical activity also may provide some benefits. Regular physical activity appears to reduce symptoms of anxiety and depression for children and adolescents. Whether physical activity improves self-esteem is not clear.

KEY LEADER PROFILE

Caroline A. Macera, PhD

Why and how did you get into this line of work? Did any one person have an overriding influence on you?

I am an epidemiologist who specializes in chronic disease etiology and control, but the decision to go into this type of work was driven by my interest in science and the scientific method. I wanted to know why things happened and how to change them. My focus on chronic diseases (e.g., heart disease, cancer, diabetes) came early on because I saw this field as a way to do what I loved (research) and apply my findings to everyday life. My interest in physical activity research developed after meeting Dr. Ralph Paffenbarger. His passion for the field and his groundbreaking work made an impression on me. I realized that not everyone can be a runner or ultra-marathoner, but everyone can incorporate a bit more activity into everyday activities and achieve substantial health benefits. My work at the Centers for Disease Control and Prevention reinforced my commitment to physical activity research.

What are your current research interests?

Teaching epidemiological methods to public health students is very important to me. Helping the next generation of researchers move the field above and beyond where things are now is really exciting as well. My research interests include the effect of physical activity on chronic conditions such as heart disease, cancer, and diabetes. You cannot be a researcher in this area without developing an interest in obesity, the major epidemic of our time.

What drives you to be a researcher and activist in the field of physical activity and health?

When you look at the field of health, so many people spend the last third of their lives with some type of disability. Many times these disabilities can be prevented by maintaining a physically active lifestyle. It is hard for young people to understand the notion that what they do now may affect their quality of life in 30 to 40 years. My goal is to search for ways to make healthy and fit lifestyles the norm and healthy choices about diet and physical activity easy.

What are one or two key issues of importance in our field that must be addressed by 2022?

Obesity is a major health problem for this generation. An unprecedented number of children and young adults are developing diabetes, primarily due to the obesity epidemic. For the first time it may be that the life span of today's children may not exceed that of their parents. There are several ways that the obesity epidemic needs to be addressed, and one of them includes increasing physical activity in schools, at home, and in worksites. Not only will there need to be individual behavioral changes, but policy and environmental changes are needed as well. Keeping children and young adults healthy will prevent a number of chronic conditions from developing as they age.

CHAPTER WRAP-UP

What You Need to Know

- Mental health conditions have been estimated to affect 26.2% of American adults in any one year, and the relative increase in medical expenditures for mental disorders rose from $35.2 billion in 1996 (in 2006 U.S. dollars) to $57.5 billion in 2006.

- Common mental health disorders include depression, bipolar or manic-depressive disorders, medical conditions related to mood changes, substance-induced mood disorders, anxiety, phobias, panic disorders, obsessive-compulsive disorder, posttraumatic stress disorders, feelings of distress, CNS dysfunctions, and addictive behaviors.

- Common risk factors for mental disorders include physical inactivity, poor self-esteem, distress, drug abuse, alcohol abuse, negative lifestyle behaviors, age, sex, suicidal thoughts, and medical treatments.

- Tests to detect mental health disorders are typically conducted by physicians (general practitioners) and specialists (psychiatrics and sleep study experts).

- Mental health benefits have been found in people who do aerobic or a combination of aerobic and muscle-strengthening activities three to five days a week for 30 to 60 minutes at a time.

Key Terms

mood disorders

dysthymia

state anxiety

trait anxiety

psychological distress

age-related decline in cognitive function

self-esteem

anorexia nervosa

bulimia

muscle dysmorphia

Study Questions

1. How does the World Health Organization define *mental health*?

2. Why is mental health an essential part of overall well-being?

3. What is the prevalence of mental disorders in the United States?

4. What are the economic costs associated with treating mental disorders in the United States?

5. What are the definitions of *dysthymia* and *cognitive function*?

6. What are the differences between state anxiety and trait anxiety?

7. What is psychological distress?

8. What are four common risk factors for mental or sleep disorders?

9. What strategies are useful for testing for mental disorders?

10. How much physical activity or exercise is necessary to reduce the risk of mental disorders and improve sleep?

E-Media

Explore issues related to physical activity, exercise, and public health at the following websites:

Human Kinetics	www.HumanKinetics.com
U.S. Department of Health and Human Services: Physical Activity Guidelines for Americans	www.health.gov/PAGuidelines
International Society for Physical Activity and Health	www.ispah.org
American College of Sports Medicine	www.acsm.org
President's Council on Fitness, Sports & Nutrition	www.fitness.gov
World Health Organization	www.who.org
National Institute of Mental Health	www.nimh.nih.gov

Bibliography

American Psychiatric Association. 2000. *Diagnostic and Statistical Manual of Mental Disorders: DSM-IV-TR*. Washington, DC: American Psychiatric Association.

Deslandes A, Moraes H, Ferreira C, et al. 2009. Exercise and mental health: Many reasons to move. *Neuropsychobiology* 59: 191-198.

Dunn AL, Jewell JS. 2010. The effect of exercise on mental health. *Current Sports Medicine Reports* 9: 202-207.

DuPont RL, Rice DP, Miller LS, Shiraki SS, Rowland CR, Harwood HJ. 1996. Economic costs of anxiety disorders, *Anxiety* 2 (4): 167-172.

Soni A. 2009. *The Five Most Costly Conditions, 1996 and 2006: Estimates for the U.S. Civilian Noninstitutionalized Population*. Statistical Brief #248. Rockville, MD: Agency for Healthcare Research and Quality. www.meps.ahrq.gov/mepsweb/data_files/publications/st248/stat248.pdf. Accessed 23 September 2011.

U.S. Department of Health and Human Services. 2008. *2008 Physical Activity Guidelines for Americans*. Washington, DC: U.S. Department of Health and Human Services. www.health.gov/PAGuidelines.

U.S. Department of Health and Human Services, National Guideline Clearinghouse. 2006. *Adult Primary Care Depression Guidelines 2006*. www.guideline.gov/summary.aspx?doc_id=6007. Accessed 20 July 2011.

U.S. Department of Health and Human Services, National Institute of Mental Health. 2011a. Statistics: Any disorder among adults. www.nimh.nih.gov/statistics/1ANYDIS_ADULT.shtml. Accessed September 22, 2011.

U.S. Department of Health and Human Services, National Institute of Mental Health. 2011b. Statistics: Major disorder among children. www.nimh.nih.gov/statistics/1MDD_CHILD.shtml. Accessed September 22, 2011.

U.S. Department of Health and Human Services, Physical Activity Guidelines Advisory Committee. 2008. *Physical Activity Guidelines Advisory Committee Report, 2008*. Washington, DC: U.S. Department of Health and Human Services. www.health.gov/PAGuidelines.

Wang PS, Simon G, Kessler RC. 2003. The economic burden of depression and the cost-effectiveness of treatment. *International Journal of Methods in Psychiatric Research* 12 (1): 22-33.

World Health Organization. 2001a. *International Classification of Functioning, Disability and Health: ICF*. Geneva: World Health Organization.

World Health Organization. 2001b. *Strengthening Mental Health Promotion*. Geneva: World Health Organization (Fact sheet no. 220).

World Health Organization. 2011. International Classification of Diseases. www.who.int/classifications/icd/en/. Accessed 22 September 2011.

PHYSICAL ACTIVITY IN PUBLIC HEALTH SPECIALIST

This chapter covers these competency areas as set forth by the National Society of Physical Activity Practitioners in Public Health:

1.4.2, 2.1.1, 2.2.2, 2.3.3, 3.2.1, 3.2.2, 6.2.1, 6.4.1, 6.4.3

CHAPTER 10

HEALTH RISKS OF EXERCISE AND PHYSICAL ACTIVITY

OBJECTIVES

After completing this chapter, you should be able to discuss the following:

» The primary health risks associated with exercise and physical activity

» Common musculoskeletal injuries associated with physical activity

» Risk factors for exercise-related musculoskeletal injuries

» Exercise-related sudden cardiac death and the factors that predict its occurrence

Opening Questions

» What are the risk factors for exercise-related musculoskeletal injuries?
» Can these injuries be prevented?
» Does exercise cause sudden cardiac death or heart attacks?
» Do the benefits of physical activity and exercise outweigh the risks?
» Must someone consult a physician prior to beginning to exercise?

Throughout part II, the emphasis has been on the role physical activity plays in improving and promoting health. Physical activity lowers the risks of heart disease, some cancers, and diabetes; improves the musculoskeletal system; and prevents bone diseases. It improves quality of life and some mental health disorders. The benefits of physical activity are remarkable, particularly its ability to lower the risks of chronic diseases, and new research continues to teach us about these, and other, positive outcomes. But does physical activity also have a downside? Can participating in physical activity and exercise actually increase the risk of certain conditions? If so, do the benefits of being physically active outweigh the risks?

In public health, it is important to understand not only the health benefits of a certain behavior or intervention you may be promoting, but also the risks. Knowledge of a downside to any program is critical to having a complete picture. Vaccination programs, environmental changes for cleaner air, educational programs for HIV prevention—all of these are examples of situations where programs with good intentions might have unintended risks to people and populations you are targeting. Understanding the risks as well as the benefits of particular programs is practicing "responsible" public health.

Two primary unintended consequences (risks) of physical activity that have been extensively studied are musculoskeletal injury and exertion-related sudden cardiac death. In this chapter, we review these risks and put them in perspective in terms of the costs and benefits of increasing physical activity.

Musculoskeletal Injuries

A physical activity–related **musculoskeletal injury** involves some type of acute disorder in a bone, muscle, joint, or connective tissue that is attributable to physical activity or exercise. Such injuries can occur suddenly, such as an ankle sprain, or over a period of **exposure**, such as a gradual pain in the shoulder of an electrician who frequently works over his head installing circuits. Ligament tears, sprains, strains, bone fractures, bruises, and joint dislocations are common musculoskeletal injuries that can result from physical activity and exercise. Clearly, such injuries can occur without physical activity (e.g., in a motor vehicle accident), but this chapter addresses only those that result from some type of body movement.

A difficult problem when studying musculoskeletal injuries involves the definition of *injury*. What qualifies as an injury? You probably know what an injury means to you, but does your mother have the same definition? How about a world-class sprinter or a heavy machine operator? Chances are that each person has a unique idea of what an injury is. Many times it involves pain, loss of function, and an inability to work or socialize. Some people have a higher threshold for pain than others do, so the same incident in two people may be classified as an injury by one person and a "bump" by the other.

Severity is also an issue. How long does the pain or loss of physical function have to last before the incident is called an injury? Thirty minutes? Thirty days? Must a person see a physician or other health professional before an incident can be classified as an injury? What if that person does not have health insurance? Surely someone without health insurance would be less likely to see a doctor or health care professional for an injury than someone with insurance and the same injury. This could result in a study or survey counting one occurrence but not the other, simply because the latter could not be "found." Does someone have to be injured during an exercise session for the injury to be considered exercise related, or does an injury caused by cumulative

DEFINITION OF INJURY

What type of exercise-related musculo-skeletal injury would be severe enough for you to report it on a survey? Would you be able to remember a sore back after doing a day's worth of gardening six months ago? Studies of musculoskeletal injury are limited because of the lack of a standardized definition of what constitutes an injury and the variability in the recall of study respondents.

exposure to exercise (e.g., arthritis in the knee) also count as an exercise-related injury?

The point is that, although the scientific literature is replete with studies that have examined musculoskeletal injuries as related to physical activity and exercise, a consistent definition has rarely been used. This makes comparing studies nearly impossible. Studies that rely on participants' self-reports of "any injury" are not comparable to those that require a doctor's diagnosis prior to being classified as an injury. What we know about the rates and risks of musculoskeletal injuries due to physical activity and exercise is therefore limited compared to the wealth of information available on the health benefits.

What is the prevalence of exercise-related injury? This simple question, unfortunately, does not have a simple answer because of many complications.

First, as discussed, the definition of *injury* is varied. Most people would count an event that was serious enough to require a trip to an emergency room, but what about something less serious? Is an event that requires you to take a few days off work, but not a trip to the emergency room, serious enough to be considered an injury? What about simply taking a few aspirin and self-treating for a few days? The main limitation to studies of physical activity and musculoskeletal injury is the use of inconsistent (and incomplete) definitions.

Another complicating factor in determining the prevalence of exercise-related injury is the fact that different types of physical activity have different participation rates and may result in different types of injury. Low-impact and noncontact exercise activities and sports are likely less risky for musculoskeletal injury than are high-impact and contact sports. These differences make it somewhat meaningless to discuss the prevalence of injuries in the same manner we talk about the prevalence of diabetes or myocardial infarction, both of which are single diagnoses. Similarly, we expect weight-bearing activities such as walking or running to be associated with injuries of the lower extremities more often than other parts of the body. Racket sports such as tennis and squash may be more likely to result in injuries of the upper extremities (shoulders and arms) or the head (from being struck by a racket or a ball).

Finally, in the United States there are no routine surveys or systems from which to generate a picture of exercise-related injuries. There have

NUMERATOR MONSTER

Studies of exercise-related injuries are particularly prone to the **numerator monster**. This problem arises when injuries are counted, but the population at risk (the denominator) is ignored. Epidemiologists rely on the number of cases (numerator) as well as the size of the population at risk (denominator) to calculate prevalence and incidence. If one or the other is not known, it is impossible to compare types of activity and their respective risks for musculoskeletal injury. For example, if 100 walkers and 100 rugby players (both numerators) showed up at an emergency room one weekend for treatment of exercise-related injuries, one might say that the activities are equally dangerous. This assumption would be incorrect, however. There are many more walkers than there are rugby players in a community. Without taking into account the difference in size between the populations at risk (the entire group of walkers and the entire group of rugby players), one would miss the fact that rugby players are much more likely to be injured than are walkers. Beware of the numerator monster.

been periodic studies (Powell et al. 1998), studies of catastrophic injuries (Mueller and Cantu 1991), and studies that focus on numerators (Gotsch et al. 2002), but none of these have tracked these problems over time. This situation obviously makes it impossible to truly understand the risks of physical activity, which limits public health professionals' ability to give an accurate risk/benefit assessment.

Although population-based exercise-related prevalence data are limited, we do know some things about what might be expected to occur for several types of physical activity in a defined time period. The data in table 10.1 are from a national survey of injury (Powell et al. 1998). By asking respondents what they were doing when they became injured, the investigators were able to compare various types of common physical activities. Obviously, each of the reported activities is fairly safe; fewer than 3 people out of 100 were injured in any 30-day period. Outdoor bicycle riding appears to be the least risky activity in terms of musculoskeletal injury, and weightlifting was the riskiest.

Although these findings may appear to be intuitive, they emphasize the need to quantify the risks of physical activity for application in the real world. For example, such information can be very useful to a program manager who is beginning a community-based walking program for sedentary adults. After reviewing the data in table 10.1, the manager now knows that she might expect one or two people in her program to be injured during the walking program in a given month (30-day period). This is useful information for program planning and evaluation. If she puts appropriate preventive strategies in place and none of her participants become injured, she can report that the participants in her program are injured less frequently than what one might expect given the literature.

Despite the problems in the scientific literature, we do know some things about the causes and risk factors for physical activity–related musculoskeletal injuries. These factors have been identified in the scientific literature from studies in epidemiology, biomechanics, physiology, and medicine. As with other health-related outcomes for physical activity and exercise, risk factors conveniently can be classified as modifiable (i.e., things that can change or be changed) and nonmodifiable (i.e., things that typically can't be changed or are difficult to change).

Modifiable Risk Factors for Musculoskeletal Injuries

- Amount and type of current physical activity
- Cigarette smoking
- Low physical fitness level
- Improper use of protective equipment
- Adverse environmental conditions

Nonmodifiable Risk Factors for Musculoskeletal Injuries

- Age
- Sex (for some types of injury)
- History of injury
- Amount of physical activity in the past (history)
- Anatomical factors
- Environmental, or external, conditions

Following are descriptions of the modifiable and nonmodifiable risk factors for musculoskeletal injury:

- *Amount and type of current physical activity.* The more physical activity a person performs, the higher the risk of musculoskeletal injury associated with

Table 10.1 Percentage of Participants Reporting a Musculoskeletal Injury by Type of Physical Activity

Type of physical activity	Percentage injured in 30 days
Aerobics or aerobic dance	1.4
Gardening or yard work	1.6
Bicycle riding (outdoors)	0.9
Walking for exercise	1.4
Weightlifting	2.4

Adapted from Powell et al. 1998.

the activity. This finding has been demonstrated repeatedly in the literature. Moreover, different types of physical activity and exercise convey different risks. For example, contact sports are more likely to be related to injury than noncontact sports.

• *Cigarette smoking.* Although exercise and cigarette smoking would appear contradictory behaviors, people whose occupations demand physical activity, such as construction workers and landscapers, may also smoke. Cigarette smoking seems to increase the risk of physical activity–related musculoskeletal injuries, possibly as a result of vasoconstriction, which restricts the amount of oxygen being delivered to the muscles or connective tissues. The structure of the site and the availability of metabolic nutrients are then altered, and the hypothesis is that this alteration makes the muscle or connective tissue more susceptible to injury.

• *Low physical fitness level.* People who have higher physical fitness levels (measured as $\dot{V}O_2$max) have been consistently shown to be at lower risk of musculoskeletal injury related to physical activity.

• *Improper use of protective equipment.* Bicycle helmets, protective padding for skateboarders, breakaway bases for baseball players, mouthguards for certain sports, shoes—all of these are examples of protective equipment that, when properly used, can prevent musculoskeletal injuries associated with physical activity.

• *Adverse environmental conditions.* Environmental conditions can be considered either nonmodifiable or modifiable. If conditions are not conducive to physical activity or could increase the risk of injury during physical activity, venues can be changed, activities can be rescheduled, or the type of physical activity can be modified (e.g., going to the gym rather than playing basketball in the rain).

• *Age.* Changes in the musculoskeletal system that occur with aging result in older people being more likely to be injured than younger people doing the same activity.

• *Sex.* Women's skeletal structure and sex hormones have been hypothesized to increase their risk of lower extremity injury (specifically, to the anterior cruciate ligament in the knee) compared to men doing the same activities.

• *History of injury.* A history of injury is one of the most consistent risk factors for injury during physical activity reported in the literature. People who have been injured previously are more likely to be injured in the future than those who have not. This is a strong rationale for efforts to prevent injuries from occurring in the first place.

• *Amount of physical activity in the past (history).* Much of what we know in the area of physical activity and musculoskeletal injury comes from studies of military recruits who participate in basic training involving substantial physical activity. Recruits who were physically active prior to the training were less likely to be injured during the training. Again, this finding makes a powerful case for injury prevention.

• *Anatomical factors.* Each human body is unique, and a person's biomechanical and anatomical characteristics may increase the risk of an (or exacerbate an existing) exercise-related musculoskeletal injury. Among the many factors that have been hypothesized are varus, or bowlegs (an abnormal inward angle of a bone); valgus, or knock-knees (an abnormal outward angle of a bone); pes cavus (an abnormally high foot arch); and pes planovalgus (flat feet). Many anatomical problems can be reversed through medical intervention.

• *Environmental, or external, conditions.* A frequently overlooked risk factor for exercise-related musculoskeletal injuries is environmental conditions. Traffic, damaged or wet playing fields or courts, and broken sidewalks are all examples of environmental, or external, conditions that could increase the risk of a musculoskeletal injury associated with physical activity.

KINESIOLOGY AND MUSCULOSKELETAL INJURIES

The scientific literature fairly consistently reports a dose-response relationship between the risk of musculoskeletal injury and the overall dose (or volume) of physical activity. The results come from studies of runners and walkers (Macera et al. 1989) and military recruits (Almeida et al. 1999). Quite simply, the more physical activity you do, the higher your risk of musculoskeletal injury. The dose of physical activity, as we learned in chapter 2, is related to frequency, intensity, and duration.

The type of physical activity also can influence the risk of injury. Low-impact weight-bearing activities (e.g., walking) or non-weight-bearing activities (e.g., swimming laps, cycling) are thought to be associated with the fewest musculoskeletal injuries related to physical activity. In contrast, running and sport participation (particularly contact sports) may carry a much higher risk of injury. This difference is thought to operate through greater stresses on connective tissue and higher-impact forces on bones and joints. In a study conducted by Hootman and colleagues in 2001, people who reported participating in sports had nearly twice the risk of activity related injury compared to nonexercisers (see figure 10.1).

Among people who play sports, collision sports (ice hockey, American football, rugby) or contact sports (basketball, soccer) carry a higher risk of injury than limited-contact sports (baseball) and noncontact sports (cycling, racewalking). These results come from multiple surveys, each with its own definition of *injury*.

An interesting line of research has emerged regarding physical activity and risk of musculoskeletal injuries. Although higher doses of physical activity appear to be related to a higher acute risk of injury associated with that activity, could it be that the overall risk of being injured (both exercise- and non-exercise-related musculoskeletal injuries) shows an overall decrease with increased physical activity? This would seem an important overall health question, particularly for older adults who are susceptible to injury as a result of falls.

Carlson and colleagues (2006) examined this question in a study of more than 96,000 adults in the United States. Survey respondents were asked about their physical activity behaviors and classi-

Physical activity can increase the risk of adverse events such as musculoskeletal injuries or sudden cardiac death. Do the benefits of participating in regular physical activity outweigh the risks?

fied into three groups: meeting physical activity guidelines (at the time of the study, 30 minutes of moderate-intensity aerobic physical activity on five or more days per week), insufficiently physically active (some activity reported, but not meeting the guidelines), and inactive (no physical activity reported). The authors studied injury patterns occurring acutely during physical activity and in times not associated with physical activity across the three groups. Key results are shown in figure 10.2.

The risk of exercise-related musculoskeletal injury was elevated during physical activity in both groups (20% in those who were insufficiently active and 53% in those who were meeting physical activity guidelines). This was expected because more active people seem to be at higher risk for injury during physical activity. The surprising finding came in the overall injury risk (combining exercise-related and non-exercise-related injuries). People who were active (–3%) or somewhat active (–12%) actually were at a lower *overall* risk of injury compared to inactive people. The finding was particularly striking for injuries not due to physical activity. These results seem to suggest that, although

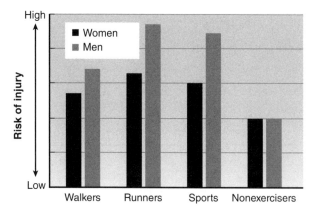

Figure 10.1 Rates of exercise-related musculoskeletal injuries for men and women.
Adapted from Hootman et al. 2001.

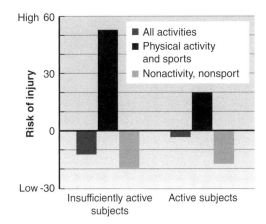

Figure 10.2   Association of habitual physical activity with risk of injury during physical activity and overall. Risk of injury is relative to inactive subjects.
Adapted from Carlson et al. 2006.

the risk of exercise-related musculoskeletal injuries is elevated during exercise, people who are habitually active have a lower overall risk of any injury.

Many people believe that stretching muscles prevents exercise-related musculoskeletal injuries. Stretching, either before or after an exercise session, feels good and increases flexibility around the joints. The thinking is that this increase in flexibility and the additional blood flow to the areas being stretched put muscles, joints, and connective tissues at a lower risk of strain or stress and subsequently a lower risk of injury. Although this is an attractive hypothesis, the results of numerous studies have failed to confirm this finding. In fact, the scientific literature is now consistent with the conclusion that stretching before or after an exercise session

does not decrease the risk of injury. Flexibility is obviously an important component of physical fitness—it just does not seem to result in a lower risk of exercise-related musculoskeletal injuries.

SUDDEN ADVERSE CARDIAC EVENTS

The images are all too familiar: A recreational runner dies during a weekend 10K race. A homeowner shoveling snow from his driveway after a winter storm falls to the ground and dies. A young, seemingly fit and healthy basketball player dies during a game. Sudden cardiac death is an unexpected death due to a dysfunction of the heart, usually within one hour of the onset of symptoms. Physical activity and exercise, particularly when performed at a vigorous intensity, are associated with a higher risk of a sudden **adverse cardiac event** (i.e., cardiac arrest, cardiac death), when compared to times of no activity or light or moderate-intensity activity.

Many types of cardiac disorders can place people at an increased risk of sudden cardiac death. As reviewed in chapter 5, the risk of atherosclerotic heart disease risk is substantially lower among people who are or who become physically active. This lower risk is primarily seen in men and women older than 30. Other disorders, including hypertrophic cardiomyopathy (a genetic disorder characterized by an overly thick wall in the left ventricle), electrical conduction disorders, and abnormalities in the cardiac arteries, are all important conditions that place people at risk of sudden cardiac death,

OVERLOAD, ADAPTATION, AND SPECIFICITY

The fundamental principles of overload, adaptation, and specificity outlined in chapter 2 are particularly relevant when discussing the prevention of exercise-related musculoskeletal injuries. Overload, or exercising beyond usual levels, stimulates bones, muscles, joints, and connective tissue to increase their function. If this overload is sustained (or repeated frequently), the body adapts to a new normal based on the physical training. Laboratory-based studies and animal studies have taught us that bone, muscles, ligaments, and tendons all adapt to increased physical activity (i.e., they get stronger with more physical activity, or overload), and that targeted activity (e.g., focusing on the lower limbs) will likely prevent injuries in that area. Large overloads with little or no time to adapt may increase the risk of exercise-related injury (particularly traumatic injury). Little or no overload does not result in any training or need to adapt and also may increase the risk of injury.

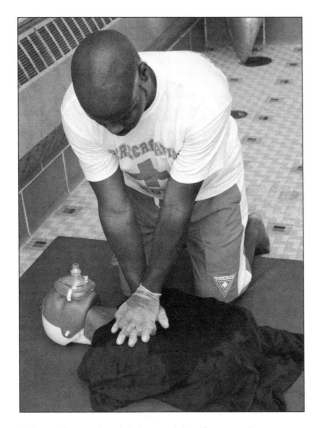

When does the highest risk of a cardiac event associated with exercise occur?

particularly during physical activity and exercise. Such conditions are mostly inherited. Although regular participation in moderate-intensity physical activity can lower the risk of chronic diseases and conditions such as obesity, atherosclerotic heart disease, diabetes, colon and breast cancer, and osteoporosis, people with conditions that increase the risk of sudden cardiac death should avoid vigorous-intensity physical activity. Moreover, people with a family history of these conditions should be examined and monitored by a physician.

Vigorous-intensity physical activity has also been shown to be a trigger for sudden cardiac death due to atherosclerotic heart disease. This process, thought to be a result of a fibrous plaque that ruptures inside a coronary artery and essentially cuts off the blood flow to a portion of the heart, can be a very dangerous situation. Evidence of this process comes from studies of people who have had adverse sudden cardiac events.

Mittleman and colleagues (1993) studied more than 1,200 men and women who had survived a myocardial infarction (i.e., a sudden loss of oxygen

to the heart muscle). They interviewed all study subjects and determined what they were doing immediately prior to or during the cardiac event and classified them based on whether they were physically active. Figure 10.3 illustrates the main results.

As figure 10.3 shows, the risk of a myocardial infarction related to exertion is largely limited to the first hour after the activity. During this time, men and women in this study were nearly six times more likely to develop a myocardial infarction. After the first hour, the risk was negligible. Although these events were nonfatal, this evidence is compelling and clearly suggests that the period of highest risk for an adverse cardiac event is during or within the first hour after cessation of exercise.

If vigorous-intensity physical activity increases the risk of adverse cardiac events, why is it recommended? Why promote physical activity if it increases the risk of death or nonfatal heart attacks? The answer to this question lies in the big picture. Multiple studies since the 1980s have shown that, although vigorous-intensity physical activity acutely increases the risk of sudden cardiac death, the *overall* risk of sudden cardiac death (throughout the rest of the day) is actually lower among people who are habitually physically active. That is, the cardiac risk that occurs as a result of an acute bout of physical activity is outweighed by an overall lower risk of sudden death when active people are compared to inactive people. This concept is perhaps best illustrated by figure 10.4.

Over a 24-hour period, the risk of cardiac arrest for someone who remains inactive (thin unbroken

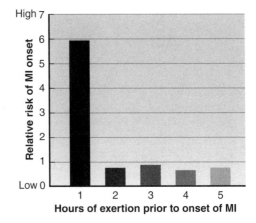

Figure 10.3 Relative risk of onset of myocardial infarction by hours of exertion prior to an event.
Adapted from Mittleman et al. 1993.

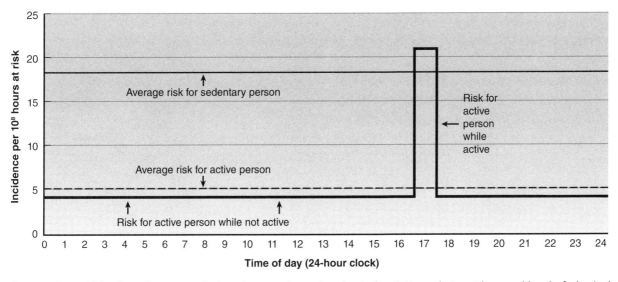

Figure 10.4   Risk of cardiac arrest during vigorous-intensity physical activity and at rest by usual level of physical activity.
Reprinted from USDHHS, PAGAC 2008.

line in figure 10.4) is far higher than the average risk for someone who is habitually active (dashed line). The risk for a cardiac arrest is far higher for that active person during or immediately after the period of physical activity (spike), but overall, the active person is at a much lower risk (even when this period of activity is taken into account) than the inactive (sedentary) person. This situation is similar to that reviewed earlier for musculoskeletal injuries. The overall benefit to being physically active far outweighs the short-term acute risk of adverse events due to physical activity.

Must someone who is sedentary consult a physician or other health care provider prior to beginning a physical activity program? This message has been around for many years and always seems like a good idea. The reality, however, is that this creates a significant barrier for many people who can use it as an excuse not to exercise.

According to the U.S. Physical Activity Guidelines Advisory Committee (U.S. Department of Health and Human Services [USDHHS], Physical Activity Guidelines Advisory Committee [PAGAC] 2008), there is no evidence that people who visit a physician or other health care provider prior to starting an exercise program are any safer than those who do not. Following the fundamental principles of exercise physiology (overload, adaptation, and specificity) by making small, comfortable increases in physical activity over one's usual behavior should minimize any acute cardiac or musculoskeletal risks associated with physical activity. In these cases, consultation with a health care provider is not necessary. People who plan to make large, high-intensity increases in physical activity without allowing for any adaptation time, who have chronic conditions that may increase acute risk, or who have general concerns about exercising should consult with a health care provider. Someone who is sedentary and wants to begin exercising with a walking program of light to moderate intensity would not need such a consultation.

KEY LEADER PROFILE

William E. Kraus, MD

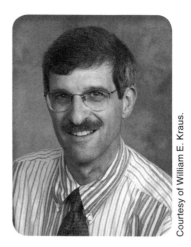

Courtesy of William E. Kraus.

Why and how did you get into this field? Did any one person have an overriding influence on you?

I got interested in exercise science during my senior year in high school. During the 1972 summer Olympics, NOVA aired a program on the East German athletics training program. Apparently they were doing muscle biopsies and targeting athletes to events and training programs for events based on their skeletal muscle characteristics, such as fiber type and enzyme content. Of course, they did not mention the use of steroids, but it was the assumption at that time that these were part of the armamentarium! I thought this was a good way to direct training regimens and that it might be a way to understand how exercise training worked.

When I arrived for medical school at Duke in the summer of 1979, knowing that I was interested in exercise science as a career goal, I sought the advice of Dr. Andy Wallace, who was then chief of cardiology at Duke. He had just started the Duke University Preventive Approach to Cardiology (DUPAC) program, which was one of the original regional cardiac rehabilitation programs in the country. Andy had just consulted with Herman Hellerstein and Ken Cooper (the father of the aerobics movement in the late 1950s) to do this, and Andy served as an inspiration and mentor to me as I began to develop my interest in exercise medicine.

I spent my medical school years assisting with cardiac rehabilitation and learning exercise physiology and exercise medicine on the job. It was a great way to learn the practical aspects of the field and how exercise provides help to patients. After clinical housestaff and cardiology fellowship training, I spent my postdoctoral research years learning molecular biology and how to apply it to animal models of exercise, while working under my other strong mentor, Dr. R. Sanders (Sandy) Williams. However, always wanting to go back to the idea of using skeletal muscle to understand the physiological and therapeutic effects of exercise, when I started my own research career in the 1990s, I developed it around human studies and finding ways to do molecular biology in human tissues. With the rapid advance of molecular techniques in the 1990s and early 2000s, the ability to perform meaningful molecular studies on small amounts of human skeletal muscle tissue became a reality. Thus, the STRRIDE studies were born, which have occupied a good deal of my attention for the last 13 years.

What are your current research interests?

We are interested in understanding several issues related to the health benefits of exercise. First, we want to better understand the dose-response effects of regular exercise on health parameters; that is, we want to find out what type, amount, and intensity of exercise is best for what health benefit. Second, we want to understand the basic underlying human physiological mechanisms that mediate these benefits. Third, we want to understand how individual genetic predisposition (for disease and physiological response to exercise in particular and lifestyle medicine in general), medical therapy, and regular exercise interact to mediate change in health in humans. These three issues require intense investigation over the next 10 years if we are to make real advances in our field.

What drives you as a researcher and activist?

I imagine that, most of all, I am driven by an underlying belief, supported ever more each day, that regular exercise is a useful and essential therapeutic tool to maintain human happiness, health, and quality of life.

CHAPTER WRAP-UP

What You Need to Know

- Musculoskeletal injuries and sudden adverse cardiac events are two important risks of physical activity participation.
- The exercise physiology principles of overload, adaptation, and specificity are important considerations in understanding the risks of exercise and physical activity.
- The definition of musculoskeletal *injury* is an important consideration for interpreting the literature.
- Studies of musculoskeletal injury prevalence and incidence must take the population at risk into account.
- The acute risks of injury and adverse cardiac events are elevated during or immediately after a physical activity bout. The overall risk of both conditions is lower among people who are habitually physically active.
- Consultation with a medical care provider may be necessary for some people, but for the majority of sedentary people who wish to become active at a moderate level, such requirements should not be a barrier to participation.
- The benefits of physical activity participation far outweigh the risks.

Key Terms

musculoskeletal injury

exposure

numerator monster

adverse cardiac event

Study Questions

1. What is a physical activity–related musculoskeletal injury?

2. Why is it difficult to compare studies that have focused on musculoskeletal injuries?

3. What is the numerator monster and how is it related to studying musculoskeletal injuries?

4. What are five modifiable and five nonmodifiable risk factors for physical activity–related musculoskeletal injuries?

5. Is there a dose-response relationship between physical activity and potential risk of injury?

6. Can regular participation in physical activity reduce the risk of musculoskeletal injuries? If so, explain how.

7. How do the basic principles of exercise physiology (overload, adaptation, and specificity) help us understand the risk of musculoskeletal injuries related to physical activity?

8. Does stretching before or after physical activity or exercise decrease the risk of musculoskeletal injuries?

9. What is a physical activity–related sudden cardiac event?

10. What are the relations among acute physical activity, habitual physical activity, and the risk of a sudden adverse cardiac event?

E-Media

Explore issues related to physical activity, exercise, and public health at the following websites:

Human Kinetics	www.HumanKinetics.com
U.S. Department of Health and Human Services: Physical Activity Guidelines for Americans	www.health.gov/PAGuidelines
International Society for Physical Activity and Health	www.ispah.org
American College of Sports Medicine	www.acsm.org
President's Council on Fitness, Sports & Nutrition	www.fitness.gov
American Heart Association	www.heart.org
World Health Organization	www.who.int
National Physical Activity Plan	www.physicalactivityplan.org

Bibliography

Almeida SA, Williams KM, Shaffer RA, Brodine SK. 1999. Epidemiological patterns of musculoskeletal injuries and physical training. *Medicine & Science in Sports & Exercise* 31: 1176-1182.

Carlson SM, Hootman JM, Powell KE, Macera CA, Heath GW, Gilchrist J, Kimsey CD Jr, Kohl HW III. 2006. Self-reported injury and physical activity levels: United States 2000-2002. *Annals of Epidemiology* 16: 712-719.

Gotsch K, Annest JL, Holmgren P, Gilchrist J. 2002. Nonfatal sports- and recreation-related injuries treated in emergency departments—United States, July 2000–June 2001. *Morbidity and Mortality Weekly Report* 51: 736-740.

Hootman JM, Macera CA, Ainsworth BE, Martin M, Addy CL, Blair SN. 2001. Association among physical activity level, cardiorespiratory fitness and risk of musculoskeletal injury. *American Journal of Epidemiology* 154: 251-258.

Macera CA, Pate RR, Powell KE, Jackson KL, Kendrick JS, Craven TE. 1989. Predicting lower-extremity injuries among habitual runners. *Archives of Internal Medicine* 149: 2565-2568.

Mittleman MA, Maclure M, Tofler GH, Sherwood JB, Goldberg RJ, Muller JE. 1993. Triggering of acute myocardial infarction by heavy physical exertion: Protection against triggering by regular exertion. *New England Journal of Medicine* 329: 1677-1683.

Mueller FO, Cantu RC. 1991. The annual survey of catastrophic football injuries: 1977-1988. *Exercise and Sport Sciences Reviews* 12: 261-312.

Powell KE, Heath GW, Kresnow MJ, Sacks JJ, Branche CM. 1998. Injury rates from walking gardening, weightlifting, outdoor bicycling and aerobics. *Medicine & Science in Sports & Exercise* 30: 1246-1249.

U.S. Department of Health and Human Services, Physical Activity Guidelines Advisory Committee. 2008. *Physical Activity Guidelines Advisory Committee Report, 2008.* Washington, DC: U.S. Department of Health and Human Services. www.health.gov/PAGuidelines.

PHYSICAL ACTIVITY IN PUBLIC HEALTH SPECIALIST

This chapter covers these competency areas as set forth by the National Society of Physical Activity Practitioners in Public Health:

2.1.1, 2.2.1, 2.2.2, 5.2.5, 6.1.1, 6.1.3, 6.1.4, 6.1.5, 6.2.1, 6.3.1, 6.3.6, 6.4.1, 6.4.4

PART III

STRATEGIES FOR EFFECTIVE PHYSICAL ACTIVITY PROMOTION

© BananaStock

INFORMATIONAL APPROACHES FOR PROMOTING PHYSICAL ACTIVITY

CHAPTER 11

OBJECTIVES

After completing this chapter, you should be able to discuss the following:

» The Guide to Community Preventive Services Task Force recommendations for physical activity promotion

» The rationale for promoting physical activity interventions through informational approaches

» Considerations for using community-wide campaigns

» Considerations for using mass media campaigns

» Characteristics of effective health education curricula

» Considerations for using classroom-based health education programs

Opening Questions

» Have you ever tried to help someone increase his or her physical activity levels or begin an exercise program by giving information on the health benefits of physical activity?

» Did it work? How successful were you over the short term and long term?

» Which types of informational approaches actually increase physical activity?

» What are the components of these approaches?

Parts I and II of this text focused on defining the field of physical activity and public health and on outlining the myriad health benefits to being physically active. In this part of the text we introduce strategies that have been proven to help people increase or maintain their physical activity levels. The information in this part, derived from behavioral, population, and environmental research, separates public health research from other forms of research. In most cases, results from public health research are translated into practice to improve the health of populations and individuals. Public health research generates new knowledge, as do virtually all other types of research. Fundamentally, though, applying the results of public health research to a particular health problem is a critical step.

The best resource for translating public health research into practice in the United States is The Guide to Community Preventive Services, or the Community Guide (www.thecommunityguide. org). The **Community Guide** contains guidance on health improvement strategies for numerous topics including physical activity. Each recommended strategy in the Community Guide is based on a rigorous review of available scientific evidence—the recommended strategies for health promotion and disease prevention that are found in the Community Guide, for example, have been thoroughly reviewed and scientifically tested.

The Physical Activity sections of the Community Guide include the following review areas, which are discussed here and in more detail in chapters 12 through 14:

• School-based methods (chapter 12)

• Behavioral and social methods (chapter 13)

• Environmental and policy approaches (chapter 14)

Understanding the Community Guide

The Community Guide is an ever-expanding resource for recommendations on evidence-based interventions to improve public health. The Task Force on Community Preventive Services (Task Force) was established by the U.S. Department of Health and Human Services (USDHHS) in 1996 to clarify which community-based health promotion and disease prevention interventions work, and which do not. The U.S. Centers for Disease Control and Prevention (CDC) is the USDHHS agency that provides the Task Force with technical and administrative support. Our review is drawn from the recommendations in the Community Guide,

Evidence-Based Public Health

The Community Guide offers practitioners and decision makers strategies for preventing or addressing health problems. To save from having to continually try various strategies that may or may not be successful, the Community Guide provides a one-stop shop for people interested in translating research into programs that have a better chance of succeeding. The Community Guide saves time and money and helps keep public health program developers from having to reinvent the wheel. When in doubt, go with the evidence-based strategy.

but we recommend that you also visit the Community Guide website and read the information contained under the General Use section (www.thecommunityguide.org/uses/general.html).

The information in the Community Guide provides support for a variety of activities related to public health, such as the following:

- *Policies.* Provides an understanding of concepts and research that can help in the development of more effective legislation and organizational policies.
- *Research.* Identifies research gaps, research priorities, and high-quality evidence-based studies.
- *Programs.* Helps with program planning and health promotion services.
- *Education:* Disseminates knowledge of effective public health strategies.
- *Funding:* Provides background information for creating grant proposals and gaining access to funding streams.
- *General:* Helps determine what works and how to make wise use of resources, and builds community support.

When reviewing research from physical activity programs, the Task Force (www.thecommunityguide.org/about/index.html) takes into account the types of activities targeted, the breadth of their impact, how programs are delivered, the target population, and the type of setting in which the programs are delivered. The Task Force also seeks the answers to the following questions about specific physical activity interventions:

- Does it work?
- If it does work, how well?
- For whom does it work?
- Under what circumstances is it appropriate?
- What does it cost?
- Does it provide value?
- Are there barriers to use?
- Are there any risks?
- Are there any unanticipated outcomes?

Reprinted from the Community Guide.

Although a review of all the criteria the Community Guide uses to judge a study or studies is beyond the scope of this text, suffice it to say that they are very stringent. Only the best studies, with adequate numbers of participants and methods, are evaluated. Once all of the studies in a particular area are assessed, the Task Force recommendations about specific physical activity programs are categorized into the following three broad groupings:

- *Recommended:* Strong or sufficient evidence that the intervention or program is effective.
- *Recommended against:* Strong or sufficient evidence that the intervention or program is not effective, or is harmful.
- *Insufficient evidence:* The available studies do not provide sufficient evidence to determine whether the intervention is effective.

Classifying a program as having insufficient evidence can mean one of several possible things. First, additional research may be needed to determine whether the intervention is effective; the number of studies may be insufficient to draw any firm conclusions; or the studies may lack sufficient quality. Second, the studies may all be of sufficient quality and size, but they may present inconsistent or contradictory findings, or both. Finally, the studies may be consistent, but the results may not be of sufficient size or intensity to confidently describe an effect—that is, the statistical significance of the findings may not support the results. In this case *insufficient evidence* does *not* mean that the intervention or program does not work.

RATIONALE FOR INFORMATIONAL APPROACHES

If knowledge is power, does more knowledge about the health benefits of physical activity translate into more power to change behavior among people who are physically inactive? **Informational approaches** may be designed to increase leisure, occupation, transportation, or at-home physical activities. These approaches are based on the idea that when people are taught about the health benefits of a certain behavior (such as physical activity), they will change their behavior for the better. Obviously, we all know that we sometimes choose to behave in ways that are not good for us; however, informational approaches can increase our knowledge and reinforce our desire to change when we are motivated to make a change.

In general, informational approaches to physical activity promotion may change behaviors through several pathways. First, an increase in knowledge about the health benefits and risks of physical activity (see chapters 5 through 10) may be sufficient for behavior change in some people. For example, a person with a substantial family history of heart disease may be more likely to become physically active if she learns that activity may lower the risk she has because of her genetic makeup.

Second, informational approaches may encourage people to be physically active by extending their knowledge about where and how to be active in their communities. For example, building a bicycle trail may not be enough to encourage physical activity; however, providing information about the trail, its entry and exit points, safety features, and other characteristics may increase the use of the trail for physical activity.

Third, informational strategies may help people identify the personal and environmental reasons they are physically inactive and help them overcome them. Finally, informational approaches can let people know when opportunities for physical activity are happening in their communities, at their worksites, or in other settings. This knowledge can then be converted to action. For example, when a neighborhood walk or other event is scheduled,

informational approaches can increase awareness, registration, and participation in the event, thereby increasing physical activity.

Informational approaches for promoting physical activity that have been evaluated by the Task Force include community-wide campaigns, mass media campaigns, and classroom-based health education curricula for youth that focus on providing information and skills development. The remainder of this chapter reviews each of these informational approaches, provides an assessment of the effectiveness of each in promoting physical activity, and provides examples of success.

COMMUNITY-WIDE CAMPAIGNS

Have you ever seen a billboard in your community that urges you to exercise more, or a late-night television advertisement or intranet site at your worksite or university that reminds you to be physically active? Have you ever received an e-mail reminding you of the health benefits and importance of being physically active? These are all examples of informational approaches to promoting physical activity. Unfortunately, these "single-stream" techniques rarely, if ever, succeed in getting people to become more active.

When single-stream techniques are combined into multiple intensive strategies targeted toward increasing physical activity, however, they can be successful. One of the recommended strategies for increasing physical activity via informational approaches is a **community-wide campaign**. These campaigns rely heavily on communication to change behavior through increased knowledge.

Community-wide informational campaigns for physical activity promotion have three defining characteristics. First, they include many community sectors. That is, they are not limited to messages from the local health department, parks and recreation department, hospital, or mayor's office. Successful programs create consistent messages and program identification (e.g., logos, tag lines) across those sectors.

Second, community-wide informational campaigns frequently include very visible, broadly targeted strategies. To succeed, they cannot be limited to one-way communication methods, such as television advertisements or billboards by the side of the road. Rather, they should be incorporated into other

TIME

Time—no one has enough of it; everyone is busy. Lack of time is consistently cited in surveys and studies as the number one reason people are not physically active. Although it may be true that people make time for things that are important to them, informational approaches to physical activity participation can help people identify why they believe they don't have time to be physically active and rearrange their schedules to make room for physical activity.

health-related events, such as health fairs, cancer screening events, and other activities where people may be. Moreover, social media strategies (mobile and desktop) can supplement one-way communication strategies and make them interactive.

Finally, community-wide informational campaigns can be successful if they are included in other activities that focus on physical activity–related health issues. For example, a heart disease prevention program at a large worksite may include physical activity promotion information. Such programs have been shown to increase physical activity in the targeted population.

Community-wide informational campaigns often include television, radio, newspaper, and other media to raise program awareness, disseminate physical activity health messages, and reinforce behavior change. Targeted mailings and communications from key influencers such as places of worship and community centers that support the informational campaign can also be very effective.

Taken together, existing studies in this area show that community-wide informational campaigns to increase physical activity do the following:

- Increase the percentage of people who report being physically active (at least in the short term) by an average of 4.2%.
- Increase caloric expenditure by an average of 16.3%.
- Increase multiple types of physical activity.
- Increase participants' knowledge about exercise and physical activity.
- Increase participants' intentions to be physically active (even though they may not actually be carrying through on those intentions).
- Reduce risk factors for cardiovascular disease that are related to physical inactivity.

Of note, however, is that these same studies reported equivocal findings on body weight. Some studies showed weight loss, but others showed no change or even slight weight gains. Thus, it is clearly possible to increase physical activity without weight loss.

Community-wide campaigns for physical activity promotion are typically not easy to organize. They require substantial planning, coordination, and evaluation efforts to determine effectiveness. Partnerships must be developed, and partners need to be counted on to assist in the campaign. An underfunded and underplanned campaign will underperform.

MASS MEDIA CAMPAIGNS

Mass media campaigns are physical activity promotion programs that rely on messaging efforts to change physical activity behavior through changes in knowledge, beliefs, and attitudes. The exposure to mass media campaigns can be measured in a variety of ways, but is usually summarized as the number of times an average target group member will view, hear, or see the message. Such campaigns can (and probably should) be part of a community-wide informational campaign, but do not constitute

© Olivier Douliery/Abaca Press/MCT/Photoshot

Community-wide campaigns, if done correctly, have been shown to be effective strategies to promote physical activity. Here, First Lady Michelle Obama leads a group of schoolchildren in jumping jacks as part of her Let's Move! initiative to reduce childhood obesity.

CASE STUDY: ACTIVE AUSTRALIA

Can the physical activity levels of an entire state be improved? Active Australia (Bauman et al. 2001) is a perfect example of a community-wide campaign that was able to show that they could. In a two-month period in 1998, a statewide campaign was initiated and conducted by the New South Wales (Australia) state health department. The campaign sought to increase physical activity levels in adults 25 to 60 years of age. Informational strategies used in the project included paid television and print advertising, marketing of campaign merchandise (branded with logos), multilingual outreach, and mailings to public health professionals and medical personnel throughout the state. Mass participation events such as community walks were scheduled in towns throughout the state to reinforce the objectives of the campaign.

At the completion of the two-month program, its effects were evaluated. A random sample of residents in New South Wales was identified for data collection before and after the program, as was a random sample of residents of outlying states who did not receive any of the materials (a control group).

The main findings from the evaluation of this program are shown in figure 11.1. Residents of New South Wales reported higher physical activity participation levels at the end of the campaign, whereas people not exposed to the campaign actually showed a statistically significant decline in hours per week of physical activity. Other markers of physical activity participation showed similar findings. Moreover, the campaign was quite successful in the target group in increasing awareness (recognition of the campaign) and knowledge of items related to physical activity (e.g., how much, how often), as well as self-efficacy to become more physically active. Although these are short-term findings (whether the gains remained past the time period of interest is unknown), they quite powerfully show that a community-based informational campaign can in fact increase the physical activity behaviors of an entire state.

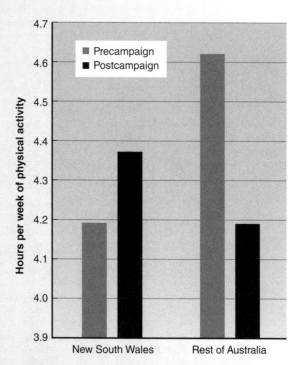

Figure 11.1 Changes in physical activity participation using a community-wide informational campaign: Active Australia.
Adapted from Bauman et al. 2001.

a community-wide effort on their own. Following are characteristics of mass media campaigns for physical activity promotion:

- Usually are large-scale efforts designed to transmit messages about physical activity to large and nonspecific audiences; anyone is considered a target.

- Are designed to increase physical activity by increasing knowledge and changing attitudes and beliefs.

- Use communication media exclusively, including newspapers, TV, radio, and billboards.

- Can rely on a single communications channel (e.g., billboards) or a combination of channels.

The Task Force found insufficient evidence to recommend mass media strategies for physical activity promotion. These kinds of programs can be very expensive and difficult to carry out, and the money spent may well be wasted. Many reasons may explain this lack of effect, including poorly produced or poorly placed media products and the lack of a defined target audience.

Although the Community Guide does not recommend mass media approaches for physical activity promotion because of insufficient evidence, research evidence has emerged that has led some authors to label the strategy as promising (Heath 2009).

Why don't mass media campaigns seem to work for physical activity promotion? Many communities and countries have used such campaigns, yet changes in behavior across a target audience have been difficult to demonstrate. Bauman and Chau (2009) reviewed mass media campaigns and provided some ideas about why they were not successful. First, the campaigns may not have had consistent, comprehensive messaging strategies. For example, some campaigns focused on increasing exercise behaviors, whereas others focused more broadly on physical activity of all kinds. Alternatively, some focused on sport participation, whereas others focused on lifestyle-related physical activity.

Second, the target behavior was inconsistent. Urging people to take a walk is not the same as urging them to meet the recommendation of 150 minutes per week of moderate-intensity physical activity. Mass media approaches may work for one type of activity outcome, but not for another.

Third, the campaign may not have been sequenced correctly. *Sequencing* here means that the messages are built on one another in a logical fashion to encouraging behavior change.

Fourth, resources may have been insufficient to reach deeply into the targeted audience or to evaluate the effectiveness of the campaign properly. As with community-wide campaigns, underfunded efforts are likely to underperform.

Finally, to have a reasonable chance at success, mass media campaigns for physical activity promotion cannot be used in isolation (Bauman and Chau 2009). They must be implemented as part of comprehensive programming with messaging supporting the policies, programs, and environment of the specific physical activity intervention. In other words, mass media campaigns that are implemented without supporting actions, policies, or places for activity are not effective.

Clearly, television, radio, and print media are giving way to electronic, web-based, and social media. Can these new media be useful in promoting

CASE STUDY: VERB

The VERB campaign was the first U.S. national mass media campaign to promote physical activity. The campaign was developed and implemented by the U.S. Centers for Disease Prevention and Control (CDC) in 2002 (Huhman et al. 2005). The purpose of the campaign was to increase and maintain physical activity among tweens (youth ages 9 to 13) throughout the United States. Paid television advertising was used on channels frequented by the target audience. Additionally, the program included radio advertising, websites, and other communications links all designed to make physical activity an attractive, "cool" behavior.

After one year, although the campaign seemed to have no overall effect on physical activity levels, several population subgroups did show increases in physical activity sessions per week. Girls, younger children, initially inactive children, and other subgroups reported becoming more active. The campaign did reach a majority of its intended audience, and most children and their parents recognized the campaign logo and materials and understood what it was about. Although not definitive, the results from VERB are promising in that a large-scale paid media campaign seemed to change physical activity behaviors in some population groups. Clearly, more research is needed in this area.

physical activity, and should they be considered mass media strategies? The question is a bit tricky because electronic media allow messages to be tailored to (and by) the people being targeted, rather than a more traditional one-size-fits-all approach that has characterized mass media strategies to date. Campaign developers can tailor the type, frequency, and appearance of the message to the characteristics of the receiver. This is very different from the "blunt instrument" approach of traditional mass media. Bauman and Chau (2009) evaluated new media physical activity interventions (primarily web-based) and concluded that these new methods of social marketing can be developed and tailored to individuals to encourage behavioral change, but more research is clearly needed in this area.

Although web-based and electronic physical activity promotion programs show promise for increasing physical activity levels, the current

How do informational approaches, such as a community campaign to raise awareness of a new playground, help promote physical activity and exercise?

evidence is inconclusive. The costs of web-based physical activity interventions are relatively low, but studies and research about their effectiveness need to be more focused on measurable outcomes as access to and the availability of web-based campaigns continue to evolve. Combinations of traditional and emerging mass media campaigns (e.g., face-to-face with electronic follow-up) are just beginning to become more readily available technologically. Examples of web-based media interventions that have had at least some success include several YouTube sites. These are found in the e-media materials at the end of this chapter, which you should find fun to explore and discuss with your colleagues.

CLASSROOM-BASED HEALTH EDUCATION PROGRAMS

Although many definitions exist, **health education** can be roughly defined as the processes through which people learn about personal health concepts and behaviors. Health education curricula have become an important part of instructional goals for elementary, middle, and high schools. Classroom-based health education programs are usually focused on providing information to help students make rational decisions about adopting healthy behaviors. Ideally, these curricula avoid an overreliance on teaching facts alone. Rather, the more effective health education curricula provide essential information and concepts and then help to shape values and norms that will result in positive health behaviors.

Clearly, variability in the quality of health education curricula is to be expected. Some may not be funded adequately, some may not address issues important to the children being taught, and in some cases the instructor may not be a helpful role model (e.g., teaching children about not smoking while being a smoker!). The Centers for Disease Control and Prevention proposed a series of criteria that, taken together, define an effective health education curriculum. These are summarized in the highlight box Characteristics of Effective Health Education Curricula.

Can information transmitted in classroom-based health education programs increase physical activity? Because so many students are exposed to such curricula, the potential for reaching a broad audience is great. Health education curricula can also supplement physical education curricula and classes. Ideally, such curricula would result in changes in self-reported or objectively measured physical activity (usually away from school), changes in BMI or adiposity, changes in physical fitness (e.g., aerobic capacity, strength), improvements in general health knowledge related to physical activity, and possibly improvements in self-confidence and self-efficacy to be physically active. Some characteristics of classroom health education programs attempting to increase physical activity are as follows:

- Provide knowledge and skills for healthy decision making.
- Work at the individual level (personal and behavioral).
- Usually include several components (e.g., tobacco use, nutrition, physical activity) and focus on reducing the risk of chronic disease.
- Teach behavioral skills, but have no added (in-class) physical activity component.

Unfortunately, the available scientific evidence does not support the use of classroom health education as a method to increase physical activity behaviors. The Community Guide has concluded that there is insufficient evidence to determine the effectiveness of classroom-based health education programs. The primary findings showed little evidence of increased student physical activity levels. As with mass media and other informational strategies, health education does seem to increase general health knowledge, knowledge related to physical activity, and even self-efficacy about exercise. Unfortunately, these increases do not translate into behavior changes.

One of the striking observations from the preceding list of characteristics of classroom-based health education programs for physical activity promotion is that very few, if any, of the programs focused solely on physical activity. Usually, the programs included physical activity information and education as part of a larger curriculum on reducing the risk factors of chronic disease. Moreover, although many programs focused on developing skills for being physically active, they did not provide stu-

CHARACTERISTICS OF EFFECTIVE HEALTH EDUCATION CURRICULA

- Have clearly defined health goals, and behaviors are linked to those goals.
- Are based on research, but rooted in theory.
- Define and describe age-appropriate peer and social norms for health behaviors, and anchor health values and beliefs.
- Help students understand their own personal risks for certain health behaviors.
- Teach skills for dealing with social pressures to engage in bad health behaviors.
- Teach skills that result in self-confidence and competence to engage in desired health behaviors.
- Provide age-appropriate and culturally appropriate learning strategies, materials, and examples.
- Provide adequate time for instruction, reinforcement of lessons, and skill and behavioral practice.
- Provide opportunities to connect with appropriate role models such as peers, family members, and community leaders.
- Include support for teachers to enhance their teaching effectiveness.

Adapted from CDC 2011.

dents with time to actually be physically active! Ideally, such a curriculum would focus solely on physical activity skills and building knowledge around the characteristics of effective health education curricula outlined earlier. It is likely that the lack of evidence of program efficacy is because programs have not gone far enough to have a reasonable chance of success.

In summary, informational approaches to physical activity promotion seem to increase knowledge, change attitudes, and even improve self-efficacy about physical activity (i.e., the belief that one can actually be more physically active). However, concurrent changes in behavior, what we are interested in, are elusive. The only strategy likely to show (short-term) increases in physical activity behavior is community-wide campaigns. More work is needed in this area to better understand it. Can you think of a research question that could help?

CASE STUDY: COLLEGE-BASED HEALTH EDUCATION

What happens to physical activity levels in late adolescence and early adulthood? Numerous researchers have reported that it declines significantly, which results in continued inactivity in most adults. Dr. James F. Sallis and colleagues (1999) designed Project GRAD (Graduate Ready for Activity Daily) to study the effectiveness of a college health education class in increasing physical activity levels. In the study, 338 students ages 18 to 29 years were randomly assigned, after baseline measurements, to intervention or control groups. Posttest data were reported for 321 students after one semester, and one and two years after baseline. The test group of students took a semester-long course that promoted the adoption and maintenance of physical activity. The lessons each week were rooted in behavioral science theory designed to support behavior change. The control students took a semester-long course on a variety of health education topics.

The GRAD intervention integrated concepts from exercise science and behavioral science based on national physical activity recommendations at the time of the study, social cognitive theory, and behavioral change theory. The intervention course included 50 minutes once a week of lecture about physical activity and behavioral topics, and a weekly 110-minute lab that included 15 minutes of physical activity (no equipment), 25 minutes of behavior group discussion, and another 45 minutes of varied physical activity with equipment as required. Primary measures included assessments of behavioral change stage and 7-day Physical Activity Recall interviews.

The main results of Project GRAD showed no significant effects on men during the semester-long intervention. Men were more active than women at baseline, at least in the maintenance stage for physical activity, which may have made increases difficult to achieve. The women in the intervention showed significant increases in physical activity levels during leisure, strengthening, and flexibility exercises.

Although the reasons for the lack of consistent results are unclear, the authors hypothesized that the participants may not have been far enough along in terms of their readiness to begin physical activity. As a result of this study, the authors recommended that starting earlier (than college) may be a better approach. Interestingly, in a follow-up study of Project GRAD, Calfas and colleagues (2000) reported no significant effects on physical activity outcomes for men or women after two years, despite excellent participation in the theoretically based intervention. College-based health education seems to improve knowledge, but not behavior.

KEY LEADER PROFILE

Russell R. Pate, PhD

Why and how did you get into this line of work? Did any one person have an overriding influence on you?

I first became interested in research on physical activity when, as an undergraduate student at Springfield College, I was invited to be a participant in a study that my exercise physiology professor, Wayne Sinning, was conducting. That experience led to my undertaking a senior thesis that gave me a taste of research. At that point I was hooked. I moved on to the University of Oregon, where I had the opportunity to do my doctoral training with a very experienced physiologist, Eugene Evonuk, and a fine cadre of fellow graduate students who shared a passion for understanding the physiological underpinnings of human performance.

Courtesy of Russell R. Pate.

What are your current research interests?

The overarching purpose of my research program is to increase our understanding of physical activity behavior in children and adolescents. The long-term goal is to provide the knowledge necessary for mounting effective public health interventions to increase physical activity in groups of young people. My interest in physical activity and fitness in youth spans my entire career. As a student I was exposed to contemporary thinking about fitness in youth and developmental influences on physiologic function.

As I was beginning my professional career at the University of South Carolina in the mid-1970s, the exercise science community was challenging traditional approaches to field measurement of physical fitness in youth. It was at that time that I began studying fitness and its measurement in young people. Over the years my research interest in children broadened to encompass issues related to physical activity behavior in kids. My research team is currently engaged in a series of studies on measurement of physical activity, factors influencing physical activity, and promotion of physical activity in children and adolescents ranging in age from preschoolers to high school students.

What drives you to be a researcher and activist in the field of physical activity and health?

I am driven by a love for the research process and by a desire to answer important questions about physical activity and its impact on public health. I enjoy the challenge of designing and conducting studies that address critical issues, and I enjoy working with co-investigators, research staff, postdoctoral fellows, and graduate students in pursuing our research goals. I believe that declining physical activity in young people is one of the great public health challenges of the 21st century, and I am motivated to do what I can to address that huge challenge.

What are one or two key issues of importance in our field that must be addressed by 2022?

Physical activity is a relatively new focus of public health, and accordingly our public health system (including both components of research and professional practice) has not yet made the investments that have been made in many other areas of public health. If we are going to stem the tide of declining physical activity in developed societies, we must do the necessary research and build the public health infrastructure to change our schools, worksites, communities, and health care systems in ways that result in many more people leading physically active lifestyles.

CHAPTER WRAP-UP

What You Need to Know

- The Community Guide is an ever-expanding resource of recommendations on evidence-based interventions to improve public health.
- The Task Force categorizes recommendations about specific physical activity interventions as recommended, recommended against, or having insufficient evidence.
- Informational approaches may be designed to increase leisure, occupation, transportation, or at-home physical activities.
- Community-wide campaigns are recommended for physical activity promotion and should include strategies to promote increased awareness and knowledge, enhance motivation and readiness to change behaviors, and teach or enhance the skills needed to establish and maintain desired behaviors.
- Mass media campaigns can be large-scale efforts that address messages about physical activity to large and undifferentiated audiences.
- Mass media campaigns seem to improve awareness and knowledge but alone should not be expected to increase physical activity.
- New media and web-based physical activity interventions show promise for increasing physical activity levels, but the current evidence is inconclusive. The costs of web-based physical activity interventions are relatively low, but studies and research about their effectiveness need to be more focused on measurable outcomes as access to and the availability of web-based campaigns continue to evolve.
- Classroom-based health education programs are usually focused on providing information to help students adopt healthier behaviors.
- Classroom-based health education programs do not seem to promote physical activity behaviors, although they seem to increase general health knowledge, exercise-related knowledge, and exercise self-efficacy.

Key Terms

Community Guide

informational approaches

community-wide campaigns

mass media campaigns

health education

Study Questions

1. What is the Community Guide, and how can you use it to develop effective physical activity interventions?

2. How does the Task Force determine whether a physical activity intervention is effective?

3. What are three ways informational approaches can increase physical activity?

4. What three categories does the Task Force use to classify physical activity interventions?

5. What are three characteristics of a community-wide program that succeeds at increasing physical activity?

6. Are mass media approaches effective at increasing physical activity? Why or why not?

7. What was the VERB program, and did it increase physical activity?

8. Are classroom-based health education approaches effective at increasing physical activity? Why or why not?

9. What are four characteristics of effective health education criteria according to the CDC?

10. What was Project GRAD? How effective was it?

E-Media

Explore issues related to physical activity, exercise, and public health at the following websites:

Human Kinetics	www.HumanKinetics.com
U.S. Department of Health and Human Services: Physical Activity Guidelines for Americans	www.health.gov/PAGuidelines
International Society for Physical Activity and Health	www.ispah.org
American College of Sports Medicine	www.acsm.org
President's Council on Fitness, Sports & Nutrition	www.fitness.gov
U.S. Centers for Disease Control and Prevention: Physical Activity	www.cdc.gov/physicalactivity/
Guide to Community Preventive Services	www.thecommunityguide.org/pa/index.html
The Community Guide Website: Information About the Task Force on Community Preventive Services and Systematic Review Process	www.thecommunityguide.org/about/task-force-members.html www.thecommunityguide.org/about/findings.html
Get Active: TV campaign to get out of your chair	http://youtu.be/AY5AILaXDdA
VERB with Hannah Montana	http://youtu.be/7AECNDfs51Q
iThrive Austin—Physical Activity :15	http://youtu.be/-Hqp-_Bo-oQ
Physical Activity—Well-Being	http://youtu.be/2gHNFj_fltc
PA Physical Activity	http://youtu.be/FDh-s3pxjRk
Get a Life, Get Active, Go Walking!	http://youtu.be/L84dSVDg5XU
Informational Approaches	http://youtu.be/781vWEFgrlE

Bibliography

Bauman A, Bellew B, Owen N, Vita P. 2001. Impact of an Australian mass media campaign targeting physical activity in 1998. *American Journal of Preventive Medicine* 21: 41-47.

Bauman A, Chau J. 2009. The role of the media in promoting physical activity. *Journal of Physical Activity Health* 6 (Suppl 2): S196-S 210.

Calfas KJ, Sallis JF, Nichols JF, Sarkin JA, Johnson MF, et al. 2000. Project GRAD: Two-year outcomes of a randomized controlled physical activity intervention among young adults. *American Journal of Preventive Medicine* 18: 28-37.

Centers for Disease Control and Prevention. 2011. Characteristics of an Effective Health Education Curriculum. www.cdc.gov/healthyyouth/SHER/characteristics/index.htm. Accessed 23 July 2011.

Heath G. 2009. The role of the public health sector in promoting physical activity: National, state, and local applications. *Journal of Physical Activity Health* 6 (Suppl 2): S159-S167.

Huhman M, Potter LD, Wong FL, Banspach SW, Duke JC, Heitzler CD. 2005. Effects of a mass media campaign to increase physical activity among children: Year-1 results of the VERB campaign. *Pediatrics* 116: 277-284.

Kahn EB, Ramsey LT, Brownson RG, Heath GW, Howze EH, Powell KE, Stone EJ, Rajab MW, Corso P, the Task Force on Community Preventive Services. 2002. The effectiveness of interventions to increase physical activity. *American Journal of Preventive Medicine* 22: 73-107.

Sallis, JF, Calfas KJ, Nichols JF, Sarkin JA, Johnson MF, et al. 1999. Evaluation of a university course to promote physical activity: Project GRAD. *Research Quarterly for Exercise and Sport* 70: 1-10.

U.S. Department of Health and Human Services. 2008. *2008 Physical Activity Guidelines for Americans.* Washington, DC: U.S. Department of Health and Human Services. www.health.gov/PAGuidelines. Accessed 23 July 2011.

U.S. Department of Health and Human Services, Public Health Service, Centers for Disease Control and Prevention, National Center for Chronic Disease Prevention and Health Promotion, Division of Nutrition and Physical Activity. 2010. *Promoting Physical Activity: A Guide for Community Action*, 2nd ed. Brown DR, Heath GW, Martin SL, eds. Champaign, IL: Human Kinetics.

PHYSICAL ACTIVITY IN PUBLIC HEALTH SPECIALIST

This chapter covers these competency areas as set forth by the National Society of Physical Activity Practitioners in Public Health:

1.1.1, 1.1.5, 1.3.1, 1.3.2, 1.3.3, 1.4.1, 1.4.2, 2.1.1, 2.1.3, 2.2.1, 2.2.2, 2.3.2, 2.3.3, 2.5.2, 3.1.1, 3.1.2, 3.2.1, 3.3.3, 3.7.1, 4.1.1, 4.1.2, 4.1.4

CHAPTER 12

SCHOOL-BASED APPROACHES TO PROMOTING PHYSICAL ACTIVITY

OBJECTIVES

After completing this chapter, you should be able to discuss the following:

» The rationale for school-based programming to increase physical activity

» The current U.S. policies and strategies for school physical activity programs along with the components of effective school-based interventions

» The potential physical activity and exercise outcomes for youth

» The evidence and guidelines for physical activity and exercise for youth

» Examples of school-based programs that increase physical activity

Opening Questions

» Can and should schools implement physical activity programs?

» What are the components of an effective school physical activity program?

» What are some successful models of school-based physical activity programs?

School-based physical activity programming, such as physical education (PE) and sport participation (athletics), have been part of school culture for over 100 years in the United States and many other countries. Physical educators and coaches promote exercise to youth and young adults in schools to help them achieve or improve their physical fitness and athletic performance. Since 2005, the national focus for PE and athletics in U.S. schools has shifted somewhat to promoting physical activity for all students (more so in PE) and promoting student health and safety (more so in athletics). Schools today are also looking to promote physical activity at all grade levels in school venues other than just PE and athletics. Enhanced school-based PE is a recommended strategy that works to increase physical activity (Kahn et al. 2002).

In addition to PE, schools are increasingly looking to integrate physical activity into other parts of students' lives. Current best practices include pre-kindergarten (pre-K) programs, kindergarten programs, before-school programs, **classroom activity breaks**, recess breaks, after-school programs, commercially sponsored programs, and summer (seasonal) programs.

Rationale for School-Based Physical Activity Programs

This chapter is based on several detailed reports (Kahn et al. 2002; Massengale 1987; Siedentop 2009) and the Guide to Community Preventive Services (Community Guide; 2011), which support the promotion of physical activity via the education sector.

In the United States, schools have been involved in the public health and safety of children and adolescents since colonial times. Initially, schools

helped combat infectious diseases; today they are being asked to lead the way in preventing and controlling the incidence of obesity, overweight, and diabetes (see chapters 5 and 6 for more). In the United States it has been estimated that over 54 million children and adolescents attend public and private elementary and secondary schools, where they spend approximately 6.5 hours per day for an average of 180 days per year. Additionally, over 1.1 million children attend pre-K schools, and over 11.5 million older adolescents attend community colleges or universities.

Because of the number of students they serve, schools at all levels are attractive public venues at which to disseminate positive physical activity messages and promote active lifestyles. Mandating increased physical activity in schools has also become popular recently, because policy makers believe that it can improve the adherence of students to varied curricular-based education to combat obesity, overweight, and diabetes, versus relying on students' voluntary compliance to lifestyle change.

National legislation in the United States, including the 2004 Child Nutrition and WIC Reauthorization Act (Lee et al. 2006), ties school funding to items such as school wellness policies that include promoting healthy eating and physical activity. National goals such as those contained in Healthy People 2010 and **Healthy People 2020** (HP2020) also encourage curricular goals in school PE that include having students work at vigorous or moderate intensities for at least 50% of class time. Further, professional associations such as the National Association for Sport and Physical Education (NASPE; 2010) have recommended that elementary students engage in 150 minutes per week of school PE, and that secondary students acquire 225 PE minutes per week.

Although PE curricula have become the main focus for increasing physical activity in schools, youth cannot realistically meet physical activity

HEALTHY PEOPLE 2020

Healthy People 2020, the most recent version of the health goals that are updated every 10 years by the U.S. Department of Health and Human Services, highlights disparities and opportunities for health improvement by setting public health targets to achieve in a 10-year period. Several objectives promote increased physical activity levels that are, or could be, school related. Following are some of the physical activity–related HP2020 objectives that target children and adolescents:

- Increase the proportion of adolescents who meet current federal guidelines for aerobic physical activity and for muscle-strengthening activity.
- Increase the proportion of the nation's public and private schools that require daily PE for all students.
- Increase the proportion of adolescents who participate in daily school PE.
- Increase regularly scheduled elementary school recess.
- Increase the proportion of school districts that require or recommend elementary school recess for an appropriate period of time.
- Increase the number of states that require licensed child care programs to provide physical activity.
- Increase the proportion of the nation's public and private schools that provide access to their physical activity spaces and facilities for all people outside of normal school hours (i.e., before and after the school day, on weekends, and during summer and other vacations).

guidelines (60 minutes per day; see the section on scientific evidence later in the chapter) by being active only in PE class. Even the very best physical education classes can offer only 20 to 30 minutes of physical activity time. Youth should be encouraged to acquire physical activity before, during, and after school in a variety of other ways, such as active transportation, play, sports, and leisure or recreation. It is also important to note that NASPE's weekly PE guidelines for children in schools address teaching structures and administrative issues as much as they do physical activity promotion.

Schools are a major source of potential influence on children and adolescents to adopt active lifestyles. Other sectors in society such as government agencies, families, and the media can also influence the adoption of behaviors that can help children and adolescents achieve caloric balance. Until recently, schools were built in neighborhoods, which often positively affected the whole community in relationship to physical activity, because they were within walking distance for most students and included

places (e.g., open green spaces, outdoor tracks, indoor and outdoor basketball courts, gyms) at which youth and adults could be active. Schools have traditionally also offered opportunities for physical activity through PE classes, recess, and organized sports. However, with the new accountability requirements such as those of the **No Child Left Behind** legislation (2001), which encourage schools to focus on academic subjects such as mathematics and reading, physical education and recess time continue to be squeezed out.

Adding to these challenges, most new schools are being built on the outskirts of cities as a result of urban sprawl. This trend reduces opportunities for physical activity for members of the school community, because students must be bused to school or driven by family members. Research shows that whether schools are new or old, parents are concerned about safety, injury, and crime, which can cause them to limit physical activity opportunities for their children if they perceive the school environment to be unsafe.

Does Physical Activity Make Children Smarter?

Balancing academics with opportunities to engage in school-time physical activity is a significant challenge for school policy makers. Although some have advocated increased school time for academics at the expense of PE, research summarized by the Robert Wood Johnson Foundation (2009) and the CDC (2010) has shown strong associations between measures of participation (e.g., physical fitness) in school-based physical activities and measures of academic performance. Based on the 2003 National Youth Risk Behavior Survey (YRBS), the CDC has also reported that students who engage in insufficient physical activity, who do not play on at least one sport team, and who watch TV or play video games more than three hours per day earn lower grades than students who are more physically active. Participation in school-based physical activities also has been found not to negatively affect academic performance.

Although a cause-and-effect relation has yet to be reported, Van Dusen and colleagues (2011) recently reported a significant dose-response relationship between physical fitness scores and academic achievement scores in a large sample (>300,000) of Texas students. There is growing evidence that participation in school-based physical activities like PE not only doesn't hurt academic scores, but actually may improve them.

Kinesiology and Physical Activity Outcomes for Youth

Selected chronic adaptations to physical activity and exercise are listed in the highlight box Adaptations to Physical Activity and Exercise Programming for Youth. The amount of physiological adaptation related to each of the identified benefits is dose dependent and influenced by the physical training principles discussed in chapter 2 and reviewed by Strong and colleagues (2005).

The exercise science–related benefits of physical activity for children, adolescents, and young adults are highly dependent on individual growth and development. Most youth experience very positive benefits associated with engaging in physical activity and exercise; however, maturation can influence the rate and timing of specific training adaptations.

Physiologically, participation in regular physical activity can help youth increase cardiorespiratory endurance ($\dot{V}O_2$max or $\dot{V}O_2$peak) by 8 to 10%. Children and adolescents can increase their strength and muscular endurance; however, strength gains prior to puberty are mostly due to neural changes (i.e., better muscle recruitment) rather than hypertrophy

(i.e., increased muscle size). There are published recommendations that pertain to youth strength and muscular endurance training (see Kenney, Wilmore, and Costill 2012 for more).

Youth can achieve significant and consistent improvements in bone health by participating in weight-bearing and muscular strength and endurance activities. The opportunity for increased bone mass in girls and boys occurs in puberty and premenarche. Their risks for metabolic dysfunction (e.g., related to HDL, triglyceride, and insulin levels; blood pressure; and percentage of body fat) and type 2 diabetes are also significantly reduced if they can get 60 minutes or more of physical activity daily.

Youth can experience many of the same biomechanical benefits reported for adults and older adults, particularly if they have the opportunity to develop a variety of motor skills by participating in a variety of physical activities. As children move through adolescence to young adulthood, they experience improved economy (i.e., lower energy cost) in activities such as running. Changes in economy have been attributed to changes in body size (i.e., filling out) and perhaps in stride frequency.

Youth engaged in organized sports have been reported to suffer more injuries that might limit future physical activity (particularly if they do not participate safely with protective equipment) than

Do children and adolescents like competition in school PE? Does it help promote physical activity? Why?

nonparticipants. However, except for information from descriptive studies of high school athletes, there is little evidence for this claim. Research on participation in school-based PE suggests that it is a very safe undertaking for the vast majority of students participating.

Behaviorally, the experiences that children (including pre-K and kindergarten students) have with physical activity, and the early behaviors that they develop as a result of participation, are thought to be critical to whether they become and remain active through adolescence and adulthood. Those

ADAPTATIONS TO PHYSICAL ACTIVITY AND EXERCISE PROGRAMMING FOR YOUTH

Physiological
Increased $\dot{V}O_2$max

Increased strength

Improved muscular endurance

Increased HDL levels

Lower triglyceride levels

Improved insulin levels

Lower blood pressure

Lower percentage of body fat

Reduced risks for metabolic dysfunction

Reduced risks for type 2 diabetes

Improved bone health

Biomechanical
Improved economy with age

Improved balance

Improved mobility

Increased motor skill and confidence to engage further in physical activity and exercise

Improved proprioception

Behavioral
Increased self-confidence

Improved self-efficacy

Improved self-esteem

Decreased depression and anxiety

Experience with behavioral change

who do not have the opportunities to develop the motor skills by participating in a variety of physical activities (as part of leisure time, recreation, PE, sports, and games) are most likely to become inactive, obese, or overweight adolescents, adults, and older adults.

SCHOOL-BASED PHYSICAL ACTIVITY AND PHYSICAL FITNESS ASSESSMENTS OF YOUTH

A discussion of all the physical activity and fitness assessments that pertain to children and adolescents is beyond the scope of this text (see Morrow 2009 for more). However, it is important to point out that the assessment techniques discussed in chapter 4 have been used extensively in school-based physical activity intervention evaluations.

Two frequently used comprehensive fitness assessment programs are **Fitnessgram** and the **President's Challenge**. They are comprehensive field assessments of physical fitness in children and adolescents, and both contain a variety of valid and reliable fitness tests (with options for teachers and individuals), age-appropriate interpretations, and strategies and programming to increase or maintain physical activity. Fitnessgram and the President's Challenge are school friendly and can be used as in-service programs for teachers and health professionals.

Fitnessgram was first developed in 1982 by the Cooper Institute in response to the need for a comprehensive set of assessment procedures for PE programs. The assessment includes health-related physical fitness field tests that assess aerobic capacity; muscular strength, muscular endurance, and flexibility; and body composition. Scores from these assessments are compared to healthy fitness zone standards to determine students' overall physical fitness levels; areas for improvement are suggested when appropriate. Activitygram is an activity assessment tool that comes with the Fitnessgram software; it enables students to record their physical activity in 30-minute increments over a three-day period. The software generates a report showing the total minutes of activity, periods of activity time each day, and types of activity.

The President's Challenge, a program managed by the President's Council on Fitness, Sports & Nutrition, encourages all Americans to make being active part of their everyday lives. The President's Challenge helps children, adolescents, adults, and older adults participate in daily physical activity to improve fitness. The program includes online challenges and allows participants to keep track of their progress.

Although many have argued that physical fitness tests should be abandoned in schools, physical fitness testing of youth seems to have gained momentum in recent years; California and Texas both mandate fitness testing of students in several grades. The idea is that testing health-related fitness parameters will help integrate behavior, fitness, motor skills, and cognition to encourage more physical activity in and out of school. Policy makers have used fitness evaluation results, at least in part, to make the case that children and adolescents score low on health-related fitness, and therefore need to become more active and fit.

McKenzie (2007), Morrow (2005), and others have reviewed the numerous pitfalls associated with youth fitness testing, particularly in school PE. For public health practitioners, the most notable point to remember is that the relationship between measures of physical fitness (i.e., physiological constructs) and physical activity (a behavior) is relatively weak. Further, youth fitness performance is influenced by genetics, growth and development, and maturation. Thus, very inactive children may do very well on school field tests of physical fitness, whereas others who may be meeting or exceeding the guideline of 60 minutes a day may not do well on some or all of the tests. It remains to be determined whether physical fitness testing of children and youth promotes more physical activity in these groups, and whether the strategy has a significant impact on the prevalence of obesity, overweight, and diabetes.

PHYSICAL ACTIVITY IN CHILDREN AND ADOLESCENTS

Important considerations concerning physical activity promotion in children and adolescents are their individual growth trajectories and the fact that older children have better motor skills than younger ones. We all know that some children mature faster than others. This results in differences in motor ability among individuals of similar ages as well as across the age range. Fourteen-year-olds have better gross motor control, more muscle mass, and generally higher fitness than younger children, for example. This difference in physical abilities due to growth and development can affect many aspects of physical activity behavior, both physical and psychological.

Changes in motor ability for a number of fitness parameters are illustrated in figure 12.1 for youth ages 6 to 17. Clearly, children at different ages differ in their results in tests that measure flexibility, muscular strength, and aerobic fitness. Further, sex differences are to be expected, particularly as children grow into adolescence. The data in this figure illustrate that many physical performance outcomes for youth are related to growth and development and maturation, and they vary by sex. Understanding the interrelationships among growth, maturation, and exercise is critical for designing and implementing meaningful physical activity programs for children and adolescents.

Scientific Evidence

Strong scientific evidence supports a consistent effect of physical activity and exercise on cardiorespiratory fitness and muscular strength in youth (U.S. Department of Health and Human Services [USDHHS], Physical Activity Guidelines Advisory Committee [PAGAC] 2008). There is also strong evidence that physical activity and exercise are positively associated with body composition, cardiorespiratory and metabolic health, and bone health; higher levels of physical activity are associated with more favorable outcomes. The evidence related to mental health benefits for youth who are active is moderate for depression, weak for anxiety, and limited for self-esteem.

Evidence of a dose-response relationship between physical activity and exercise and cardiorespiratory and metabolic health in youth has not been specifically determined, and more dose-response studies are needed for evaluating other health outcomes in youth. There is strong evidence that physical activity and exercise has positively affected youth fitness levels for boys and girls. Evidence of effects of age, sex, and race or ethnicity on body composi-

tion, cardiorespiratory and metabolic health, and mental health in youth is unclear. A strong association has been shown between physical activity and exercise and bone health for both boys and girls, and it is influenced by growth, development, and maturation. Children should initiate physical activity at least by the early teen years to maximize physiological benefits.

Many studies have demonstrated significant gains in physical fitness measures (cardiorespiratory and muscular) in children and adolescents who participated in vigorous-intensity aerobic activities three or more days per week, and muscle-strengthening and bone-strengthening exercises two or three days per week. Cardiorespiratory and metabolic health are also significantly improved by taking part at least three days per week in vigorous-intensity aerobic activities, and bone health responds positively to weight-bearing activities performed at least three days per week.

Guidelines

The guidelines for physical activity for youth were highlighted in chapter 6. In addition to understanding that youth should engage in at least 60 minutes each day of vigorous- or moderate-intensity physical activity (including aerobic, muscle-strengthening, and bone-strengthening activities), it is important to know some strategies for replacing inactivity with activity. Following are some suggestions for getting youth active and encouraging them to stay that way (USDHHS, PAGAC 2008):

- Children and adolescents who are inactive or doing less physical activity than the guidelines suggest should slowly increase their activity in ways they enjoy. A gradual increase in the number of days and the time spent being active will reduce the risk of injury and is consistent with the basic principles of exercise physiology covered in chapter 2.

- Children and adolescents who meet the recommended guidelines should continue to be active on a daily basis and, if appropriate, become even more active. Evidence suggests that more than 60 minutes of activity every day may provide additional health benefits.

- Children and adolescents who exceed the guidelines should maintain their activity level

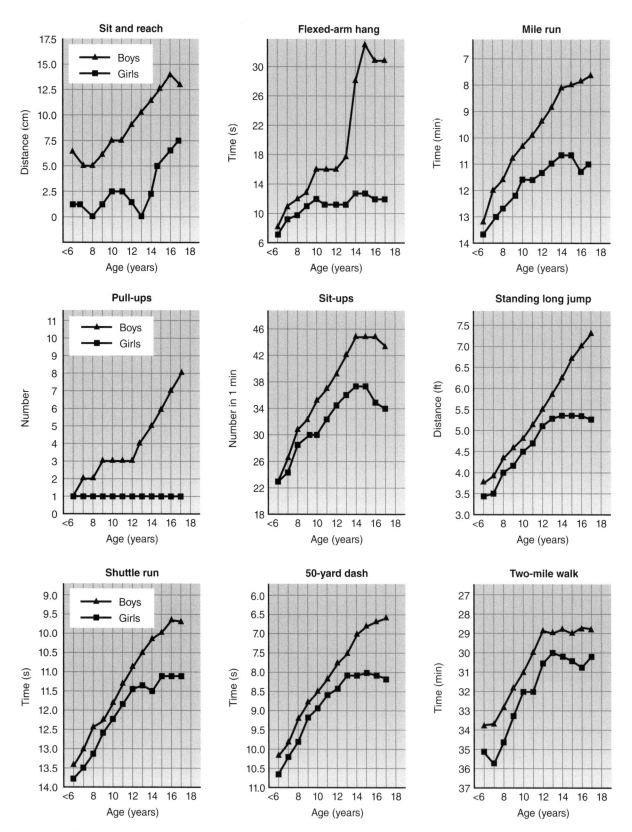

Figure 12.1 Changes in motor ability from the ages of 6 years to 17 years.

Reprinted, by permission, from J.H. Wilmore, D.L. Costill, and W.L. Kenney, 2008, *Physiology of sport and exercise,* 4th ed. (Champaign, IL: Human Kinetics), 397. Data from the President's Council on Fitness and Sports 1985.

and vary the kinds of activities they do to reduce the risk of overtraining or injury.

- Youth should have opportunities to participate in a wide range of developmentally appropriate activities.

- Youth with disabilities will most likely be less active than those without disabilities and should specifically be encouraged to incorporate physical activity into their lifestyles, using adjustments to accommodate their physical and mental challenges.

SCHOOL-BASED PHYSICAL EDUCATION

Since the early 1990s, school PE programs have focused on **health-related PE (HRPE)** (McKenzie 2007). The HRPE concept promotes public health objectives and focuses on the health benefits of physical activity. This focus on behavior rather than physiological status is important, because for many years, school PE was primarily focused on sport performance and yearly fitness testing of students. Students usually were not prepared for the fitness tests administered, scored poorly, and became turned off to becoming fit. In fact, many of today's public policy makers experienced PE themselves in "a setting in which embarrassment, humiliation, anger, discomfort, noninvolvement, apathy, rebellion, compliance, and irrelevant behavior appear to be the norm" (Massengale 1987, p. 56). Perhaps it is not too hard to understand why resistance to promoting school PE through HRPE continues, both inside and outside the PE profession.

As covered in chapter 11, the Community Guide (Kahn et al. 2002) is a resource for evidence-based disease prevention and health promotion strategies. One of the strategies that work is high-quality physical education. Outcomes of interest in studies of PE include the amount of time (days per week and minutes per period) as well as the amount of time per period that students spend being physically active in PE class. Additionally, physiological outcomes such as improvements in physical fitness and changes in body composition have been investigated. The studies of the effectiveness of school PE in increasing physical activity suggest the following:

- The *amount* of time spent in vigorous- or moderate-intensity physical activity during PE class can be increased by an average of 50.3%.

- Increases in the *percentage* of time spent in vigorous- or moderate-intensity physical activity in PE can range from 3.3% to over 15% (average increase is 10%).

- Significant increases in *aerobic capacity* (on average, 8.4%) can be obtained with such programs as well.

These increases have been demonstrated in studies that have involved lengthening the time period of existing PE classes, adding new PE classes, and increasing the amount of time dedicated to vigorous- or moderate-intensity physical activity during class without lengthening class time. These increases are most often achieved by exchanging active lessons for inactive ones or modifying existing lessons to increase the intensity of the physical activities students perform.

Crucial to the Community Guide recommendations for PE is the use of high-quality studies. The

Can participating in youth sports promote physical activity and exercise for a lifetime?

studies reviewed were not those in which a ball was rolled out during a PE period or students were lined up to shoot basketballs one at a time. Instead, the Community Guide recommendations are based on studies of high-quality PE programs that include all students, have appropriate class sizes for instruction, use developmentally (age-) appropriate curricula delivered by trained PE specialists, have adequate equipment to meet educational goals, focus on an appropriate mix of motor skills, increase student understanding, and provide active opportunities to practice. These characteristics (and others) are encouraged by NASPE, the primary association for physical education professionals in the United States. PE works when it is high-quality PE.

Several commercial school-based PE programs are available to enhance the PE curricula of primary, middle, and secondary schools. Most of these programs have been tested with positive results. A list of these resources is found at the end of the chapter.

Physical education is not the only aspect of health that schools can affect. Because such a large proportion of children are enrolled in schools, it is a logical place to help them learn and practice healthy behaviors. The primary public health framework, developed by the CDC, for promoting school health, including increased physical activity, is the **coordinated school health** (CSH) model, which consists of the following eight interactive components:

- Health education
- Physical education
- Health services
- Nutrition services
- Counseling, psychological, and social services
- Healthy and safe school environment
- Health promotion for staff
- Family and community involvement

From CDC, www.cdc.gov/healthyyouth/cshp/components.htm.

CASE STUDY: THE SPARK PROGRAM

The Sports, Play, and Active Recreation for Kids (SPARK) program is a comprehensive health-related PE program for elementary school children. The program promotes the use of enjoyable physical activities during PE classes coupled with teaching movement skills. Children are active in vigorous- or moderate-intensity physical activity for most of a PE class period; approximately 50% of the time is devoted to health-related physical activity, and 50% of the time is devoted to skill building. In addition to the weekly modules in health-related physical activity and skill-related sports, the program includes an emphasis on self-management skills and also homework designed to engage parents.

In their evaluation of the program, Sallis and colleagues (1997) randomly assigned seven schools (nearly 1,000 elementary school students) to the SPARK program or the usual PE program (control). The SPARK schools were further randomly assigned to one of two conditions: one in which PE classes were led by trained PE specialists, and the other in which PE was led by classroom teachers with no special training in PE. After two years of implementation, schools with the SPARK program achieved significantly more minutes per week of physical activity during PE classes than did the control schools (see figure 12.2). Further, students in classes that were led by trained PE specialists spent the most time being physically active. High-quality physical education promotes physical activity.

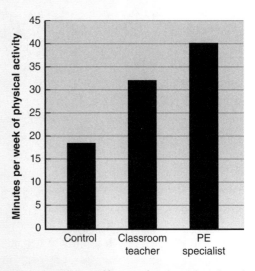

Figure 12.2 Effects of a health-related physical education program (SPARK) on minutes per week of physical activity in elementary school children.
Adapted from Sallis et al. 1997.

OPPORTUNITIES FOR SCHOOL-RELATED PHYSICAL ACTIVITY

Despite the fact that schools are a helpful place to promote physical activity because of the sheer numbers of students there, even the best PE classes cannot provide the entire recommended 60 minutes per day of physical activity that children and adolescents need. This realization has led to the emergence of ongoing studies of the usefulness of other school-related settings that may help children get closer to the 60-minutes-per-day guidelines. Examples include before- and after-school programs; sports and intramurals; in-class physical activity breaks; active, structured recess breaks; seasonal break (summer) programs; and active transportation options to and from school such as walking and biking. Many of these strategies are showing promise in adding physical activity opportunities to the day for children and adolescents, but none has as yet specifically been recommended as an evidence-based strategy.

DEVELOPMENTAL CONSIDERATIONS FOR PHYSICAL ACTIVITY IN YOUTH

The maturation, growth, and development factors (roughly, although not perfectly, estimated by the child's age) that affect physical fitness and were discussed earlier have implications for strategies to promote physical activity among children and adolescents. Clearly, different approaches are needed depending on age, sex, and developmental stage. High-quality school PE curricula should take such differences into account. Figure 12.3 shows how the emphasis of PE programs should change as children mature.

During the preschool and elementary school years, the emphasis of PE programs should be on engaging in general physical activity that develops and improves motor skills (locomotor, such as traveling, fleeing, and dodging; nonmanipulative, such as jumping, landing, and balancing; and manipulative, such as kicking, throwing, and catching). Participation in activities that develop physical competence, offer choices, maximize fun and enjoyment, and minimize anxiety (e.g., games and lead-up sports) can positively influence children's activity levels.

Once students reach the ages of 10 to 14, the emphasis on physical activity should shift toward individual and group activities, including school and club sports. Because time for school PE usually decreases for this age range, teachers should focus

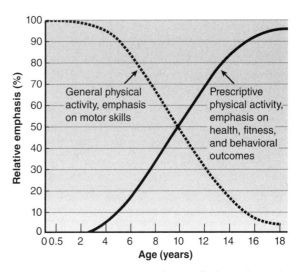

Figure 12.3 Changing emphasis of physical activity during childhood and adolescence.
Adapted, by permission, from American Academy of Physical Education, 1991, *New possibilities, new paradigms?*, edited by R.J. Park and M.H. Eckert, American Academy of Physical Education Papers, No. 24. (Champaign, IL: Human Kinetics Publishers), 30-38.

on factors that encourage youth to adopt and maintain physical activity levels, such as the following:

- Being active with friends
- Having fun and developing the skills needed for performing activities of choice
- Having time for physical activities that they feel competent at
- Sharing time with adults who are positive role models for physical activity
- Access to neighborhood facilities to engage in leisure-time activities

KEY LEADER PROFILE

Adrian Bauman, MD, MPH, PhD, FAFPHM

Courtesy of Adrian Bauman.

Why and how did you get into this line of work? Did any one person have an overriding influence on you?

Self-reflection is not something I have time to do very often, given my many academic research, teaching, and service roles. I have no background in exercise science, although I think I attended one exercise physiology class in medical school in 1973. My background was in clinical work, and I moved into public health in the early 1980s, really enjoyed it, and specialized in both health promotion and traditional epidemiologic research. My subsequent MPH and PhD dissertations were on patient education program evaluation, but even then I had started concurrently to work on physical activity epidemiology.

Why was I interested in that area? I saw physical inactivity as a classic public health risk factor for which individual behavior change and multisectoral societal responses would be needed. It was distinctly different from other cardiovascular risk factors at that time, such as hypertension and dyslipidemia, which required pharmacological therapy. As a historical curiosity, the obesity epidemic had hardly started at the time, and no one in public health was concerned about it as a population problem.

I had no real overriding mentors in an ongoing sense, but I visited two people through a World Health Organization Travelling Fellowship in 1987, and they both were particularly kind and acutely influential on me. Both of them have had an impact on the whole field of physical activity and public health. My first visit in 1987 was to Ken Powell, who then ran the CDC branch that was pioneering physical activity surveillance and also public health frameworks for addressing inactivity. My second visit was to Steve Blair, whose energy and enthusiasm combined with good epidemiologic evidence moved the field forward through advocacy, by encouraging everyone working in public health to take physical inactivity seriously.

What are your current research interests?

I currently spend a lot of my time in diverse areas of physical activity and public health research. My more specific interests include research into sitting time and health, developing integrated measurement and surveillance systems for physical activity, and evaluating social marketing and mass media efforts to promote physical activity to whole populations.

Physical activity and public health is a challenging area, but my interest is maintained because of its breadth. I work in physical activity policy, behavior change, epidemiology, translational research, and occasionally interventions in clinical settings. I work with policy makers, urban planners, environmental scientists, and many other kinds of professionals; this keeps me thinking about new issues and new methods.

What are one or two key issues of importance in our field that must be addressed by 2022?

Key issues to be solved or implemented by 2022 include working out how to really implement well-resourced, multisectoral, broad physical activity programs across whole communities and countries. This means implementing many national physical activity plans in developed and developing countries. Societal as well as health goals would be decreased inactivity time, more incidental physical activity, and less sitting time.

- Good parent–adolescent communication
- Self-esteem
- Access to equipment

For older adolescents (ages 15 to 18), more structured physical activity programs that help them transition toward adulthood are recommended. This age group is quite vulnerable to becoming sedentary, losing caloric balance, and gaining weight rapidly as they become more independent. The emphasis for young adults should be on health, fitness, and behaviors that can be adopted and maintained in adulthood. Those attending community colleges or universities should be encouraged to participate in physical activity classes to establish or maintain skills for active living.

In summary, numerous methods have been, or can be, implemented in schools to increase students' physical activity levels. Aerobic fitness levels increase with school-based PE programs, although other physiological measures do not respond similarly. Although high-quality PE is the only evidence-based strategy that has been shown to work to increase physical activity, others are still being investigated and are showing promise. High-quality PE programs delivered by trained physical educators seem to have the greatest impact on physical activity in schoolchildren. The evidence that physical activity and physical fitness can positively affect academic achievement is tantalizing. Finally, the cost-effectiveness of school-based PE and physical activity interventions has not been evaluated.

CHAPTER WRAP-UP

What You Need to Know

- In the United States, over 54 million children and adolescents attend public and private elementary and secondary schools, where they spend approximately 6.5 hours per day for an average of 180 days per year. Additionally, over 1.1 million children attend pre-K schools, and more than 11.5 million older adolescents attend community colleges or universities. This makes schools very attractive environments in which to promote physical activity.
- There is no evidence that time spent in physical activity or PE has negative influences on academic achievement, and emerging evidence suggests that physical activity positively influences academic achievement.
- The primary coordinated school health (CSH) components that U.S. educators have used to promote increased school physical activity are health education; physical education; health services; nutrition services; counseling, psychological, and social services; healthy and safe school environment programs; health promotion for staff; and family and community involvement.
- The Community Guide recommends high-quality school PE as an evidence-based strategy to promote physical activity.
- The health-related PE (HRPE) concept promotes public health objectives and encourages students to become more physically active, rather than just focusing on physical fitness, which was the desired product or outcome of PE in the past.
- In addition to PE, schools can increase youth physical activity levels through before-school programs, classroom activity breaks, recess breaks, after-school programs, sports and intramural programs, and summer (seasonal) programming.
- During the preschool and elementary school years, school-based programs to increase physical activity should emphasize general physical activity and improving the motor skills (locomotor, such as traveling, fleeing, and dodging; nonmanipulative, such as jumping, landing, and balancing; and manipulative, such as kicking, throwing, and catching).
- Once students reach the ages of 10 to 14, the emphasis should shift toward a variety of individual and group activities, including school and club sports.

- The emphasis for young adults should be on health, fitness, and behaviors that can be adopted and maintained into adulthood.
- The exercise-related benefits for children, adolescents, and young adults are highly dependent on individual growth and development. Most youth can experience the very positive benefits of engaging in physical activity and exercise; however, maturation can influence the rate and timing of these training adaptations.
- The guideline for physical activity participation for children and adolescents above five years of age is 60 minutes daily of vigorous- or moderate-intensity activity. Significant gains in youth physical fitness (cardiorespiratory and muscular strength) can be obtained by participating at least three days per week in vigorous-intensity aerobic activities, and by performing resistance training two or three days per week.

Key Terms

classroom activity breaks
Healthy People 2020
No Child Left Behind
Fitnessgram

President's Challenge
health-related PE (HRPE)
coordinated school health (CSH)

Study Questions

1. What are three reasons physical activity should be promoted in school-based settings?
2. What are three goals of Healthy People 2020 related to increasing school-based physical activity?
3. What are four of the eight components associated with coordinated school health programs?
4. How is health-related PE different from school PE programs in the past?
5. How might school-based fitness testing promote physical activity?
6. What are four positive outcomes of school-based PE related to increasing student physical activity levels?
7. What are two common strategies to increase vigorous- or moderate-intensity physical activity in school PE programs based on the Community Guide recommendations?
8. What two or three factors do research-based school physical activity interventions have in common?
9. How do growth and development issues influence youth fitness and physical activity levels?
10. Were your own personal experiences in school PE positive or negative with regard to the promotion of lifetime fitness and physical activity? Why or why not?

E-Media

Explore issues related to physical activity and youth at the following websites:

U.S. Department of Health and Human Services: Physical Activity Guidelines for Americans	www.health.gov/PAGuidelines
President's Council on Fitness, Sports & Nutrition	www.fitness.gov
U.S. Centers for Disease Control and Prevention: Adolescent and School Health	www.cdc.gov/healthyyouth
National Association of Sports and Physical Education	www.aahperd.org/naspe
Guide for Community Preventive Services	www.thecommunityguide.org/index.html
Fitnessgram and Activitygram	www.fitnessgram.net
School-based approaches	http://youtu.be/s7YX2giUJ84

Following are school-based physical education curricula (no endorsement implied):

Bienestar	www.sahrc.org
CATCH (Coordinated Approach to Child Health)	www.catchinfo.org
Eat Well & Keep Moving	www.eatwellandkeepmoving.org
Middle School Physical Activity and Nutrition (MSPAN)	http://rtips.cancer.gov/rtips/programDetails.do?programId=285123
Sports, Play, and Active Recreation for Kids (SPARK)	www.sparkpe.org
Planet Health	www.planet-health.org
Youth Fit for Life	http://cbpp-pcpe.phac-aspc.gc.ca/intervention.pdf/en/389.pdf

Bibliography

Centers for Disease Control and Prevention, National Center for Chronic Disease Prevention and Health Promotion Division of Adolescent and School Health. 2010. The Association Between School-Based Physical Activity, Including Physical Education, and Academic Performance—Revised Version. www.cdc.gov/HealthyYouth. www.cdc.gov/healthyyouth/health_and_academics/pdf/pa-pe_paper.pdf. Accessed 26 September 2011.

Community Guide. 2011. The Guide to Community Preventive Services www.thecommunityguide.org/pa/index.html. Accessed 26 September 2011.

Healthy People 2020. 2010. www.healthypeople.gov/2020/default.aspx. Accessed 30 July 2011.

Kahn EB, Ramsey LT, Brownson RG, et al. 2002. The effectiveness of interventions to increase physical activity. *American Journal of Preventive Medicine* 22: 73-107.

Kenney L, Wilmore J, Costill D. 2012. *Physiology of Sport and Exercise,* 5th ed. Champaign, IL: Human Kinetics.

Lee S, Wechsler H, Balling A. 2006. The role of schools in preventing childhood obesity. *Research Digest.*7(3):1–8.

Massengale JD, ed. 1987. *Trends Towards the Future in Physical Education.* Champaign, IL: Human Kinetics.

McKenzie T. 2007. The preparation of physical educators: A public health perspective. *Quest* 259 (4): 345-357.

Morrow JR Jr. 2005. Are American children and youth fit? It's time we learned. *Research Quarterly for Exercise and Sport* 76: 377-388.

Morrow JR, Zhu W, Franks D, Meredith M, Spain C. 2009. 1958-2008: 50 years of youth fitness tests in the United States. *Research Quarterly for Exercise and Sport* 80: 1-11.

National Association of Sport and Physical Education. www.aahperd.org/naspe. Accessed 30 July 2011.

Park RJ, Eckert MH, eds. 1991. New possibilities, new paradigms? *American Academy of Physical Education Papers* 24. Champaign IL: Human Kinetics, pp 30-38.

Robert Wood Johnson Foundation. 2009. Active Living Research, Active Education, Physical Activity, and Academic Performance. www.activelivingresearch.org. Accessed 14 July 2011.

Sallis JF, McKenzie TL, Alcarez JE, Kolody B, Faucette N, Hovell MF. 1997. The effects of a 2-year physical education program (SPARK) on physical activity and fitness in elementary school students. *American Journal of Public Health* 87: 1328-1334.

Siedentop D. 2009. National plan for physical activity: Education sector. *Journal of Physical Activity and Health* 6 (Suppl): S168-S180.

Strong, W.B., Malina, R.M., et al. 2005. Evidence based physical activity for school-age youth, *Journal of Pediatrics*, 146, pp. 732-737.

U.S. Department of Health and Human Services, Physical Activity Guidelines Advisory Committee 2008. *Physical Activity Guidelines Advisory Committee Report, 2008.* Washington, DC: U.S. Department of Health and Human Services. www.health.gov/PAGuidelines.

Van Dusen DP, Kelder SH, Kohl HW III, Ranjit N, Perry C. 2011. Associations of physical fitness and academic performance among school children. *Journal of School Health* 81:733-740.

PHYSICAL ACTIVITY AND PUBLIC HEALTH SPECIALIST

This chapter covers these competency areas as set forth by the National Society of Physical Activity Practitioners in Public Health:

1.1.1, 1.4.1, 1.4.2, 2.1.1, 2.1.3, 2.2.1, 2.2.2, 2.2.3, 3.4.2, 4.2.2, 4.5.3, 4.5.4, 6.2.1, 6.2.3, 6.4.1, 6.4.4

CHAPTER 13

BEHAVIORAL AND SOCIAL APPROACHES TO PROMOTING PHYSICAL ACTIVITY

OBJECTIVES

After completing this chapter, you should be able to discuss the following:

» The key behavioral theories and theoretical models used to explain physical activity behavior

» The definitions of behavioral and social approaches to physical activity promotion

» The rationale for promoting physical activity using behavioral and social approaches

» Theoretical models underlying behavioral and social approaches to physical activity promotion

» Evidence-based strategies for behavioral and social approaches to physical activity promotion

» Examples of each kind of strategy

OPENING QUESTIONS

Almost everyone cycles in and out of being physically active. Some weeks are good; others are not. Weeks turn into months and then into years. Life circumstances and demands have a way of shifting priorities.

» Has this ever happened to you? Have you had to change or stop an exercise routine because of life changes? How did you work through it? What strategies did you use to continue to make exercise and physical activity a priority?

» Has this ever happened to someone in your family, or a close friend? What were the barriers for this person continuing to be physically active?

» Alternatively, have you ever wondered how someone can be physically active day after day, year after year? What do these people do to be successful?

This chapter provides answers to these and other questions.

Behavioral and social approaches to physical activity promotion rely on individual strategies, skills for behavior change and maintenance, and structuring the social (as opposed to physical) environment to support physical activity. Historically, much of the early scientific work in physical activity promotion used behavioral and social approaches. These strategies, with their roots in psychology, behavioral science, and social psychology, have helped identify the ways people improve many health behaviors, not just physical activity.

BEHAVIORAL THEORIES AND THEORETICAL MODELS OF BEHAVIOR CHANGE

Behavioral scientists rely on theories and theoretical models to explain and predict health behaviors and changes in and the maintenance of those behaviors. These theories and theoretical models are very important for guiding individual-focused programs and promotion projects for physical activity. Such theoretical models typically are used to determine what can be expected from helping individuals adopt the skills they need to begin or continue physical activity.

Five popular behavior theories and theoretical models that have been used to explain and predict the physical activity behavior of individuals are depicted in table 13.1. Shown are the name of the

theory or model, key behavioral constructs that the theory or model seeks to modify or support, and a brief explanation of the theory or model.

Although a complete treatment of each aspect of these theories and models is beyond the scope of this textbook, it is important to be familiar with them and with the constructs they are designed to influence. One popular example, the transtheoretical model, is presented in here. Different scientists prefer different behavior change theories or models. Unfortunately, no single theory or model has proven to excel in its ability to predict behavior change and maintenance. Each has strengths and weaknesses based on, many believe, the varying effects of the environment on individual behavior. Chapter 14 provides a more in-depth discussion of environmental influences on physical activity behavior.

The transtheoretical model presents a behavioral continuum and addresses how people may move along that continuum to aid in the development of ideas about how to change physical activity behavior. Fundamentally, a person is classified into one of five behavioral stage categories—from not being ready to change or begin physical activity at all to maintaining that behavior for an extended period of time. The idea behind the transtheoretical model is that knowing where a person is along the continuum of behavior change makes it theoretically easier to understand how that person could increase physical activity behavior and the processes through which those changes could occur. The transtheoretical model was first developed to

Table 13.1 Popular Behavioral Theories and Theoretical Models for Physical Activity Behavior

Theory or model	Key behavioral constructs	Explanation
Health belief model (Janz et al. 2002)	Perceived susceptibility, perceived severity, perceived benefits, perceived barriers, cues to action, self-efficacy	People become physically active if they feel at risk for a negative health outcome (i.e., perceived susceptibility and severity), expect that by being physically active they will prevent that negative health outcome (i.e., perceived benefits), and believe that they can initiate and maintain the physical activity (i.e., perceived barriers, self-efficacy).
Theory of reasoned action/theory of planned behavior (Azjen 1991)	Attitudes, subjective norms, behavioral control, behavioral intention	People's intentions to be physically active depend on their beliefs about physical activity weighted by evaluations of these beliefs (i.e., attitudes), the beliefs of other people about physical activity weighted by the value attributed to these opinions (i.e., subjective norms), and the perceived ease or difficulty of being physically active (i.e., perceived behavioral control). People who intend to become physically active are likely to do so.
Social cognitive theory (Bandura 1986)	Reciprocal determinism, environment, outcome expectancies, observational learning, reinforcement, self-efficacy	Reciprocal relationships exist among the environment, personal factors (e.g., beliefs), and physical activity. Beliefs (i.e., outcome expectancies, self-efficacy) can influence actions, and vice versa. Beliefs are molded by structures within the social and physical environment. Physical activity can influence the environment and is determined by that environment. These processes occur through observational learning and reinforcement.
Self-determination theory (Ryan and Deci 2000)	External regulation, introjected regulation, identified regulation, integrated regulation, intrinsic motivation	Actions vary in the degree to which they are volitional, without any external influence. Motivation to be physically active occurs along a continuum from external regulation (i.e., rewards, others' demands), to introjected regulation (i.e., moral reasons), to identified regulation (i.e., useful outcomes), to integrated regulation (i.e., important for personal growth), to intrinsic motivation (i.e., mastery, enjoyment).
Transtheoretical model (Prochaska and DiClemente 1983)	Stages of motivational readiness for change, processes of change	People progress through five stages of change on the way to being physically active: precontemplation, contemplation, preparation, action, and maintenance. Processes of change are activities that people use to move through the stages: consciousness raising (increasing awareness), dramatic relief (emotional arousal), environmental reevaluation (social reappraisal), social liberation (environmental opportunities), self-reevaluation (self-reappraisal), stimulus control (reengineering), helping relationship (supporting), counter conditioning (substituting), reinforcement management (rewarding), and self-liberation (committing).

Adapted from Dunton et al. 2010.

FROM INDIVIDUALS TO POPULATIONS

Social and behavioral approaches to physical activity promotion have historically and necessarily focused on changing or maintaining the exercise behaviors of individuals. As we learned in the first part of this textbook, public health focuses on populations and to a lesser degree on individuals. Why then are behavioral and social approaches of interest if they focus on a person-by-person strategy? The answer is that we can learn a substantial amount about how to change populations by understanding how individuals can (and do) change. Although behavioral and social approaches may not strictly be recommended for population-wide changes in physical activity, they do provide a useful point at which to begin because populations are composed of individuals. Understanding the systems dynamics that influence individual behavior changes on a population level becomes the missing link.

explain smoking cessation behavior, but it has been extrapolated to other health behaviors. It is one of the more frequently used models in physical activity intervention studies.

A graphic description of the **transtheoretical model** as it relates to physical activity is shown in figure 13.1. The continuum of behavior has five categories that are used to classify how prepared a person is to change his behavior, or his *motivational readiness*. The first is **precontemplation,** in which a person has not even thought about becoming physically active or may be unaware of the importance of being physically active. The second category, or stage, is **contemplation**. Here, a person may be thinking about making a change to be physically active a short time in the future. She may be aware of the health benefits, but has not yet reached a tipping point to make the behavior change. The third stage, **preparation**, is when a person has reached that tipping point and is making small changes in behavior (e.g., taking the stairs rather than the elevator). In the preparation stage, a person may be looking for support from friends and family as he begins to be active. **Action** is the fourth stage of the transtheoretical model. Here, a previously sedentary person has become physically active recently and perhaps is meeting the U.S. physical activity guidelines of 150 minutes per week of moderate-intensity physical activity. Finally, the **maintenance** stage is when a person has been consistently active for at least six months.

Knowing people's stages of readiness helps practitioners tailor behavioral intervention programs to match those stages. For example, people in the preparation stage should be encouraged to move to action by giving them more reasons to exercise and fewer reasons to not exercise. Strategies that are useful in the precontemplation stage include teaching about the health benefits of exercise, the reasons for being physically active, and how to meet the minimal physical activity guidelines. People at this stage should also be educated about the risks

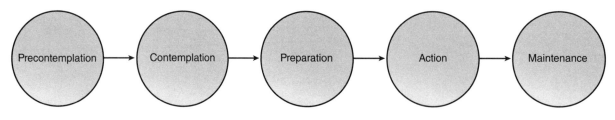

Figure 13.1 The transtheoretical model of health behavior change for physical activity.
Adapted from Marcus et al. 1992.

of being inactive. Clearly, these strategies would be of little use to those in the preparation or action stage because they would have already acquired this knowledge.

Although strategies for behavior change may differ depending on the stage, two overriding concepts that are relevant at each stage of motivational readiness are **decisional balance** and **self-efficacy**. Decisional balance refers to a person's ability to weigh the pros and cons of being physically active and to take action based on that balance. Typically, the cons outweigh the pros in the precontemplation and contemplation stages, whereas the pros should outweigh the cons in the action and maintenance stages.

Self-efficacy in physical activity behavior refers to the confidence or perceived ability that a person may have to be physically active and deal with the external threats and barriers that could result in slowing, stopping, or reverting progress. Research has identified self-efficacy as a key construct of several of the health behavior theories in table 13.1. Strategies at each stage of motivational readiness incorporate some form of skills training to grow self-efficacy for people to become physically active and to maintain that behavior.

Of course, a person may remain in one stage or another for extended periods of time, or may experience relapse as a result of illness, lack of time, or other priorities. However, identifying the person's stage and implementing strategies for dealing with the reasons for relapse can prove successful.

What keeps us from being physically active? Time is the number one reason people don't exercise at levels recommended for health enhancement. Individually adapted behavior change programs for physical activity promotion frequently incorporate strategies for helping people identify and overcome barriers. Barriers are the real or perceived things that a person believes are in the way of or preventing increases in physical activity. Identifying barriers can help people prioritize physical activity. Once the barriers are identified, strategies can be developed to overcome them.

A short quiz that is useful in identifying barriers to physical activity is shown in form 13.1. Do any of these barriers apply to you? What strategies might be useful in overcoming the barriers to help you make physical activity a priority in your life?

SOCIAL SUPPORT FOR HEALTH BEHAVIOR CHANGE

Related to, but distinct from, theories of behavior change is the concept of *social support*. Social support in public health refers to the degree to which people perceive that they are receiving assistance to overcome health challenges. Social support is thought to be key to promoting physical activity, either by being integrated into the existing health behavior theories and models reviewed earlier, or as a stand-alone strategy.

Social support has been a foundation of attempts to understand health behavior change and maintenance for years. It stems from observations that people with shared experiences and goals benefit from the support they receive from others. This support could be in many forms: two people who want to become more physically active and begin to walk together in the mornings before work; a husband supported by his wife to restart a physical activity program; or a mother who supports her child to be physically active by enabling participation in a sport league after school. Each of these examples (and others) constitutes some form of social support for physical activity.

The three basic types of social support are perceived, received, and connected (Barrera, 1986). *Perceived support* refers to the perceptions that one is adequately supported. For physical activity, an example would be a woman who knows she can count on a church friend or friends to walk with her when she needs company. *Received support* is more direct and measurable. It refers to the amount of direct support a person can count on for physical activity. An example of this is a basketball team that must have five members to play. Each member counts on the others to be at the playground so the team can play. Finally, *connected support* refers to the degree to which a person is socially integrated. Social integration provides implicit social support as a result of the connections made through participation. Examples are clubs, communities and community events such as fun runs, the workplace, and family and friends. Connected support is thought to be helpful in physical activity promotion because of the experiences that can be shared through a social network.

Form 13.1 Barriers to Being Active Quiz

How likely are you to say?	Very likely	Somewhat likely	Somewhat unlikely	Very unlikely
1. My day is so busy now, I just don't think I can make the time to include physical activity in my regular schedule.	3	2	1	0
2. None of my family members or friends like to do anything active, so I don't have a chance to exercise.	3	2	1	0
3. I'm just too tired after work to get any exercise.	3	2	1	0
4. I've been thinking about getting more exercise, but I just can't seem to get started.	3	2	1	0
5. I'm getting older so exercise can be risky.	3	2	1	0
6. I don't get enough exercise because I have never learned the skills for any sport.	3	2	1	0
7. I don't have access to jogging trails, swimming pools, bike paths, etc.	3	2	1	0
8. Physical activity takes too much time away from other commitments—time, work, family, etc.	3	2	1	0
9. I'm embarrassed about how I will look when I exercise with others.	3	2	1	0
10. I don't get enough sleep as it is. I just couldn't get up early or stay up late to get some exercise.	3	2	1	0
11. It's easier for me to find excuses not to exercise than to go out to do something.	3	2	1	0
12. I know of too many people who have hurt themselves by overdoing it with exercise.	3	2	1	0
13. I really can't see learning a new sport at my age.	3	2	1	0
14. It's just too expensive. You have to take a class or join a club or buy the right equipment.	3	2	1	0
15. My free times during the day are too short to include exercise.	3	2	1	0
16. My usual social activities with family or friends do not include physical activity.	3	2	1	0
17. I'm too tired during the week, and I need the weekend to catch up on my rest.	3	2	1	0
18. I want to get more exercise, but I just can't seem to make myself stick to anything.	3	2	1	0
19. I'm afraid I might injure myself or have a heart attack.	3	2	1	0
20. I'm not good enough at any physical activity to make it fun.	3	2	1	0
21. If we had exercise facilities and showers at work, then I would be more likely to exercise.	3	2	1	0

Follow these instructions to score yourself:

1. Enter the circled number from each of the 21 items in the Barriers to Being Active Quiz above in the numbered spaces provided below, putting together the number for statement 1 on line 1, statement 2 on line 2, and so on.

2. Add the three scores on each line. Your barriers to physical activity fall into one or more of seven categories: lack of time, social influences, lack of energy, lack of willpower, fear of injury, lack of skill, and lack of resources. A score of 5 or above in any category shows that this is an important barrier for you to overcome.

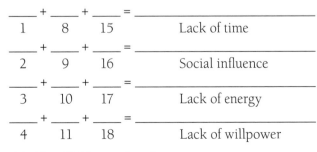

____ + ____ + ____ = _____			____ + ____ + ____ = _____		
1	8	15 Lack of time	5	12	19 Fear of injury
____ + ____ + ____ = _____			____ + ____ + ____ = _____		
2	9	16 Social influence	6	13	20 Lack of skill
____ + ____ + ____ = _____			____ + ____ + ____ = _____		
3	10	17 Lack of energy	7	14	21 Lack of resources
____ + ____ + ____ = _____					
4	11	18 Lack of willpower			

Reprinted from CDC (www.cdc.gov).

How physically active are your family members? Why are they active or inactive?

Behavioral and social approaches for physical activity promotion increase physical activity either by enabling people with behavior change and management skills according to one of the health behavior theories or theoretical models in table 13.1 or by structuring the social environment so that it is conducive to physical activity behavior. Behavioral approaches give people skills to be physically active and to overcome barriers to implement those skills. Social approaches are one way to make it easier to use those skills.

Behavioral and social approaches increase physical activity participation as part of leisure, occupation, transportation, or at-home activities. Like informational approaches (discussed in chapter 11), behavioral and social approaches to increasing physical activity are based on the theory that when people are told to engage in specific health behaviors that are generally perceived as being good for them (e.g., improving health), they will change their behavior.

INDIVIDUALLY ADAPTED HEALTH BEHAVIOR CHANGE PROGRAMS

Individually adapted behavior change strategies integrate key components of the health behavior theories and theoretical models to help people change and maintain physical activity behaviors. The evidence base behind these types of programs is strong, and these strategies work when implemented appropriately.

What kind of improvements can be expected in individually adapted behavior change programs? According to the Guide to Community Preventive Services Task Force (Community Guide) recommendations on physical activity (Kahn et al. 2002), the studies that led to their recommendations showed an average 35% increase in physical activity behavior (pretest to posttest measures). Increases in physical fitness ($\dot{V}O_2$max) and caloric expenditure have been documented as well using these kinds of programs. Although studies measure physical activity in a variety of ways and vary in length, such increases are impressive. For the physical activity outcomes, it means that if someone reported being somewhat active each week (e.g., walking about two hours, or 120 minutes, each week), participation in such a program could be expected to help the person achieve the 150-minutes-per-week goal recommended in the *2008 Physical Activity Guidelines for Americans* (USDHHS 2008). Because they are not one-size-fits-all approaches, individually adapted behavior change strategies seem to work for both women and men. Moreover, they can be adapted to a variety of settings, including worksites, communities, schools, and possibly families.

To be most successful, individually adapted behavior change programs tailor the type and dose of interventions to the individual's needs. As discussed earlier, this could be a stage-matched education curriculum for someone in the precontemplation or contemplation stage if the transtheoretical model is being used to guide the program development. Alternatively, such a strategy could include a keen focus on how the person interacts with his or her environment to become more physically active, if social cognitive theory is guiding the program. The key point is that there are models and theories, but

POINT-OF-DECISION PROMPTING

Although prompting (reminders) is a technique used to increase physical activity in an individually adapted behavior change program, it can be aimed at larger groups of people. The Community Guide recommends **point-of-decision prompting** as an evidence-based strategy for physical activity promotion. The idea is similar to other individually based prompting strategies. At a point of decision (e.g., to make an active choice or a sedentary choice), the prompt is placed to encourage selection of the active choice. These strategies have most often been used at elevators and escalators in buildings where choosing to walk the stairs is a reasonable and convenient physical activity alternative.

Even though the Community Guide classifies point-of-decision prompting as an informational approach, prompts could have different effects on different people, thus also making it a more individualized approach. Although point-of-decision prompting can influence the acute decisions of people to be physically active, the extent of lasting changes in physical activity behavior is unknown. Other than stairs and elevators, can you think of other situations in which point-of-decision prompting might be useful in promoting the active choice?

each type of strategy in individually adapted behavior change is guided by an emphasis on the needs of the individual and the individual's needs for success.

How do people change from being sedentary to being physically active? Although there is no one way for everyone, behavioral scientists rely on strategies that increase self-efficacy and move a person's decision balance from more cons to more pros. Five strategies that are part of many individually adapted behavior change programs for physical activity promotion are shown in table 13.2.

The art in this kind of work is finding the appropriate mix of strategies that will be most effective for the individual. Whereas some people need only simple reminders, others need several (or all) strategies to move into a physically active lifestyle.

Table 13.2 Behavioral Strategies Useful in Individually Adapted Behavior Change Programs to Increase Physical Activity

Strategy	Intended consequence	Example
Substitution	Stay physically active when you may not be even thinking about being active.	Push back from a desk during work and take several two-minute walks during the day.
Social support	Find a partner or partners to help you stay active.	Join a walking club or exercise with family members.
Self-reward	Provide positive feedback to yourself for being physically active.	Set pedometer goals and reward yourself with a gift when short- and long-term goals are reached.
Commitment	Encourage tangible commitments for yourself to being physically active.	Sign a self-contract; become a physical activity support to someone who is trying to become more active.
Reminders	Use prompting tools to remind yourself and others about activity.	Place exercise shoes and equipment where they are visible (e.g., exercise machines in the middle of the house or apartment instead of hidden in a back room).

One of the first long-term studies to examine the role of individually adapted behavior change in increasing physical activity levels compared the transtheoretical model combined with aspects of social cognitive theory to a more traditional model of exercise promotion (Dunn et al. 1999). Project Active researchers recruited 235 sedentary women and men to participate in a two-year individually adapted behavior change program. These participants were between the ages of 35 and 60, were not obese, and were otherwise healthy (except for being sedentary).

The 235 participants were randomized into two groups. The structured exercise group received a standard exercise prescription encouraging them to become active at a certain intensity (based on their baseline tests) for 20 to 60 minutes on three to five days each week. These participants were also given complimentary access to a health club gymnasium for the first six months of the study. Finally, contact was maintained with them for the full 24 months through newsletters and periodic mailings.

Participants in the lifestyle exercise group received an individually adapted behavior change program that used the transtheoretical model to match their stages of motivational readiness with appropriate behavioral intervention strategies and processes. They attended weekly meetings for the first four months of the program and then biweekly meetings for the next two months. These meetings focused on the cognitive and behavioral strategies thought to help people make positive, lasting changes. Each participant's stage of motivational readiness was identified, and the strategies and processes used to increase knowledge and change behavior were matched to these stages. Topics covered in these meetings (among others) included how to set a goal and monitor progress, how to reinforce positive behavior, how to overcome barriers that get in the way of being physically active (problem solving), how to build social support, and how to prevent relapse.

The main outcome of interest in Project Active was physical activity (measured by estimated energy expenditure) in the two groups. Would those in the lifestyle group increase their physical activity as a result of the individually adapted behavior change program? How would any change in the lifestyle group compare to that in the structured exercise group?

The main findings from Project Active are illustrated in figure 13.2. People in both groups significantly increased their physical activity levels from baseline through 24 months. Perhaps most interesting, the lifestyle group showed a statistically *equivalent* increase in energy expenditure from physical activity compared to the structured exercise group. Similar findings were seen for increases in aerobic fitness as well. These results showed that not only is an individually adapted behavior change program feasible for increasing physical activity among previously sedentary people, but also, it can be a useful alternative for people for whom the more structured, traditional exercise prescription model may not work or be appropriate. Can you think of some examples in which such a model may not be the best choice? Project Active confirmed the utility of behavior change programs for physical activity promotion.

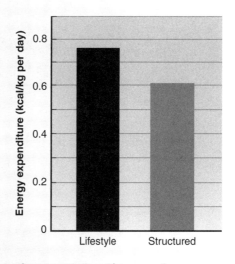

Figure 13.2 Changes in energy expenditure (in kcal/kg per day) over 24 months due to physical activity in Project Active.

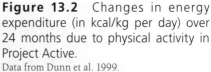
Data from Dunn et al. 1999.

SOCIAL SUPPORT INTERVENTIONS IN COMMUNITY SETTINGS

Another recommended strategy for increasing physical activity via behavioral and social approaches is **social support interventions in community settings.** As mentioned earlier, social support is a broad concept, but generally refers to any strategy for developing or strengthening the social environment to encourage (or overcome barriers to) physical activity. Social support can be an important strategy in individually adapted behavior change programs, but the type of social support we refer to here is at the community, rather than individual, level.

Community social support interventions focus on changing physical activity behavior by building, strengthening, and maintaining social networks in the community that provide supportive relationships for behavior change (e.g., a buddy system, contracts with others to complete specified levels of physical activity, and walking groups or other groups to provide friendship and support).

Studies of community-based social support programs to promote physical activity generally report the following findings (Kahn et al. 2002):

- Social support interventions in community settings increase physical activity as measured in a variety of ways (e.g., blocks walked or flights of stairs climbed daily, frequency of attending exercise sessions, minutes spent in physical activity).

- Time spent in physical activity can be expected to increase by an average of 44% above baseline levels.

- Frequency (days per week) of physical activity can be expected to increase on average nearly 20%.

- More support (greater participation in social networks for physical activity) may be associated with a higher level of physical activity. The potential for a dose-response relationship between the two has not yet been demonstrated, however.

- Social support for physical activity promotion seems to be effective in a variety of settings (e.g., communities, worksites, universities),

Does socialization promote participation in regular physical activities?

CASE STUDY: HEART AND SOUL PHYSICAL ACTIVITY PROGRAM

Peterson and colleagues (2005) developed a unique physical activity program targeted at women in midlife (35 to 65 years of age) using churches as the target community and focusing on developing a social support system. Physical activity was promoted through the church-based support network. The goal of the project was to increase moderate-intensity physical activity among program participants compared to those in churches not receiving the program. Over a 12-week period, participants met for one hour each week and worked during this time to identify and develop ways to support each other to be more physically active. During the meetings, participants were asked to share their physical activity goals, barriers, and successes with other members and group leaders. Strategies for finding outside support (e.g., people to exercise with) were included. Participants in the control churches received an exercise brochure, information regarding physical activity recommendations, and several follow-up telephone calls during the 12-week session.

Although the study was fairly small, the participants in the social support group demonstrated a 63% increase in the minutes per week spent in physical activity that was at least of moderate intensity. Those in the information-only control group churches increased their minutes per week of moderate-intensity physical activity also, but not nearly as much at the social support group (43%). The authors of this study concluded that this type of social support–based physical activity promotion showed some promise as a strategy to promote physical activity.

for both sexes, across the adult age range, and among people with various baseline physical activity levels (both sedentary people and those who were already active).

As with any community-based work, developing community-based social support systems can be challenging. Whether it is a faith-based exercise program for one specific place of worship, a neighborhood walking club across several blocks, or a sport league in a parks and recreation department in an entire town or city, the most important first step in developing such programs is always to know the target audience. The most well-intentioned program will never get off the ground unless leaders and stakeholders are consulted and their ideas are considered prior to program development. Chapter 15 introduces the logic model and program evaluation strategies. A key first step is defining the problem through census data, health survey data, or other sources. Perhaps most important is to leverage existing networks and other sources of social support rather than trying to create new ones.

A final note: although individually adapted behavior change programs and community-based social support systems have been scientifically shown to increase physical activity, other types of social and behavioral strategies have not. This may be because of a lack of evidence (i.e., the strategy has not been rigorously tested) or because the strategy does not actually work. It is important to know what works, but also what doesn't work in physical activity promotion. Knowing the evidence base allows for more informed decisions about physical activity promotion programs. Clearly, programs for which there is evidence of effectiveness should take precedence.

According to the Community Guide (Kahn et al. 2002), social and behavioral approaches to physical activity promotion for which there is not enough evidence to recommend their implementation include college-based health and physical education courses; classroom-based health education curricula for elementary, middle, and high school students; and family-based social support programs.

KEY LEADER PROFILE

Antronette (Toni) K. Yancey, MD, MPH

Courtesy of Antronette K. Yancey.

Why and how did you get into this field?

Social justice advocacy has been in my blood since my junior high school was desegregated in 1969. I was drawn to preventive medicine and public health because, as a general practitioner in New York City in the late 1980s, I was regularly reminded of the need to work more upstream—to change organizational, regulatory, and legislative policies. The health of many of the teens I treated in foster care was already compromised by some combination of smoking, fast food, STDs, substance abuse, and pregnancy, the last of which feeds the cycle of poverty and poor parenting. Many of the adults I saw in community clinics were walking train wrecks, debilitated by diabetes, hypertension, and cardiovascular disease by age 50. I felt as though I were treating massive hemorrhaging with Band-Aids.

Did any one person have an overriding influence on you?

I believe that it's very important to have a mentor, preferably a visionary, like mine, Lester Breslow, MD, MPH. He read every memo, report, article, and video script I wrote while precepting my preventive residency and beyond, and got back to me in a few days with extensive comments and recommendations. This from a man who is revered in the field as a pioneer, the modern "father of public health." He provided a concrete example of how to get it done, and, sometimes more important, what to get done, when to do it, and whom to or not to engage in the effort.

What are your current research interests?

My primary research interest is in the design, implementation, and evaluation of population-based physical activity promotion interventions. I've worked most actively in recent years on changing the social and cultural environments in organizational settings to make the active choice the default option, and on teasing apart the elements required for successful intervention, including leadership and role modeling.

What drives you as a researcher and activist?

In 1996, when I was making the transition from academia to the public sector, directing Richmond, Virginia's, city health department, I spent a lot of time imagining how I could improve the fitness of an entire city. I knew I wanted to bring the department into the 21st century by focusing on chronic disease prevention and particularly physical activity promotion. But my challenge was to figure out how to inspire or induce people to do something that people have tried to avoid throughout most of human history, and which is still shunned in most forms by all but an elite sliver of the population. Sitting at my desk one day, I started thinking about the times in life when being active is strictly "fun and games," when it is play, not work. My book *Instant Recess: Building a Fit Nation 10 Minutes at a Time* is a synthesis of my 20 years of work in the field.

What are one or two key issues to be addressed by 2022?

We must find ways to insert physical activity into the national dialogue about health policy, and I think that the way to do it is in small increments, delivering value along the way. I believe that short recess breaks for adults and kids could be the smoking bans of the fitness movement, creating an entry point for organizations to get involved, because organizational change has been the driving force behind the major public health successes of the late 20th century, such as decimating tobacco use and drunk driving.

CHAPTER WRAP-UP

What You Need to Know

- Behavioral and social approaches can improve readiness for individual and community behavior change.
- The behavior theories and theoretical models highlighted in table 13.1, as well as others, help with an understanding of why people are physically active or inactive, and also help to develop approaches for behavior change programs.
- Behavioral and social approaches to increasing physical activity are based on the theory that when people are told to engage in specific health behaviors that are generally perceived as good for them (e.g., improving health), they will change their behavior.
- Individually adapted health behavior change programs teach behavioral skills such as goal setting and self-monitoring of progress toward those goals, building social support for new behaviors, reinforcing behavior through self-reward and positive talk, structured problem solving to maintain behavior change, and preventing relapse into sedentary behavior.
- Individually adapted health behavior change programs are recommended to increase physical activity.
- Social support interventions focus on changing physical activity behavior through building, strengthening, and maintaining social networks that provide supportive relationships for behavior change (e.g., a buddy system, contracts with others to complete specified levels of physical activity, and walking groups or other groups to provide friendship and support).

Key Terms

transtheoretical model

precontemplation

contemplation

preparation

action

maintenance

decisional balance

self-efficacy

point-of-decision prompting

individually adapted health behavior change programs

social support interventions in community settings

Study Questions

1. How do the five theories or models of behavior change differ?
2. How can a person's stage of motivational readiness be measured?
3. What is the goal of behavioral and social approaches to increase physical activity?
4. What are five common barriers to being physically active?
5. What are three basic types of social support?
6. Why are social support interventions in community settings recommended by the Community Guide as physical activity programs?
7. What are three ways individually adapted health behavior approaches can increase physical activity?
8. What are two community-based social support strategies for increasing physical activity?
9. What is point-of-decision prompting, and how does it work?
10. What was Project Active? Was it successful or unsuccessful?

E-Media

Explore issues related to physical activity, exercise, and public health at the following websites:

U.S. Department of Health and Human Services: Physical Activity Guidelines for Americans	www.health.gov/PAGuidelines
President's Council on Fitness, Sports & Nutrition	www.fitness.gov
U.S. Centers for Disease Control and Prevention: Physical Activity	www.cdc.gov/physicalactivity
Guide to Community Preventive Services	www.thecommunityguide.org/pa/index.html
Healthy and Active Australia	www.healthyactive.gov.au
Behavioral and Social Approaches	http://youtu.be/8xmilUPtqiM
	http://youtu.be/2P6G0LjPbco

Bibliography

Ajzen I. 1991. The theory of planned behavior. *Organizational Behavior and Human Decision Processes* 50: 179-211.

Bandura A. 1986. *Social Foundations of Thought and Action*. Englewood Cliffs, NJ: Prentice Hall.

Barrera M. 1986. Distinctions between social support concepts, measures, and models. *American Journal of Community Psychology* 14, 413-445.

Dunn AL, Marcus BH, Kampert JB, Garcia ME, Kohl HW III, Blair SN. 1999. Comparison of lifestyle and structured interventions to increase physical activity and cardio-respiratory fitness: A randomized trial. *Journal of the American Medical Association* 281, 327-334.

Dunton GF, Cousineau M, Reynolds KD. 2010. The intersection of public policy and health behavior theory in the physical activity arena. *Journal of Physical Activity and Health* 7 (Suppl 21): S91-S96.

Janz NK, Champion VL, Strecher VJ. 2002. The Health Belief Model. In Glanz K, Lewis FM, Rimer BK, eds. *Health Behavior and Health Education*. San Francisco: Jossey-Bass, 45-66.

Kahn EB, Ramsey LT, Brownson RG, et al. 2002. The effectiveness of interventions to increase physical activity. *American Journal of Preventive Medicine* 22: 73-107.

Kohl HW III, Dunn AL, Marcus BH, Blair SN. 1998. A randomized trial of physical activity interventions. *Medicine & Science in Sports & Exercise* 30: 275-283.

Marcus BH, Banspach SW, Lefebvre RC, Rossi JS, Carleton RA, Abrams DB. 1992. Using the stages of change model to increase the adoption of physical activity among community participants. *American Journal of Health Promotion* 6: 424-429.

Peterson JA, Yates BC, Atwood JR, Hertzog M. 2005. Effects of a physical activity intervention for women. *Western Journal of Nursing Research* 27: 93-110.

Prochaska JO, DiClemente CC. 1983. Stages and processes of self-change of smoking: Toward an integrative model of change. *Journal of Consulting Clinical Psychology* 51: 390-395.

Ryan RM, Deci EL. 2000. Self-determination theory and the facilitation of intrinsic motivation, social development, and well-being. *American Psychology* 55: 68-78.

U.S. Department of Health and Human Services. 2008. *2008 Physical Activity Guidelines for Americans*. Washington, DC: U.S. Department of Health and Human Services. www.health.gov/PAGuidelines.

PHYSICAL ACTIVITY IN PUBLIC HEALTH SPECIALIST

This chapter covers these competency areas as set forth by the National Society of Physical Activity Practitioners in Public Health:

1.1.1, 1.1.5, 1.4.1, 1.4.2, 2.1.3, 2.3.3, 3.1.3, 3.1.4, 3.3.1, 3.3.2, 3.3.3, 4.1.1, 4.1.2, 4.1.5, 4.2.5

© Stephen Finn

CHAPTER 14

ENVIRONMENTAL AND POLICY APPROACHES TO PROMOTING PHYSICAL ACTIVITY

OBJECTIVES

After completing this chapter, you should be able to discuss the following:

» How aspects of the physical and built environment can encourage or inhibit physical activity

» How enhancing access to opportunities for physical activity, and places to be physically active, works

» The role of urban design in physical activity promotion

» The differences between street-scale urban design and land use policies and community-scale urban design and land use policies, and how each type can be used for physical activity promotion

» How built and physical environmental influencers on physical activity are measured

» Physical activity policy concepts and practical examples of how to create or enhance places in which to be physically active

OPENING QUESTIONS

» Can the built environment influence physical activity? If so, which aspects help? Which aspects hurt?

» How can the local physical environment be changed to increase opportunities for physical activity?

» What strategies aid in the development and implementation of policies to promote physical activity by changing or enhancing the built environment?

What do you think of when you hear the term **built environment**? The apartment or home in which you live? The gym or fitness center downstairs? The buildings at your university? A hike and bike trail in the center of town? Although many people have proposed different various definitions, for our purposes, the built environment refers to the set of constructed structures that influence opportunities for physical activity. These structures can be positive (e.g., sidewalks on a busy street), or negative (e.g., a six-lane road with no crosswalks that prevents pedestrians from getting to one side from the other) in their influences on physical activity. **Physical environment** is a broader term encompassing the built environment as well as other physical supports or barriers. Walking on hilly terrain can be more difficult for older people or people who are of low fitness levels than walking on flat terrain such as the beach. Wild animals (e.g., stray dogs) can be a part of the physical environment that makes playing outdoors difficult for children.

We have known for centuries that health is related to the physical environment. Where you live matters. Although this can be (and is) related to social factors, physical factors also play an incredible role in determining health. The quality of the water and air, the type of housing, roads, safety issues, and other physical and environmental factors are critical to overall health, health care, and disease prevention.

As covered in chapter 13, strategies for individually adapted behavior change and enhancing social support can increase physical activity among individuals. Although these strategies may be sufficient for some people, public health is also interested in changes that affect population health. Thus, physical activity and public health should also focus on changes that may affect an entire group of people such as the citizens of a town, students enrolled in an urban school district, members of a neighborhood association, or residents of an apartment complex.

Environmental and policy change initiatives to improve public health seek to augment the approaches focused on individuals. Actions at the environmental and policy levels directly affect organizations and physical structures rather than people in an attempt to achieve longer-term and more sustainable results. To maximize success, environmental and policy changes must involve many sectors of influence outside of health. For

POLICY INFLUENCES ON HEALTH

Changing or maintaining a behavior is difficult when external forces continue to work against the behavior. Effective health and health behavior policies include tobacco taxation and purchase age restrictions to reduce smoking, water supply fluoridation to reduce dental caries, speed limits to reduce motor vehicle fatalities, and breakaway bases in Little League Baseball to reduce musculoskeletal injuries in players. The field of public health differs from the medical and health care fields in that it acts at the population level to reduce disease and disability. Can you think of other examples of public health policies?

example, building a bicycle lane on a major street must involve the transportation department, the city government, and public safety experts.

Major changes in public health rarely, if ever, depend solely on behavior change in individuals. Policy initiatives, environmental initiatives, or both, usually facilitate individual behavior change.

Environmental and policy approaches to physical activity promotion create or enhance opportunities, support, and cues to help people be more physically active. These approaches are often combined with informational outreach activities to enhance their effectiveness. Environmental and policy approaches may involve making changes to the built or physical environment, making changes in organizational norms and policies, or enacting legislation that improves health.

Environmental approaches to physical activity promotion are found in two broad areas: access and urban design and land use. This chapter describes these areas and offers examples.

ACCESS

Broadly defined, the term **access** refers to the right to approach or use something. In physical activity and public health, creating or enhancing access to places for physical activity is an evidence-based strategy for increasing physical activity and exercise. Studies that have examined the issue of increased access have focused on strategies for changing the physical environment. Converting an old rail bed to a hike and bike trail, reducing the cost of fitness center memberships for employees in a company (economic access), and unlocking the school playground basketball court so that it can be used on weekends are all simple strategies for increasing access. The highlight box Increasing Access to Physical Activity at a Worksite offers examples of ways to increase access to physical activity in a work setting.

Although intuitive to some, the idea of creating or enhancing access to physical activity is not as straightforward as it seems. Simply building a new trail, cleaning up a park and building a new ball field, or providing more exercise equipment in a fitness center may not be enough. Most of the studies that have pointed to higher physical activity participation with increased access to places in which to be active have also included some form of informational outreach. Essentially, it does not appear to be enough to build or make it easier to get to physical activity resources. Appropriate (and targeted) information is needed to let people know about the existence of these resources. Signs, texting, announcements on websites, mailings, and e-mails are all ways to let people know that access exists.

In general, the physiological and behavioral results of studies examining increased access to places in which to be physically active support it as a recommended strategy (Kahn et al. 2002). Caloric expenditure among participants may be expected to increase an average of just over 8% above baseline. The percentage of people in a defined population exposed to increased access to physical activity

INCREASING ACCESS TO PHYSICAL ACTIVITY AT A WORKSITE

- Provide secure and covered parking for bicycles to encourage bicycling to and from work.
- Install employee shower facilities and changing rooms.
- Provide financial incentives for active commuting to work.
- Create a culture of physical activity by encouraging brief exercise breaks throughout the workday.
- Provide on-site fitness facilities or buildings that are conducive to physical activity, as well as easy access to walking and running routes; distribute walking maps.
- Offer reduced-fee fitness or recreation center memberships to employees and their families.
- Encourage employees to participate in community-based worksite exercise competitions or community-based mass participation events.

might be expected to increase physical activity participation around 3%. Cardiorespiratory fitness (aerobic capacity) should increase more than 5% on average. Some participants will increase much more than that, however, and others may not increase at all (or may actually decrease). Results of effectiveness (increased participation in physical activity) have been seen in various types of settings, in low-income communities, and in various racial or ethnic groups. Men and women seem to respond equally well to increasing access to places in which to be physically active.

URBAN DESIGN AND LAND USE POLICIES

The second way physical activity can be influenced by the built or physical environment is through urban design and land use policies. **Urban design and land use** are separate constructs but are fre-

quently combined in discussions of the physical and built environment and physical activity. Many definitions seem to exist for each. For the purposes of this text, urban design refers to the form, function, and outward appearance of the physical environment in defined entities, such as neighborhoods, towns, cities, and communities. Examples of urban design for physical activity are landscape design in municipal parks, street design for pedestrian safety, and recreation center design and placement within a community.

Land use, on the other hand, refers to the management, planning, and development of land in defined jurisdictions. Land use policies most frequently occur at the local level to advance the well-being of the communities that control the land. **Zoning** is another commonly used term that is synonymous with legislatively determined land use policy. Examples of land use strategies for physical activity include setting density targets (i.e., creating guidelines about the people living per square mile)

CASE STUDY: DEVELOPMENT AND PROMOTION OF WALKING TRAILS

Brownson and colleagues (2004) reported the results of an effort to increase access to walking trails in the state of Missouri in the United States. Adults 18 years of age and older were targeted. The study group was mostly women (75%), most had less than a high school education (60%), and 70% were Caucasian. Control groups were selected from similar communities in the neighboring states of Arkansas and Tennessee. The program was designed to increase the physical activity levels in rural communities, in part, by creating walking trails. Participants received eight individually tailored newsletters to promote interpersonal activities and social support, while advertising community-wide events such as walk-a-thons and walking clubs. The walking trails in the intervention were equipped with tracking systems to help people acquire individually tailored walking reports. The cost of trail development was approximately $3,000 (USD) per trail (six trails from 0.13 to 2.38 miles, or 0.3 to 3.8 kilometers, in length). Of particular interest in this study were changes in the use of the walking trails and walking behaviors in general (minutes per week).

After a one-year promotion effort in six communities, changes in walking trail use and walking behavior were observed in this study. However, these changes were limited to increases in trail use among those who already had been using the trails at the baseline or as they were built. In other words, there was an increase in use among people who had already reported using the trails. Although this is a positive result, the program seemed to miss affecting those who hadn't used the trails prior to the study. Certain demographic subgroups showed increases in weekly walking participation, but a change was not observed in the population as a whole. The authors concluded that this way of increasing access in rural communities should be used to understand how to design future studies in this area.

CHANGEABILITY?

Clearly, community-scale changes in the physical and built environment cannot happen overnight, or even in the span of a few years. These types of changes are structural and require a long-term view and strong leadership given that physical activity and health benefits may not be seen for many years or decades. Because these are expensive propositions, community leaders must consider the financial impacts. Political challenges can slow such changes as well—not all stakeholders may be supportive of broad community-level changes. Despite such difficulties, the long view is necessary, and the question perhaps most relevant is, Can we afford to *not* make these changes for the sake of the health of our citizens?

for a neighborhood, green space mandates (e.g., preservation of undeveloped land) within town or village limits, and urban mixed-use developments that provide destinations within walkable distances to living spaces.

Although no evidence-based recipe for the perfect, or even optimal, urban design or land use elements that will maximize physical activity exists, many strategies have been shown to work in communities. Ongoing research will help us to identify specific approaches. For now, urban design and land use policies and practices can be separated into two categories: community scale and street scale.

COMMUNITY-SCALE URBAN DESIGN

Community-scale urban design land use policies and practices involve changes and enhancements to the physical and built environment of urban areas of several square miles (or kilometers) or larger. Community-scale policies can involve a small town or village or larger cities or towns. These kinds of approaches strive to make entire communities more amenable to physical activity, whether that activity is transportation related or exercise purposefully done in recreational or discretionary time. Strategies at this level of influence include connecting

transportation arteries; creating landscaping and lighting to enhance the aesthetics and perceived safety of the community; providing tax incentives for developers to build sidewalks and trails in new developments; implementing community-wide programs that encourage bicycling; and designing residential areas so that destinations such as workplaces, schools, and areas for leisure and recreations are within safe walking or bicycling distances.

Studies examining community-scale changes for physical activity promotion have used a wide variety of outcome measures (Heath et al. 2006). Some have studied the absolute number of walking trips in a community in a given time period; others, the distance of those trips, minutes of walking per week, and number of pedestrians in a certain area. Although these outcomes are not entirely comparable, the general interpretation of the science base is that the physical and built environment can improve levels of physical activity (regardless of how it is measured) an average of more than 160%. Clearly, this is a major influencer of physical activity habits in a community.

Often overlooked, but very relevant, are the effects community-scale environmental changes can have on factors outside of physical activity. Various studies have reported higher amounts of green space in communities that encourage physical activity. People in these communities have a greater sense of community, or belonging, compared to those other communities, and some studies even reported lower crime rates. Although these other benefits are not absolute (or guaranteed), they are useful additional positive outcomes that may help advocacy efforts for more physical activity–friendly communities.

STREET-SCALE URBAN DESIGN

Street-scale urban design and land use policies and practices include changes to the built and physical environment in smaller geographic areas, generally limited to a few blocks. These kinds of approaches seek to make neighborhoods or similar areas more livable and amenable to a variety of physical activity opportunities. Strategies include enhancements for pedestrians such as marked street crossing areas or pedestrian bridges over multilane highways; traffic-calming strategies such as traffic circles, stop lights, signs, and speed bumps; bicycle lanes; improved lighting; landscaping; and the repair of

CASE STUDY: BOGOTÁ, COLOMBIA

 Gomez and colleagues (2010) reported on efforts in the capital city of Bogotá, Colombia, to remake the city center and outlying areas to increase the mobility of citizens and recover public space to enhance the quality of life in the city. Many of these changes involved community-scale improvements in land use and urban design. A large, modern mass transit system was created to reduce motor vehicle traffic and air pollution, green space and parks were created, and bicycle paths (*ciclorutas*) were constructed to connect the parks. The remaking of Bogotá has been one of the most substantial urban redesign efforts ever undertaken.

The authors studied the role these changes throughout the city may have had on physical activity participation. Using geographic information system (GIS) techniques (these are described in more detail later in this chapter), the study mapped the neighborhoods in which participants lived and analyzed their proximity to major community-level built or physical environment elements that may be related to physical activity participation. They considered neighborhood density, land use mix, the density of parks, the completeness of *ciclorutas*, the proximity to transit stations, and the topographic slope of neighborhoods (Bogotá is situated on a plateau in a mountainous area). Men and women aged 18 to 65 made up the study population.

Results suggested that active people were more than twice as likely to live in neighborhoods with the highest density of parks compared to inactive people, and those who lived closer to a new transit station were 27% more likely to be physically active than similar residents who lived farther away. Although causality is difficult to determine with this study design, it is clear that community-level built and physical environmental variables are associated with increased levels of physical activity.

eyesores such as broken windows and graffiti to increase safety and aesthetics. Although street-scale approaches can be translated community-wide, they are most often initiated in neighborhoods in which leaders are looking to improve residents' standard of living.

Like community-scale studies, studies examining street-scale characteristics and their relation to physical activity have used a variety of outcome measures. Researchers have looked at increases in the number of walkers or bicyclists in a given area, the prevalence and change in prevalence of people

CAUSE AND EFFECT?

One of the major challenges facing researchers in the area of environmental and policy change for physical activity promotion is the lack of true experiments. It is impossible to randomly assign people (well, not impossible, but certainly not practical) to a neighborhood that has street-scale supports for physical activity or to a neighborhood that doesn't. People are free to live where they wish. So researchers are stuck with measuring what happens without experimentation. They cannot know whether characteristics of the neighborhood make people more active or whether people who are already physically active or are looking for those characteristics are more likely to choose that neighborhood. This issue can confound studies looking to assess whether the cause (a neighborhood more conducive to physical activity) results in the effect (higher levels of physical activity participation). How would you design a study to tackle this problem?

who are physically active, and the number of users of a walking and jogging path. This variety makes summarizing the effects of street-scale changes difficult. All things considered, a 35% average increase in physical activity might be anticipated with appropriate street-level changes. The types of changes necessary clearly depend on the neighborhood that is targeted. Because no two are alike, the general recommendation can encompass many specific strategies.

CASE STUDY: MUELLER

The Mueller neighborhood in Austin, Texas, was built on the site of a municipal airport that dated back to the early years of aviation and was closed in 1999. As part of a redevelopment effort, the 800+ acres were transformed into a new urbanist–inspired neighborhood. New urbanism is a neighborhood planning approach that creates sustainable communities by adhering to six principles: walkability (pedestrian-friendly street design), connectivity (interconnected street grid), a diverse and mixed-use built environment (shops, offices, and residences all in a geographically convenient area), mixed housing (a variety of housing choices), aesthetically pleasing and high-quality architecture, and public spaces in the neighborhood. Walking trails, parks, traffic-calming devices, sidewalks, bicycle lanes, and other strategies are hallmarks of the Mueller neighborhood design.

The development of Mueller offered a unique opportunity to study the effects of the street-scale design on physical activity participation. Although it was not possible to experimentally assign people to the neighborhood, a natural experiment was possible. Calise and colleagues (2012) surveyed newly arrived residents in Mueller about their physical activity habits before and after their move to the neighborhood. The hypothesis was that people who were less active in their previous neighborhoods would have higher reported physical activity levels in Mueller. The design of the neighborhood might be one explanation for any changes observed.

Key results of the study are shown in figure 14.1. After moving to Mueller, study participants reported an average of nearly 30 minutes more of moderate-intensity and vigorous-intensity physical activity each week compared to when they lived in their previous neighborhoods. Interestingly, there was an increase in reported recreational walking (about 45 minutes each week) after the move to Mueller, but there was actually a *decrease* in reported walking outside the neighborhood (approximately 20 fewer minutes each week). The vast majority of the increase was in minutes per week of walking *inside* Mueller. The authors concluded that the design characteristics of the neighborhood seemed to influence physical activity behaviors.

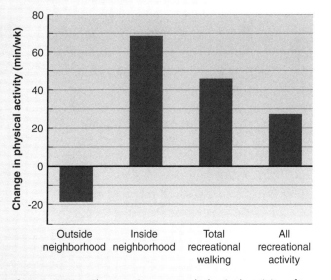

Figure 14.1 Changes in reported physical activity after moving to a neighborhood supportive of physical activity.
Adapted from Calise et al. 2012.

MEASURING THE BUILT AND PHYSICAL ENVIRONMENT

Measuring the built and physical environment is a challenge. As covered in chapter 4, there are a variety of ways to assess physical activity behaviors and energy expenditure in individuals. Despite the limitations discussed there, we are able to reasonably accurately assess physical activity in individuals and reliably separate those who are most active from those who are inactive or somewhat active. We can reasonably assess and classify those who are meeting or exceeding physical activity guidelines (U.S. Department of Health and Human Services [USDHHS], Physical Activity Guidelines Advisory Committee [PAGAC] 2008). We can also identify the context and types of physical activity being done. Although there is much room for improvement, these methods have evolved over the years and have improved.

Because of the importance of the built and physical environment in physical activity promotion, it must be measured, and measured well, if we are going to further our understanding in this important area. Unfortunately, this is a much less developed area of research than that of physical activity in individuals. Given the relative recency of attention to the built and physical environment, this is to be expected; new methods and techniques are sure to continue to be developed.

Brownson and colleagues (2009) published a comprehensive review of tools and techniques that have been used to measure the built and physical environment. They developed a categorization scheme that conveniently separates existing tools into three broad categories: self-reported measures of perceptions of the environment, direct observation techniques (audits), and secondary analysis techniques using existing datasets and geographic information systems (GIS).

Self-report techniques are used to query study participants regarding their perceptions of environmental supports of or barriers to physical activity. Most of the existing scientific evidence of an environmental effect on physical activity behaviors comes from studies relying on self-reports of perceptions of the environment. Such tools are typically administered in an interview (on the telephone or face-to-face), on the Internet, or via a mail survey. These questionnaires typically assess perceived aspects of access; community-scale and street-scale characteristics such as traffic, aesthetics, urban design, safety and crime; and the availability of local physical activity resources.

Self-report tools can be useful because most are relatively inexpensive to administer and thus can be used in large population studies. A respondent's perceptions of the environment may or may not be consistent with reality, however. Respondents may over- or underestimate aspects of the environment based on their personal situation or health status. For example, someone who is rarely outside may not have an idea of the environmental supports of and barriers to physical activity. Or less healthy people may not see the same things that their healthier neighbors do. The perception of crime is also a key issue. Studies suggest that perceptions of crime in neighborhoods do not match the reality from objective crime statistics.

To minimize the biases that can result from self-reports, scientists and planners often use more objective measures of assessment called **audits**. Audits quantify aspects of the built and physical environments at the community and street levels

WHAT IS A NEIGHBORHOOD?

Studies relying on self-report often ask respondents to describe the physical activity resources in their neighborhoods. Unfortunately, there is not uniform agreement among scientists as to what exactly constitutes a neighborhood. A half-mile radius? A one-mile radius? Ten blocks in all directions? Further, people's perceptions of what constitutes their neighborhoods vary. To some, it is the area between their workplace and home. To others it is a specific street. Still others consider their apartment building their neighborhood. Obviously, this variability injects substantial uncertainty into assessment techniques and can make comparisons among studies very difficult. What do you consider your neighborhood to be? Do you know what kinds of physical activity opportunities are available there?

that can be observed, such as the completeness of sidewalks, noise and traffic levels, the presence of abandoned or unsafe buildings, and the cleanliness and usability of parks and park equipment. The results of these audits can be used for research, to change and improve the environment, or both.

Audit tools can be particularly useful when collecting data that are more standardized and not prone to respondent bias. Audits are usually more expensive to administer than self-report tools because of the personnel needed to collect and enter data for analysis and interpretation. Training personnel to collect audit data is crucial, as is reducing the variability of observations by standardizing terms and observation techniques. If a study is large, ongoing training and error checking of observers may be also required.

Geographic information systems (GIS) offer a third technique for assessing the environment. GIS allows the analysis of geographic and social data (e.g., distances, landmarks, density, traffic, crime, resources, green space) by overlaying data in map format from multiple sources. Although the details of this technique are beyond the scope of this text, suffice it to say that GIS is an objective technique that allows multiple inputs from secondary data sources. With available computing power, GIS techniques have grown in popularity in a few short years, allowing desktop analyses of complicated geospatial datasets.

By using objectively derived datasets containing measures of the built environment, researchers and planners can use GIS to identify barriers to and supports of physical activity and relate the presence of those barriers and supports to physical activity levels in a community. GIS is most useful when large areas are under consideration and audits are not feasible because of costs and logistical difficulties.

Because GIS techniques rely on existing datasets that have been assembled for purposes other than research, the datasets may suffer from incompleteness or errors. These problems can present challenges when analysts attempt to relate the exposure data (e.g., neighborhood density) to the prevalence of physical activity in a geographically defined area. These challenges notwithstanding, GIS offers a tremendous resource that has yet to be fully explored for assessing aspects of the built and physical environments as they may relate to physical activity behaviors.

How walkable is your neighborhood? Would you walk more if it was easier?

PHYSICAL ACTIVITY POLICY

Physical activity policy can be defined as a legislative or regulatory action, including formal and informal rules that are implied or explicit, that is instituted by an organization with the power to support or inhibit physical activity participation. Such organizations include governments at all levels, nongovernmental agencies (including employers, schools, and places of worship), and less defined groups such as neighborhood associations and social groups. Most, if not all, environmental approaches to physical activity promotion must result from some kind of policy change. The environmental approaches (as reviewed in this chapter) then influence physical activity behavior, which in turn affects the health of individuals and populations.

According to Schmid and colleagues (2006), three broad classifications of policies can influence physical activity positively (i.e., encourage physical activity through environmental changes and supports) or negatively (i.e., discourage physical activity through environmental changes and barriers).

First, policies can include formal written codes, regulations, or court decisions that carry legal authority. Municipal zoning regulations that regulate the types of businesses that can operate in a given area are examples of formal written codes. City codes that require a certain distance of building setback from a street to accommodate sidewalks and pedestrian traffic are another example. Finally, a state legislative mandate for providing daily physical education to elementary school children is another example of a formal written regulation.

Nonlegal, but accepted and written, standards are a second type of policy that may influence physical activity participation. The key difference between these types of policies and the legally binding ones are that these are not usually mandated. As a result of tradition or professional standards (e.g., urban planning, architecture, or engineering), these types of policies become standards of practice and can be used to encourage or impede physical activity participation.

Finally, unwritten social norms can also influence physical activity policy. More difficult to quantify, these are usually based in the culture of an organization or society and have the ability to influence physical activity in a variety of ways. Social groups in which most members value exercise and regularly participate may encourage new members to be physically active (i.e., social support). A worksite culture that encourages physical activity participation through informal walking groups is another example of unwritten social norms.

A stepped approach is necessary for understanding the effectiveness of physical activity policy. Further, physical activity policy can occur at multiple levels, in multiple types of organizations, and in multiple geographic settings. Figure 14.2 illustrates this three-dimensional matrix.

On the left side of the matrix is the stepped approach. Beginning at the bottom, a policy is first identified as encouraging or inhibiting physical activity. Next, the determinants of that policy are identified. For example, why are certain zoning codes in place that require culs-de-sac in an area of a city instead of grid-type connected streets? Next, the degree of implementation of the policy is of interest. How completely is the mandate being fol-

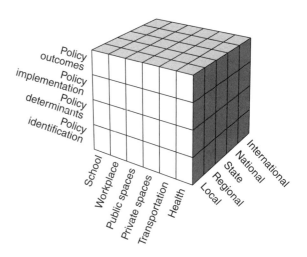

Figure 14.2 Physical activity policy matrix.
Adapted from Schmid et al. 2006.

How many parks are within 1-2 miles of your home? Are they maintained well enough to encourage their regular use for physical activity? How could they be improved to support more physical activity?

KEY LEADER PROFILE

James F. Sallis, PhD

What are your current research interests?

After getting frustrated with interventions targeting individual behavior change that had small or no effects, I decided we needed to try something different. Over several years, I developed ideas, collected data, and made connections outside the health field, to lay the foundation for understanding how environments might affect physical activity. I came to believe that changing environments and policies is required if we are going to overcome the epidemic of sedentary living. This was a relatively novel idea that required new models of behavior, new measures, new research designs, and new collaborators. I have been working on these topics for about 15 years, but in that time environmental and policy approaches to physical activity promotion have gone from "What's that?" to the U.S. government and others investing hundreds of millions of dollars to making it happen. This makes me believe that our research can make a big difference.

Courtesy of James F. Sallis.

What drives you to be a researcher and activist in the field of physical activity and health?

The evidence is absolutely compelling that being physically active is one of the most important things everyone needs to do to stay healthy. It affects most of the leading diseases as well as seemingly every bodily organ and system. However, there is a huge gap between what we know and what we do, as individuals and as a society. It is unacceptable to me that we know that helping people be active will greatly improve their health, yet we invest little in making it happen. PE and recess are declining, over 95% of federal transportation funds support roads for driving, and zoning laws make it illegal to build walkable neighborhoods in most cities. There is so much we need to do. That's why I keep generating evidence and do more each year to take that evidence to people who can use it.

What are one or two key issues of importance in our field that must be addressed by 2022?

We need to implement what we know. We now have a National Physical Activity Plan for the United States. It is an excellent blueprint for what needs to be done, and most of the actions are consistent with research. We have to find a way to implement many of the strategies, but it will not be easy. We need everyone's help. This course is preparing you to be a force for healthy change in your city, state, and country. Be active, and use what you learn.

lowed? Are there consequences for not following the policy? For example, does a school district conduct annual fitness testing of all students if a state law requires it, but there are no consequences if it is not completed? Finally, what are the outcomes of the policy? Does it influence physical activity behavior? Does it impede it?

Along the bottom of the three-dimensional policy matrix are the various sectors in which physical activity policies can be implemented. Each policy type (legal, written or unwritten practices or standards, and unwritten norms) has roles in various sectors. For example, legal policies may address public spaces, transportation, and schools. Unwritten practices can be prominent in worksites and the health sector (e.g., physicians' offices and hospitals). Social norms can be prominent in private spaces such as places of worship.

Finally, each policy type can operate at various levels of influence—local, state, regional, national, or international. Any cell in this 4-by-6-by-5 matrix can (theoretically) be isolated and examined for policies and policy determinants that may be used to affect the environmental supports of or impediments to physical activity. Can you think of examples in one of the cells?

CHAPTER WRAP-UP

What You Need to Know

- Environmental and policy approaches may involve changing the physical environment; developing social networks; changing organizational norms and policies; or enacting laws or legislation that affect public health professionals, community organizations, legislators, departments of parks and recreation, transportation departments, planning commissions, and the media.
- Creating or enhancing access to places for physical activity, combined with informational outreach activities, involves the efforts of worksites, coalitions, agencies, and communities.
- Urban planners, architects, engineers, developers, and public health professionals use community-scale urban design land use policies and practices to change the physical environment of urban areas of several square miles or more to promote physical activity.
- Urban planners, architects, engineers, developers, and public health professionals use street-scale urban design and land use policies and practices to change the physical environment of small geographic areas, generally limited to a few blocks, to promote physical activity.
- The three categories of measures of the built and physical environment are self-reported perceptions, objective audits, and geographic information systems (GIS). Each has strengths and weaknesses, and all are evolving.
- Understanding the determinants and outcomes related to policies that may support or inhibit environmental influences on physical activity is important.
- Policies can be formal (i.e., legal) or informal (i.e., based on tradition or social norms).
- Policies that support or impede physical activity can occur in multiple sectors and at multiple levels of influence. Understanding the intersection is important in understanding how to change policies to support physical activity.

Key Terms

built environment

physical environment

access

urban design

land use

zoning

community-scale urban design

street-scale urban design

audits

geographic information systems (GIS)

Study Questions

1. How does the built environment influence participation in physical activity?

2. What are three ways to increase access to physical activity at a worksite?

3. How can signs, texting, announcements on websites, mailings, and e-mails influence access to physical activity opportunities?

4. Which environmental and policy approaches are effective physical activity interventions?

5. What is urban design, according to this text?

6. What is community-scale urban design land use? Give an example.

7. What percentage of increase in physical activity can be expected with appropriate street-level changes in urban design?

8. How can audits be used to evaluate the physical environment of a community for walkability?

9. How can GIS be used to assess the built environment in relation to physical activity access as well as barriers?

10. What are three broad classifications of policies that can positively influence physical activity participation?

E-Media

Explore issues related to physical activity, exercise, and public health at the following websites:

Human Kinetics	www.HumanKinetics.com
U.S. Centers for Disease Control and Prevention: Physical Activity	www.cdc.gov/physicalactivity/
Guide for Community Preventive Services	www.thecommunityguide.org/index.html
Robert Wood Johnson Foundation: Active Living Research	www.activelivingresearch.org
Journal of Physical Activity and Health: Special supplement on active living environments	http://journals.humankinetics.com/JPAH-supplements-special-issues/jpah-volume-8-supplement-january
Journal of Physical Activity and Health: Special supplement on physical activity policy	http://journals.humankinetics.com/JPAH-supplements-special-issues/jpah-volume-7-supplement-march-supp
Physical Activity Policy Research Network (PAPRN)	http://paprn.wustl.edu
New Urbanism	www.newurbanism.org
International Physical Activity and the Environment Network (IPEN)	www.ipenproject.org
Environmental and Policy Approaches	http://youtu.be/pEtrIzklUiY
	http://youtu.be/azP6runT9AI
	http://youtu.be/xLDqHnp61hc

Bibliography

Brownson R, Baker EA, Boyd RL, Calto NM, Duggan K, Housemann RA, Kreuter MW, Mitchell T, Motton F, Pulley C, Schmid T, Walton D. 2004. A community-based approach to promoting walking in rural areas. *American Journal of Preventive Medicine* 27: 28-34.

Brownson R, Hoehner CM, Day K, Forsyth A, Sallis JF. 2009. Measuring the built environment for physical activity: State of the science. *American Journal of Preventive Medicine* 36 (4 Suppl): S99-S123.

Calise TV, Dumith SC, Dejong W, Kohl HW III. 2012. The effect of a neighborhood built environment on physical activity behaviors. *Journal of Physical Activity and Health* (in press).

Gomez LF, Sarmiento OL, Parra DC, Schmid TL, Pratt M, Jacoby E, Neiman A, Cervero R, Mosquera J, Rutt C, Ardila M, Pinzon JD. 2010. Characteristics of the built environ-ment associated with leisure-time physical activity among adults in Bogotá, Colombia: A multilevel study. *Journal of Physical Activity and Health* 7 (Suppl): S196-S203.

Heath GW, Brownson RC, Kruger J, Miles R, Powell K, Ramsey LT, and the Task Force on Community Preventive Services. 2006. The effectiveness of urban design and land use and transport policies and practices to increase physical activity: A systematic review. *Journal of Physical Activity and Health* 3 (Suppl): S55-S76.

Kahn EB, Ramsey LT, Brownson RG, et al. 2002. The effectiveness of interventions to increase physical activity. *American Journal of Preventive Medicine* 22: 73-107.

Schmid TL, Pratt M, Witmer L. 2006. A frame-work for physical activity policy research. *Journal of Physical Activity and Health* 3 (Suppl 1): S20-S29.

PHYSICAL ACTIVITY IN PUBLIC HEALTH SPECIALIST

This chapter covers these competency areas as set forth by the National Society of Physical Activity Practitioners in Public Health:

1.1.1, 1.1.2, 1.4.2, 2.1.3, 2.2.1, 2.2.2, 2.2.3, 2.4.6, 2.5.2, 3.7.1, 3.8.1, 3.8.2, 3.8.3, 4.1.1, 4.1.2, 4.5.3, 5.2.1

CHAPTER 15

Program and Policy Evaluation for Physical Activity and Public Health

Objectives

After completing this chapter, you should be able to discuss the following:

» The importance of program evaluation for physical activity promotion

» The types of evaluation and when and how they should be used

» The role logic models can play in program evaluation

» A six-step approach for effective evaluation

OPENING QUESTIONS

» Why should physical activity programs be evaluated?

» Does your program work?

» Does a school district policy that mandates daily physical education classes for elementary school children result in uniform implementation across the district?

» When should an evaluation of a physical activity promotion project begin?

» What constitutes success in physical activity programming?

Previous chapters highlighted effective evidence-based strategies for physical activity promotion. These strategies have been tested in research or practice settings and are designed for both individual and community levels. Interventions as fundamental as reminder signs at elevators help people choose to take the stairs instead. Projects such as improving sidewalks and neighborhood design also help people become physically active.

We know that these types of approaches, if not specific programs, are effective because they have been evaluated. Evaluations allow us to make quantifiable conclusions about what a program has been able to change (or not) and by how much. This chapter covers fundamental aspects of evaluation and how it is critical for advancing our knowledge of what works in physical activity and public health.

The frequently asked question, When should an evaluation of a physical activity promotion project begin? has a simple answer: Evaluation plans should be intertwined into all aspects of program planning from the program's inception. If you wait until the project is underway to start thinking about evaluation, you've waited too long!

The research offers many ideas for programs to increase physical activity. The translation from science to practice is an important aspect of public health—not only in physical activity, but in all fields. A new program at a senior center designed to reduce the risk of falls by promoting balance exercises, a walk-to-school program that encourages and rewards children and their families for getting to school without the means of a car, and a worksite walking program that uses pedometers and gives employees breaks on the cost of their health insurance if they remain physically active are all examples of programs to promote physical activity.

Despite the contributions of science and its continuing evolution, the importance of measuring the effectiveness of a particular physical activity program cannot be overstated. Using the example of the senior center balance program, we would want to know the following: How many adults in the senior center attended the balance exercise classes? How frequently? Were the instructors trained according to protocols? Were the instructors effective in delivering the program? Were there any injuries? Was the number of falls lower than expected among participants? Was there a dose-response effect—that is, did people who participated more frequently in the program experience a lower risk for falls than infrequent participants did? Each of these questions (and more) addresses a critical aspect of program evaluation.

Program evaluation is important for many reasons. Obviously, you first want to know whether your program has had the desired effect; therefore, quantification of the effects on participants is important. You have spent much effort putting the program together and delivering it—did it work the way you wanted? Did participants increase their physical activity? By how much? Was the increase enough to elicit changes in health status?

In addition to answering these fundamental questions, program evaluations also serve other important functions. For example, a good program evaluation can be used to plan for resources—funding, personnel, and program materials—which can help to ensure overall success. Funders usually want a well-designed program evaluation that will reassure them that their investment helped participants. Evaluations can also identify program strengths and weaknesses—after all, some things will go right, some will be unexpected, and some will go wrong.

What policies in your community or neighborhood support or inhibit physical activity? How can they be changed?

If the program was successful, what aspects likely led to the success? What aspects need improvement? If the program is an ongoing one—say, a walk-to-school program—an evaluation can shed light on participation trends and suggest reasons for any shifts observed from year to year. For example, did the hiring of a new program coordinator have noticeable effects on participation? What did the new program coordinator do that resulted in the increase in participation? Did the timing of these new activities coincide with the observed increases in participation? Program evaluation, in short, is a set of strategies to help us understand what happened as a result of our efforts.

Evaluation strategies can also be used in the larger context of public health policy development and implementation. For example, does a new municipal policy designed to increase bicycle use through the construction of bicycle lanes in a certain part of the city actually result in more bicycling? Is there evidence of less motor vehicle traffic in the same area? If there are increases in bicycle lane use and bicycling, when did the increases occur? Are they limited to certain times of the day or days of the week? Are the increases transient, or can they be sustained over the course of months and years?

Are there any unexpected outcomes or increases in adverse events (such as more bicycling-related injuries) as a result of the new policy? As with programs to increase individual physical activity levels, a good evaluation design can demonstrate the impact of policies aimed at groups of people, including those in cities, states, and larger geographic regions.

The U.S. Centers for Disease Control and Prevention (CDC) has proposed a Physical Activity Evaluation Framework to help in the development and follow-through of evaluation strategies for physical activity programming (USDHHS 2002). The framework has six crucial steps:

1. Engage stakeholders.
2. Describe and plan the program.
3. Define the evaluation.
4. Gather data.
5. Develop conclusions from the evaluation.
6. Communicate findings to ensure use.

Stakeholders are people and organizations with direct or indirect interests in the project. They can include funding agencies, target populations, policy leaders and decision makers, project staff, and the evaluation team. Stakeholders should be included in the program and its evaluation early in the process. Although the interest of each stakeholder is unique (e.g., target populations have a different interest than, say, decision makers), engaging all of them at multiple levels is important. This engagement process should involve communications strategies and briefings to ensure that all have an adequate understanding of the project. Evaluation efforts are usually a good meeting ground for project stakeholders because all have a common interest at some or all points along the formative, process, outcome, or cost-effectiveness framework for evaluation. Chapter 16 provides more information on developing and maintaining productive partnerships for public health programming.

WAYS TO MEASURE PROGRAM AND POLICY EFFECTIVENESS

As soon as the program has been defined, thoughts should turn to defining the evaluation. This definition is an interactive process that can result in changes to the program plan. Because the program implementation plan and the evaluation plan inform each other, they should be addressed simultaneously. Asking questions about formative, process, outcome, and cost-effectiveness evaluations helps to clarify aspects of the program that may not have been part of the original idea. The logic model, discussed later, helps to create this interactive experience.

Although the evaluation questions one can ask about a physical activity program may seem endless, they can be conveniently categorized into four categories (see figure 15.1).

FORMATIVE EVALUATION

Formative evaluation is the first level of a physical activity program evaluation. In formative evaluation, the fundamental questions focus on the needs, utility, and design features of a physical activity promotion program or policy and its individual components. Questions asked during the formative evaluation stage do not focus on the outcome (e.g., changes in physical activity behaviors), but rather on the overall design of the program. Formative evaluation questions for the senior balance exercise program discussed earlier could include the following: What is the extent of the problem of falling in the group? How frequently will participants be willing to come to the exercise classes? Will any incentives be needed to increase participation? Is there a leader in the target population who can assist with outreach? What kind of equipment will be needed? How should instructors be trained? Clearly, this information can help the program manager or policy decision maker craft components that will have the best chance of hitting the intended target.

The primary information sources of data during formative evaluation are (1) expert opinion and previous work and (2) the target population of the program. Expert opinion, including that of the people implementing the program, comes from previous experience, similar work in other settings, published and unpublished examples, and other sources. The experiences of people who have done something similar can be invaluable to a program manager looking to build a physical activity promotion program. Often, these experiences are not published and are available only through networking with people who have similar intentions. A quick search

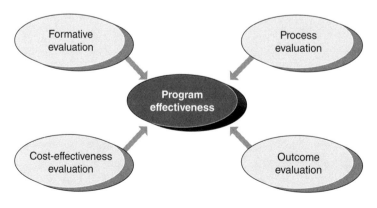

Figure 15.1 Four evaluation categories for a physical activity promotion program.

of the Internet or published literature indexing services (such as MEDLINE), or both, can begin the journey down the road to gathering expert opinion.

Needs assessments provide the second source of data for formative evaluations. Needs assessments can be formal, standardized surveys or interviews, or more qualitative discussions with key people knowledgeable about effective strategies. However it is done, the general purpose of a needs assessment is to gather information from the target population (or one that is similar). For the municipal bicycle lane example, good needs assessment questions would be, How great is the need? Would people even be interested in using the bicycle lane at all? If not, why not? If so, what will keep users using the bicycle lane over time? Will incentives to participate be helpful? How about disincentives to driving in the same area? What type of outreach should be used? In other words, what might work to fill the need?

Formative evaluation should be used throughout the program. Although it is necessary before the program starts to determine the best targets and strategies for implementation, it should not stop once the program begins. A well-designed formative evaluation provides feedback throughout program implementation so that changes and adjustments can be made. For example, interviews with participants and nonparticipants in a balance exercise class for seniors can help program planners understand the characteristics of each group, why some continue to come, and why others dropped out or never came. They can then adjust the program accordingly.

PROCESS EVALUATION

Unlike formative evaluation, **process evaluation** focuses on program or policy implementation. Of interest is how well the physical activity program or policy is operating and what can be done to improve those operations. Process evaluations help program managers assess the quality of program or policy delivery. They examine delivery strategies that appear to be successful and those that may not have worked as planned. Both are important not only for evaluating the program or policy, but also for informing future efforts.

Process evaluation should include the assessment of **delivery alternatives**. Program directors and policy decision makers can learn how best to deliver a program or implement a policy by testing their strategy against an alternative. For example, in a walk-to-school program, a pilot test could be designed to determine whether leader-supported walking groups (in which a parent volunteer leads children to school) result in more children walking to school than individual groups of children and their families. Such a systematic approach makes for a much stronger evaluation. Alternatives can be identified in the formative evaluation stage (discussed earlier), with input from potential program participants or experts. They can then be tested during a process evaluation.

Using the balance exercise class for seniors example, process evaluation questions may include the following: Are all exercise classes being offered as scheduled? Is instructor training and certification consistently being offered according to predetermined protocols? What is the attendance at the classes? How do these data compare to prestated goals? Are there trends in class attendance over the weeks and months that the program is being offered? Are other systems that have been put in place working as intended? Why or why not? What program adaptations seem to affect attendance?

The information gleaned from these process evaluation questions can be useful for monitoring the fidelity of the program or policy implementation, but also for understanding the program or policy outcome evaluation data (discussed next). For example, if outcome data for the balance training class show one or two sites that have reduced the risk of fall-related injuries among participants by 60 to 80%, the process evaluation data can be used to determine whether differences in implementation exist between the sites that are doing well and those that are not. If class attendance was routinely very high at the sites where falls have become less prevalent (and conversely very low or intermittent at sites that have seen no change in the risk of falling among participants), it could be that the dose of physical activity was higher among those participants who now are at lower risk. Clearly, this process is not linear, but requires an understanding of evaluation outcomes on all levels.

OUTCOME EVALUATION

Outcome evaluation, also sometimes referred to as *impact evaluation,* focuses on cause and effect. Did the program or policy have the intended effect?

Questions about the design and implementation of the program or policy have been answered in the formative and process evaluation stages. Now the question is, Has the program or policy increased physical activity behavior? Outcome evaluation seeks to answer this question while also assessing, in concert with process and formative evaluations, how the intended effects were reached (or not). Outcome evaluation metrics for physical activity promotion programs or policies could include minutes per week that the target population engaged in moderate- or vigorous-intensity physical activity, changes in their physical fitness levels, the percentage of the target population meeting physical activity guidelines, or the number of people using a new bicycle lane.

Because cause and effect is of central interest in outcome evaluation, changes in parameters such as these are often used as outcome measures. This, of course, assumes that a baseline assessment was done prior to the start of the program or implementation of the policy that will allow the calculation of change in the outcome of interest.

Outcomes of physical activity promotion programs and policies can be assessed using any of the measures discussed in chapter 4. As discussed, each has its strengths and weaknesses. In the walk-to-school example, the number of steps randomly selected students take a day, both before and after program implementation, as measured using pedometers, may be an important outcome of interest.

The underlying reasons for observing an outcome can be as important as measuring the outcome itself.

EVALUATION QUESTIONS

Physical activity evaluations come in all shapes and sizes. One characteristic of successful evaluations is that they have a very clearly conceptualized evaluation question (or questions). This helps define for the target population as well as the program staff the *what*, the *how*, and the *who*. Before starting any physical activity evaluation, program planners should work hard to state the question as clearly as possible.

To this end, physical activity program and policy evaluations also focus on hypothesized "upstream" determinants as well. For example, self-efficacy (i.e., a person's belief in his ability to become more physically active and sustain that behavior) is a likely mediator in participation in physical activity. If the exercise program for seniors shows an increase in participants' self-efficacy and no such increase in those who did not participate, then the program evaluator might reasonably conclude that increases in participants' beliefs that they can do the exercise program are one of the reasons physical activity increased in the group. A thorough outcome evaluation of a physical activity program not only measures the effects of the program or policy on physical activity behavior, but also suggests possible determinants of that behavior. This gives program staff a better understanding of the mechanisms through which the program may be working.

COST-EFFECTIVENESS EVALUATION

A final category of evaluation, and one that is often overlooked in public health evaluations, is related to the economics of the program or policy. The purpose of a **cost-effectiveness evaluation** is to assess not only the overall costs of delivery, but also how these costs compare with those of alternative program delivery options. Of interest is also how the costs of program or policy delivery and implementation compare to costs (real or estimated) of not delivering the program or policy.

In public health, the costs related to health issues (short term as well as lifetime) must be balanced against program delivery costs. These health costs include the money spent to treat a disease, illness, or medical condition per person as well as indirect costs such as loss in productivity. Although a complete treatment of methods of cost-effectiveness evaluation is beyond the scope of this textbook, the basic message is that this form of evaluation should not be overlooked. In the balance class for seniors example, costs of implementation include staff salaries, physical resources, evaluation costs, participation time (the cost of doing one thing at the expense of something else). These costs are then measured against any medical care expenditures (savings) that may be due to the balance exercise program (e.g., fewer emergency room visits, bone fractures, joint replacement surgeries).

Roux and colleagues published a cost-effectiveness analysis for physical activity and public health in 2008. Using information on physical activity program costs as well as gains in health outcomes to be expected from increases in physical activity (e.g., lower risks of heart disease, some cancers, and diabetes), the authors estimated that the evidence-based physical activity promotion strategies highlighted in chapters 11 through 14 of this text provided a good value for the money needed to implement them. In short, the health gains associated with increased physical activity far outweighed the costs of the programs. Not only do the interventions promote physical activity, but they are also cost-effective in terms of their effect on the direct and indirect costs related to disease.

Of the four types of evaluation covered in this chapter, the cost-effectiveness evaluation often has the most influence on policy makers and leaders. If a physical activity promotion program or policy is shown to actually save money by reducing the burden (and costs) of disease and disability, leaders and policy makers often consider it a wise investment.

LOGIC MODELS FOR PHYSICAL ACTIVITY PROMOTION AND POLICIES

The first steps of evaluation are interlaced with the overall program development and delivery. The *who, what, where, when, why,* and *how* questions are answered in this phase. Who is your target audience? What do you want the target audience to do? When do you want them to do it? How will it get done? Who is in charge of what aspects? In these early steps, the **logic model** for evaluation should be developed and refined because it will assist in evaluation as well as overall program development.

Program evaluation in physical activity and public health can get very complicated. From the initial framing of the evaluation questions and understanding antecedents and outcome relationships, program managers must keep track of many inputs, outputs, and the relationships among them. Logic models provide the hub around which all evaluation activities for a physical activity promotion evaluation can be linked. Logic models describe and define the relationships among resources, the

target population, short-term and long-term effects, and the ultimate desired outcomes.

A logic model is fundamentally a description of the processes and interrelationships that can lead to a result (cause and effect). A logic model can be graphic or narrative in its depiction of these relationships. The most fundamental use of a logic model is to ensure consistent communication among project personnel and stakeholders. A well-constructed logic model helps everyone involved or interested in a project work from a consistent set of terminology and gain a common understanding of the intended and unintended (the expected and unexpected) consequences of a program or policy. Finally, the logic model helps to drive program and policy evaluations in a timely manner when coupled with a project timeline. A schematic of a generic logic model for program and policy evaluation is found in figure 15.2.

The interrelationships of interest actually begin in the lower right-hand corner of the model in the area labeled "Ultimate goal." In fact, it is sometimes easier to build a logic model from right to left rather than from the more natural left to right. The ultimate goal of most physical activity promotion programs and policies—although not often measured in the context of specific programs or policies—is improved health for the participants. This could be determined by examining risk and rates of disease or mortality. Although such outcomes are central to the reasons for promoting physical activity, clearly, waiting for some of them to happen, or not to happen, can take years and is beyond the scope of measurement for many programs and policies, particularly those interested in more short-term behavior changes. Although ultimate goals are beyond the horizon for many physical activity promotion programs, program managers should nonetheless keep them in mind because they are a reminder of why the programs and policies exist.

Of note, each of the four ways to evaluate a physical activity program or policy (formative, process, outcome, and cost-effectiveness) detailed in the preceding section is part of the evaluation framework and is overlayed onto the logic model schematic.

Outcomes, which are clearly of interest in a physical activity promotion program, are presented schematically in figure 15.2 immediately above (and before) the ultimate goal. Outcomes can be conceptualized in terms of time. What can be expected

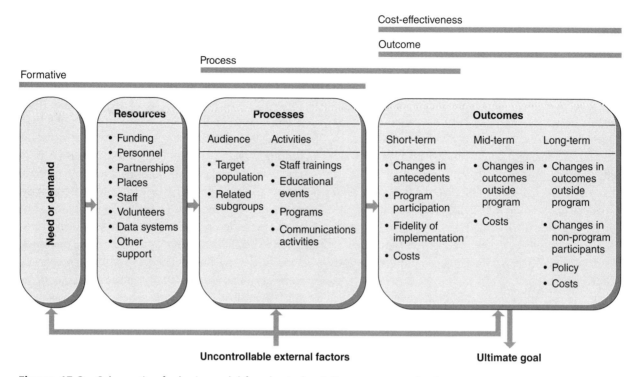

Figure 15.2 Schematic of a logic model for physical activity program evaluation.

(and should be measured) shortly after program implementation (short-term outcomes)? What can be expected after short-term outcomes (mid-term and long-term)? Short-term outcomes of a physical activity promotion program could include improvements in knowledge, skills, initial program participation, and physical activity behaviors. Mid-term outcomes could include ongoing program participation and costs associated with continuing program delivery. Long-term outcomes could include ongoing participation in physical activity after the program has ended, changes in physical activity participation among the family members of program participants, direct and indirect costs, and possibly policy changes to maintain the program. As is shown in figure 15.2, outcome, cost, and some process evaluation activities can occur in each of these outcome boxes.

Moving to the left in figure 15.2, the logic model is also used to define the processes of both the program and the target population(s) (audience). The audience for a program should be fairly well defined, but could also include subgroups such as family members or neighbors not directly participating in the program. Activities of the program should be defined as processes as well. These include training sessions for the staff members who will actually deliver the program and specific activities (e.g., exercise classes, educational activities or products,

special events such as mass participation events, stakeholder meetings, communications activities). Formative and process evaluation strategies also occur in the processes phase.

Next in figure 15.2 (to the left) is resources. A thorough understanding of resources (and how to acquire and manage them) is critical to include in a program evaluation. Without resources, programs don't exist. Resources can certainly include funding,

HOW MUCH SHOULD AN EVALUATION COST?

Program planners frequently neglect to budget for an adequate evaluation, particularly at the proposal stage. Even the best, most innovative physical activity promotion program will be unable to demonstrate effectiveness without an adequate evaluation budget. How much money is needed? Although there are no clearly developed guidelines, a good rule of thumb is to dedicate 10% of an overall project budget to evaluation activities (personnel, data collection, analysis, and reporting).

Why does changing public policy about physical activity promotion require many steps over time?

but they also include paid personnel, volunteers, the time needed from all staff, partnerships (e.g., other stakeholders in the program), and the physical plant (e.g., places for the exercise classes). Items in the "Resources" box in figure 15.2 are limited to those that a program evaluation staff can control or measure. Formative evaluation is used to assess and define the resources needed for a program.

The formative evaluation should involve defining and clarifying the need for the program before initiating it. This process could include analyzing health data (linking to the ultimate goal discussed early), gathering input and determining demand from stakeholders in the target population (e.g., bicyclists in the case of constructing bicycle lanes), or conducting surveys. This needs assessment is used to justify program development (and consequently, evaluation). Needs assessments can be standardized (e.g., questionnaires) or more interactive (e.g., expert opinion). Physical activity programs should not be developed or delivered without a clear understanding of the need for them and the ultimate goal.

The final consideration when developing a physical activity program is uncontrollable external fac-

tors. Factors outside of the control of program staff and participants can negatively affect any step along the logic model. Although program staff may not be able to control outside influences (e.g., competing programs or changes in the weather that may influence physical activity participation), it is helpful to recognize these possibilities and plan accordingly to minimize their effects.

Each logic model is unique because each program is unique, as are their evaluation processes and needs. The generic framework in figure 15.2 should be modified according to needs of the specific program. Because logic models can get very complex, program managers should start with a basic structure and then continue to develop it as program planning and a work plan emerge.

A real-life example of a logic model for a physical activity promotion program is shown in figure 15.3. The Walk a Hound, Lose a Pound project seeks to increase physical activity participation by promoting dog walking in a community (USDHHS 2010). Can you identify unique aspects of this logic model that would not show up in others?

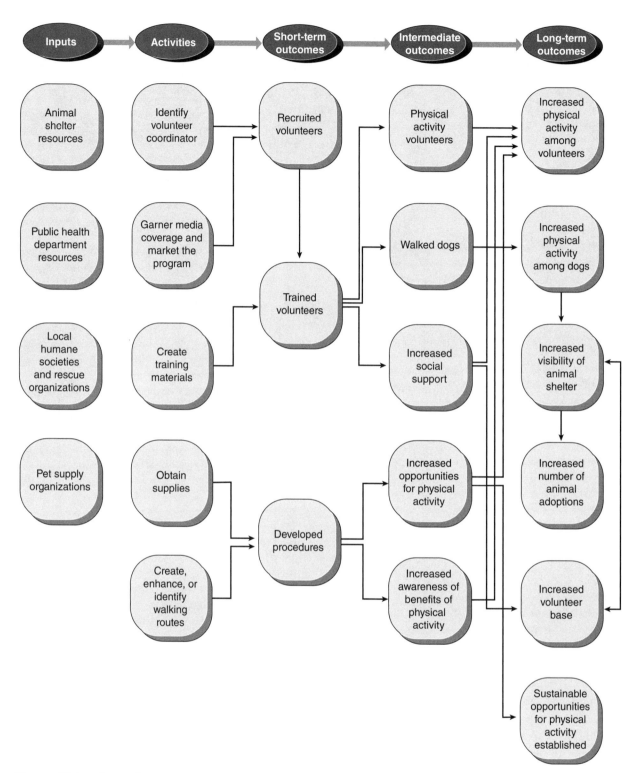

Figure 15.3 The Walk a Hound, Lose a Pound logic model.

EVALUATION DESIGNS

The success of a physical activity program or policy often rests largely on its evaluation design. These generally fall into two broad categories: experimental and observational. **Experimental evaluation studies**, considered to be the gold standard, are characterized by one important aspect that observational studies do not have: **randomization**. With randomization, people, groups of people, or places are randomly assigned to the program treatment or to a control (i.e., not receiving the treatment). Randomization has many benefits, but its most important is that it distributes any unknown errors randomly among the treatment and control groups. This is important because unknown errors can affect the results of a program evaluation, thereby affecting the inferences drawn from that evaluation.

Unfortunately, randomization is rarely possible in the real world of physical activity and public health. People must live their lives, and assigning people to a no-physical-activity control group is not entirely ethical, particularly because we know the health benefits of physical activity. Experimental evaluation studies have proven useful for physical activity and public health promotion programs, most often when the programs were short and focused on specific (smaller) groups of people.

When experimentation is not possible or impractical because of costs or other barriers, **observational evaluation studies** are a useful alternative. Often referred to as *quasi-experimental designs,* these programs differ from experimental ones in that they are not randomized. Groups are divided into treatment and control based on convenience, receptivity, targeting, and other factors. Control participants or populations (i.e., those not receiving the program) are selected based on their comparability to the treatment group on factors such as age, sex, race or ethnicity, and baseline physical activity participation. What the randomization procedure does during experimental studies—spreading the unknown error equally among groups—statistical techniques do during observational studies. These techniques minimize sources of unknown variability during data analysis.

Frequently, a clearly delineated control group is unattainable in public health practice. For example, if a policy mandates daily physical education classes

DOSE-RESPONSE ANALYSIS

Advances in physical activity research have made the one-treatment, one-control design somewhat antiquated. More often, a key question is, Does the response to a physical activity program or policy (i.e., participation in physical activity) depend on the dose of the program or policy to which a person is exposed? Three or more evaluation groups are needed to create a **dose-response analysis**. These can be conceptualized in a variety of ways, but most often they consist of high-dose, medium-dose, and no-dose groups. If the results suggest that the outcomes correlate with the dose received, cause and effect is easier to argue.

for elementary school students in an entire school district, no control group would be available. In this case, a delayed implementation design could be used. In such a design, all children would receive the benefit of increasing physical activity through quality physical education classes, but in some schools, the policy would be delayed. This would give the evaluators an opportunity to estimate program effects without withholding the program or having an "untouched" control group.

DATA COLLECTION AND ANALYSIS

Without data, there is no evaluation, only opinions. Each question at each phase in a physical activity program evaluation must have data sources to assess the effect of the program. These data sources can be quantitative (e.g., questionnaires, attendance sheets, delivery logs, production schedules, training manuals) or qualitative (e.g., interviews). The data sources used in an evaluation must help answer specific questions. For example, if a process evaluation question for the balance exercise class for seniors involves the fidelity of class delivery (i.e., whether the balance exercise class was delivered according to schedule each week), a data system must be in place to capture that information. If changes in balance

CASE STUDY: WEST VIRGINIA WALKS

 Reger-Nash and colleagues (2010) published an evaluation of a community-based physical activity promotion program called West Virginia Walks. This program was a follow-up to an earlier social marketing program designed to increase walking behavior among residents in a 12-county area of north-central West Virginia. At the time, more than 360,000 people resided in the target area. The program goal was to increase walking behavior among 40- to 65-year-old residents who were insufficiently physically active at the start of the program. The program was a community-wide approach, using mass media outreach (see chapter 11).

The evaluation design for West Virginia Walks focused on identifying any behavioral outcome changes in physical activity (specifically, reported walking behaviors) that could be attributed to the program. The treatment group consisted of people living in the target area. The control group consisted of residents of another county in West Virginia who were not exposed to the program or the mass media outreach from the program. Also of interest were process evaluation markers—in this case, self-enrollment statistics, participation in a web-based tracking tool to help people log their behaviors, and most important, media attention to the program. Media attention was an important process evaluation measure because the program relied heavily on the mass media (i.e., television, radio, and print) for promotion and awareness. Other process evaluation measures included documentation of any municipal-level changes in policies or in the environment designed to support physical activity that could be tied to the program.

Evaluation results showed several positive findings. First, there was very high recognition of the West Virginia Walks program in the treatment area. Residents in the area targeted by the program were significantly more likely to recognize and know about the program than were residents in the control area. This process evaluation indicator suggested that the program reached its intended audience.

Short-term changes in walking behavior were used as the primary outcome evaluation indicator. Households in the program area and in the control area were randomly selected prior to the campaign. Walking and other physical activity behaviors of residents in these randomly selected households were assessed by telephone interview prior to the start of the campaign and following its completion. Differences in participation rates between the two groups were interpreted as being due to the campaign.

After the eight-week community-wide campaign ended, the outcome evaluation results showed that 12% of the target population became more active by moving from being insufficiently active to being active at least 30 minutes a day on five or more days each week. This change translated to an average additional 30 minutes each week of total walking time in the targeted community. Similar changes were not observed in the control community. These results showed that the West Virginia Walks program increased physical activity throughout a large geographic region.

indicators and a reduction in falls are desired mid-term and long-term outcomes of the project, data systems must be able to capture that information. Data system development and data collection can be thought of as a process evaluation question as well.

Conclusions from the evaluation are drawn from the analysis and interpretation of the data. Data analyses should be closely tied to the questions asked in the evaluation and represented in the logic model. For example, if changes in class participation are of interest, then preprogram and postprogram data are needed. Changes in these data can then be interpreted as due to program participation.

Changes in outcomes of interest that are due to a physical activity program or policy can be measured in all participants or in a smaller, randomly selected group that (hopefully) is representative of the full group. If the program is small enough and the evaluation budget is large enough, all participants can be measured before and after the program (and sometimes along the way) for outcome evaluation. Using our balance classes for seniors example, physical activity behaviors and incidence of falls could conceivably be measured in all participants over the course of the program. Larger physical activity programs and policies that affect large groups of people or entire communities require carefully designed sampling techniques that identify a relevant subgroup of participants. Any changes (or lack thereof) observed in these subgroups are meant to be representative of the situation in the larger target population.

Conclusions from data analyses are also closely tied to the evaluation design. Control groups are especially important when evaluating program effects. A change in those who participated in the program cannot be assumed to be the result of a physical activity program without measuring similar outcomes in the control or comparison group. This principle applies to both quantitative and qualitative data. Analyses of results are much less ambiguous when a comparison group or groups are included.

An important consideration when interpreting evaluation findings is how the results compare with those of other similar programs. The similarities could be in the target population, the type of program, the type of outcomes assessed, other variables, or some combination of these. Consulting someone who has expertise and interest in data analysis and statistics (such as a local partner in a college or university) can be invaluable.

PUBLISHING AND COMMUNICATING RESULTS

Evaluation plans and data analyses are useless unless the findings are communicated. Communication targets for evaluation results should include immediate stakeholders, with the first priority being those who provided the funding. Often overlooked, but equally important, is the communication of key findings to program participants. As key stakeholders in a physical activity promotion program, participants (or targets) should be informed all along the way, including receiving a full briefing on the outcomes of the project. Although many may not have an interest in the overall results, some will. Such communications are critical to ensure future support, if necessary.

The format of communications will differ depending on the needs of the audiences. For example, the funding agency will need a detailed final report with all associated materials, including the formative, process, outcome, and cost-effectiveness evaluation results. Project staff will also need such detail. Program participants and partners may be satisfied with top-level summary findings, and other interested parties may be satisfied with a fact sheet posted to a website or printed, which summarizes the project and key findings. Policy makers will likely be interested only in the key recommendations and the cost-effectiveness results. The important message here is not to underestimate the importance of communicating the results of an evaluation.

DO SOMETHING!

Translating research into practice is often what separates public health from other scientific disciplines. New knowledge is generated from well-designed research studies in all fields. Because public health is action oriented, new research findings need to be translated into action to improve health. New vaccines, new policies to reduce workplace injuries, and new strategies to improve the quality of drinking water are examples of research-to-practice translations. Can you think of others?

KEY LEADER PROFILE

Willem van Mechelen, MD, PhD

Why and how did you get into this field?

At the age of 18 years I started training to become a physical education teacher at the Amsterdam Institute for Higher Vocational Education. During the five years of this training I was a participant in experiments for a PhD thesis on the muscle-heart reflex. By the end of the third year, I decided that I wanted to learn more about physical activity, exercise, and health than my college could offer. So in 1975 at the age of 23 I started working as a physical education teacher to pay for my medical studies, which I had started one year earlier. In 1982 I passed my medical exams and became a medical doctor. From 1983 onward I started working as a research assistant in the broad field of exercise physiology and health at the faculty of human movement sciences at VU University in Amsterdam and as a

Courtesy of Willem van Mechelen.

specialist in training to become an occupational physician. My main drive for this combined track was the notion that an ounce of prevention through physical activity and exercise was worth at least a pound of medical cure. This track led to a specialty registration as board-certified occupational physician in 1988, a PhD in human movement sciences in 1992, and registration as a board-certified epidemiologist in 1993. I maintained this double career track until I moved full time to the VU University Medical Center in Amsterdam, where I was appointed as a full professor of occupational and sports medicine in 1999. From there I started to build a group that is devoted to studying all aspects of physical activity, exercise, fitness, work, and health. The rest is history.

Did any one person have an overriding influence on you?

A number of people have had an influence on my career. I was a participant in the PhD studies of Professor A. Peter Hollander. I also became familiar with the work of Professor Han Kemper, who allowed me to start working as his research assistant from 1983 onward. In 1992 he was the supervisor at my PhD exam. As a relatively fresh researcher in the field, I was lucky to become a member of the scientific advisory board of Polar Electro Oy. This board contained Steve Blair, Neil Armstrong, Tim Noakes, Barry Franklin, and Ilkka Vuori, all of whom have inspired me. Pekka Oja, one of the former directors of UKK Institute in Tampere, Finland, also has inspired me.

What are your current research interests?

We know about the physiology of physical activity and exercise, we know about the health effects of both, and we know that we can get people more active in the short term. But what we don't know very well is how to keep the population physically active for the long term.

What drives you as a researcher and activist?

My main drive is to preserve and maintain health through physical activity because it adds quality to life, not only for individuals but for society as a whole.

What are one or two key issues to be addressed by 2022?

How do we get more people, more often, more active? How can we best tailor our physical activity and exercise interventions? That is, what works best for whom? Where, why, and how will it work?

CHAPTER WRAP-UP

What You Need to Know

- Four kinds of evaluation strategies are important to physical activity programming. Formative, process, outcome, and cost-effectiveness evaluations are concerned with different types of questions and should be conceptualized accordingly.
- Physical activity program evaluations can be experimental or observational in design. Each has strengths and weaknesses.
- Logic models are useful tools to guide evaluation strategies as well as to help overall planning for program design and implementation.
- Following six critical steps to program evaluation will facilitate its success.
- Data from well-designed physical activity evaluation projects strengthen our understanding of cause and effect.

Key Terms

program evaluation

stakeholders

formative evaluation

needs assessments

process evaluation

delivery alternatives

outcome evaluation

cost-effectiveness evaluation

logic models

experimental evaluation studies

randomization

observational evaluation studies

dose-response analysis

Study Questions

1. When should the evaluation of a physical activity promotion program begin?
2. What are the six steps to the CDC's Physical Activity Evaluation Framework?
3. What is the definition of the term *stakeholders*?
4. What are two primary data sources used during formative evaluation?
5. How does process evaluation differ from formative evaluation?
6. Why is outcome evaluation often referred to as *impact evaluation*?
7. What two considerations can be used to determine the cost-effectiveness of a physical activity intervention?
8. What is a logic model, and does it apply to physical activity interventions?
9. What is one fundamental difference between experimental evaluation studies and observational evaluation studies?
10. What is a dose-response analysis? What type of data might be collected for both quantitative and qualitative sources?

E-Media

Explore issues related to physical activity, exercise, and public health at the following websites:

Human Kinetics	www.HumanKinetics.com
U.S. Department of Health and Human Services: Physical Activity Guidelines for Americans	www.health.gov/PAGuidelines
World Health Organization	www.who.int
International Society for Physical Activity and Health	www.ispah.org
U.S. Centers for Disease Control and Prevention	www.cdc.gov/physicalactivity/
American College of Sports Medicine	www.acsm.org
President's Council on Fitness, Sports & Nutrition	www.fitness.gov
West Virginia Walks	www.wvwalks.org

Bibliography

Reger-Nash B, Bauman A, Cooper L, Chey T, Simon KJ, Brann M, Leyden KM. 2008. WV Walks: Replication with expanded reach. *Journal of Physical Activity and Health* 5: 19-27.

Roux L, Pratt M, Tengs TO, et al. 2008. Cost effectiveness of community-based physical activity interventions. *American Journal of Preventive Medicine* 35 (6): 578-588.

U.S. Department of Health and Human Services. 2002. *Physical Activity Evaluation Handbook.* Atlanta, GA: US Department of Health and Human Services, Centers for Disease Control and Prevention. www.cdc.gov/nccdphp/dnpa/physical/handbook/pdf/handbook.pdf. Accessed 1 June 2010.

U.S. Department of Health and Human Services, Public Health Service, Centers for Disease Control and Prevention, National Center for Chronic Disease Prevention and Health Promotion, Division of Nutrition and Physical Activity. 2010. *Promoting Physical Activity: A Guide for Community Action*, 2nd ed. Brown DR, Heath GW, Martin SL, eds. Champaign, IL: Human Kinetics.

PHYSICAL ACTIVITY IN PUBLIC HEALTH SPECIALIST

This chapter covers these competency areas as set forth by the National Society of Physical Activity Practitioners in Public Health:

2.1.1, 2.1.3, 2.3.1, 2.3.2, 2.3.3, 3.1.1, 3.1.2, 3.1.3, 3.2.1, 3.2.2, 3.4.1, 3.4.2, 3.7.1, 3.8.1, 3.8.2, 3.8.3, 5.5.1, 5.5.5, 5.5.6

© PhotoDisc

PARTNERSHIP DEVELOPMENT AND ADVOCACY

OBJECTIVES

After completing this chapter, you should be able to discuss the following:

» What partnerships are and how they advance public health

» The importance of developing public health partnerships to promote physical activity and exercise

» Nine key questions for developing effective partnerships

» How to educate, collaborate, and engage with external partners

» What health advocacy is and its importance in public health

» Strategies for global advocacy of physical activity and exercise

» A detailed example of partnership development for promoting physical activity and exercise in a large population

Opening Questions

» What do you think of when you hear the word *partnership*?
» Are there different kinds of partnerships?
» How could, or would, forming collaborative public health partnerships help in the promotion of physical activity and exercise?
» What strategies would you develop or implement to develop effective collaborative partnerships to promote physical activity and exercise?

Partnerships. The word conjures many images: partnerships in business, dancing, life, government (such as the United Nations). Partnerships involve the joining of two or more persons or organizations for the purpose of achieving a common goal. Implied in that definition is that both the rewards and the risks of a joint venture are shared between the partners.

As the field of public health has evolved, it has become apparent that population-wide health improvement, including disease prevention and health promotion, must focus at a level broader than that of the individual. The medical model of a health care system that treats sick people is inadequate to address larger population and societal health problems. In public health, contrary to the medical model, health (and disease) are considered in relation to systems, environments, social forces and norms, and regulations. Each of these factors, in turn, influences health behavior and disease outcomes at the population and individual levels. Because of these overarching influences, public health relies on partnerships to work across sectors of influence to effect systemic change. The goal is a shared responsibility (risks and rewards) for improving population health.

Partnerships are necessary in the field of physical activity and public health. Physical activity, and barriers to physical activity, occurs throughout the day during leisure (discretionary) time, occupational physical activity, transportation-related physical activity, and school physical education. These areas are all opportunities in which to engage in physical activity, yet they rarely join together to achieve a common goal. The private health club industry (with gymnasiums and fitness centers) does not see a role for itself in school-based physical education to promote physical activity in children. State departments of transportation, with their emphasis on increasing capacity for motor vehicle travel (and reducing traffic congestion) pay little attention to increasing opportunities for active transportation with bicycle lanes and sidewalks.

Because of the broad societal implications of increased physical activity, partnerships are critical to the emerging field of physical activity and public health. Partnership development refers more to working with organizations than individuals via collaboration. Calise, Moeti, and Epping (2010) offer a detailed explanation of how to develop successful partnerships. The basic concepts are provided here.

All partnerships are not created equal. Generally, they fall into one of three categories: cooperation, coordination, or collaboration. **Cooperation** partnerships are usually less formal and less structured than the other forms. For example, two organizations can agree to partner to promote physical activity in a community, yet each continues with its respective work. They may share resources and communicate on a regular basis, but not much else happens.

Coordination partnerships are more formal and structured around the compatibility of the mission statements of all the partners. An example would be two or more groups united to promote physical activity by sharing resources (e.g., personnel, finances, other tangible support) toward a common goal (e.g., a community physical activity promotion campaign). Coordination partnerships are more in-depth and formalized than cooperation partnerships.

Finally, **collaboration** partnerships are the most organized and structured form of physical activity partnering, uniting groups or individuals (or both)

What types of partnerships have you been involved with? Why were they successful or unsuccessful?

toward a common cause or mission (e.g., the global promotion of physical activity and exercise). Collaboration partnerships are the most formalized type of partnership, involve some written understanding among the partners, and require a shared long-term goal. Collaboration partnerships are characterized by substantial resource sharing (e.g., personnel, facilities, equipment, funding), statements of shared objectives, and mutual accountability for shared work. Clearly, the different types of partnerships have different expectations and levels of effectiveness. However, not all organizations have the ability or the resources to support collaboration partnerships. The key is that partners unite toward a common cause—promoting physical activity.

KEY FACTORS IN BUILDING PARTNERSHIPS

How are partnerships built? What should you look for in potential partners? How will you know that the partnership is working? There are no easy answers to these questions, and they can be different for different types of partnerships. Nine key questions that can increase the chance of successful partnerships for public health and physical activity promotion are shown in table 16.1. Although answering each of these questions will not guarantee an effective partnership, it should increase the probability that the partnership will work as intended.

DEVELOPING A PARTNERSHIP FOR PHYSICAL ACTIVITY PROMOTION— ACTIVE TEXAS 2020

Active Texas 2020 is a contemporary physical activity partnership designed to increase levels of physical activity in a large population. The plan involves the State of Texas, which has a population of over 24 million people (2009), and the development of a state-level plan to make physical activity a health priority throughout Texas by 2020.

In January 2008, a partnership was initiated among the state governor's office, a local municipal mayor's office, and a local university. This partnership had the explicit goal of developing a state plan for physical activity promotion. The partnership was collaborative, and each partner was able to bring special expertise and resources to the partnership. A work plan was established. In January 2009, the Texas Fit City Summit: Strategies for Health and Fitness gathered elected officials, city management organizations, and health and fitness professionals from across the state who shared the goals of encouraging healthy lifestyles and creating healthier communities through physical fitness.

During the Texas Fit City Summit, Texas municipal and local leaders and health professionals participated in focus groups that used a structured format to answer these questions: What tools and resources would help us increase physical activity? What barriers do we face in promoting physical

Table 16.1 Key Questions for Effective Public Health Partnerships

Key Question	Importance
Do I need a partnership to accomplish my physical activity and public health objectives?	If the answer is no, then a partnership may not be effective and, in the worst case, could get in the way.
Who should I recruit?	Effective partnerships are most frequently helpful when each partner brings unique skills and resources. Partners with overlapping resources can be less helpful. Partners with no resources are not partners.
Once the partnership is created, who should lead?	Strong, collaborative leadership, and a leadership plan, are critical to successful public health partnerships. A clear understanding of the overall goal of the partnership as well as the goals of the individual partners is critical.
What are the goals of the partnership?	Clearly defined, agreed-upon, and communicated goals are critical to minimize miscommunication and different expectations among partners.
What is the level of involvement and cooperation of each partner?	As detailed in the text, partnerships can be at the level of cooperation, coordination, or collaboration depending on the interest and resources that partners bring.
How will the partnership operate?	Successful partnerships most often are those with a clear organizational structure. All partners should be clear on their roles and expectations for bringing or helping to locate resources to achieve the desired goal.
On what should the partnership focus?	Although day-to-day decisions and short-term objectives require energy, it is important to keep an eye on the long-term, overarching goal—that is, the reason the partnership was established.
How do we get to our long-term goal?	Successful partnerships, although always needing to focus on the long-term goal, must set a plan of achievable short-term objectives. When attained, the sum of the short-term objectives should help attain the long-term goal.
Is the partnership working?	Evaluation is critical to an effective partnership. Is progress toward the common goal being made? What parts of the partnership are effective? What needs improvement? Are other partners needed? Evaluation in public health partnerships should be ongoing and integrated into the fabric of the partnerships.

Adapted from Calise, Moeti, and Epping 2010.

activity? What strategies have proven successful for us in increasing physical activity in our communities?

As an outgrowth of the summit, Active Texas 2020: Moving Communities Toward Health was developed by a coalition of Texas public health leaders, agencies, and organizations interested in improving the health of all Texans by promoting physical activity.

Active Texas 2020 is a state-level plan that provides measurable goals and a framework; it is an adaptable set of tools that local leaders can use to design their own community plans based on their unique priorities, assets, needs, risks, opportunities, and challenges. The hope is that focusing on and enabling physical activity behavior changes at the local level will help all Texans become more physically active and healthier.

Active Texas 2020 (2011) was built on the following eight principles that appeal to a broad range of partners:

1. Physical activity improves health.
2. Public health approaches to increasing physical activity are needed to improve populations' health.
3. Make healthier choices easier choices.
4. All health is local.
5. Health is everyone's business.
6. Prioritize leadership, collaboration, and partnerships.
7. Work from the evidence base.
8. Evaluate effectiveness.

Active Texas 2020's leadership strategies are shown in the highlight box Leadership Strategies to Increase Physical Activity.

LEADERSHIP STRATEGIES TO INCREASE PHYSICAL ACTIVITY FROM ACTIVE TEXAS 2020

1. The Imperative for Leadership and Collaboration

The critical first step to effectively improve the health of a community through increased physical activity is to establish a coordinated leadership effort that can ensure collaborative investment of resources within that community. The approach selected by any given community should be shaped by the local community's size, resources, needs, and interests.

2. Community Goals and the Tools to Achieve and Measure Them

Having established a leadership strategy that provides effective coordination of resources and efforts, a community will be poised to assess local priorities, set goals and targets, and to define the measures they will use to track progress.

3. Strategies and Resources to Reach Community Goals and Targets

Evidence-Based Strategies. Both for the purposes of effectively improving health and for supporting good proposals for funding, implementing evidence-based interventions is the most effective way to use limited resources. Many proven and promising interventions are available to achieve better health and to get the greatest return for the investment of resources and effort.

Reaching Priority Populations. A key reason for using population-based approaches is to increase community leaders' ability to reach beyond cultural differences and socioeconomic barriers that adversely affect some areas of their community more than others. The planning process that includes assessment, goal and objective setting, strategy selection, and evaluation must include a continuous focus on needs of priority populations.

Getting the Resources and Sustaining the Initiative. With clear targets and a plan to achieve them, a community leadership team and the respective sector representatives working to implement programs will be more fully prepared to articulate how requests for funding will be used.

Return on Investment. Community leaders promoting physical activity initiatives are in a uniquely strong position when it comes to advocating for their initiatives. That is because the science tying increased physical activity to health improvements is solid; and the evidence is strong that investments in improving health through physical activity results in cost savings for individuals and employers and more.

4. Implementing and Evaluating Plans to Increase Physical Activity

After the leadership team has assessed needs and priorities, set goals, objectives, and measures, chosen evidence-based strategies, and procured resources necessary to begin working toward the objectives, it is time to begin implementing the plans established.

Reprinted from Active Texas 2020 (2011), pp. 3-4.

Active Texas 2020 is a good example of a state and local partnership involving partners that came together for a common goal: developing and disseminating a Texas state plan for physical activity promotion. The plan was published in September 2010.

U.S. National Physical Activity Plan

The U.S. National Physical Activity Plan (Pate 2009) is an excellent example of a partnership with the goal of increasing physical activity for the sake of public health. Although the research indicating the health consequences of physical inactivity is substantial (and growing), no unified approach to improving the physical activity levels of Americans was attempted prior to 2008. National action plans had existed for tobacco control, heart disease control, and other health issues, but not for physical activity. Moreover, six other countries had already established plans that focused specifically on physical activity (Australia, Northern Ireland, Norway, Scotland, Sweden, and the United Kingdom). The extent of inactivity in the United States combined with the overall health benefits that can be realized if a higher proportion of Americans were more physically active made it clear that a comprehensive set of strategies, including policies, practices, and initiatives, to increase physical activity participation in the United States was needed. A partnership was born.

With leadership from the U.S. Centers for Disease Control and Prevention (CDC), a coordinating committee (leadership) was established. Because physical activity (and inactivity) is pervasive in the United States, a multisector partnership was established.

The partnership brought together representatives from transportation and community planning, educational, nonprofit, mass media, health care, public health and recreation, and fitness and sport organizations.

The long-term goal of the partnership was to create a social movement that would dramatically increase the level of physical activity participation throughout the country. Broad-based input was solicited from stakeholders and others who might contribute to the overall goal. Finally, evaluation strategies were built into the partnership's work from the start. This would allow for monitoring the effectiveness of the partnership, its partners, and the progress toward its goal.

The U.S. National Physical Activity Plan was launched in May 2010. The partnership was the only way such an effort could be undertaken and brought to a positive conclusion. The partnership has now moved to developing strategies to implement the plan. More details and updates can be found on the National Physical Activity Plan website (www. physicalactivityplan.org).

Strategies for Physical Activity Advocacy

Public advocacy involves recommending, advancing, and supporting a particular cause or policy. Physical activity advocacy, therefore, involves publicly advancing and supporting improved health through increasing physical activity. Clearly, such an enormous task could take a variety of directions,

LEADERSHIP AND PERSISTENCE

The importance of strong leadership in developing public health partnerships cannot be underestimated. Leadership does not have to consist of just one person; it can come from a team or executive committee (as with the U.S. National Physical Activity Plan just discussed). By definition, partnerships are inclusive—the voices of all interested partners are heard. The contributions and interest of partners can wane without strong leadership to help all partners focus on the long-term goal—whether that goal is to increase physical activity participation at the state or national level, at the local community level, or at a worksite. The leadership team needs to be tenacious and persistent to reach the long-term goal and to continue past it with cycles of reevaluation, reassessment, reprioritizing, choosing new strategies to reach new priorities, and then starting the cycle over again.

PARTNERSHIP DEVELOPMENT AND ADVOCACY

FUNDING

Funding public health partnerships is always a challenge. Where does the financial support come from? How do we apply for support? How should I budget for a project? The answers to these and other questions differ with each project and partnership. A complete treatment of grant writing is beyond the scope of this text. However, a good place to begin looking for initial financial support is members of the partnership organizations. When collaborative partners bring start-up money with them, early work is easier to accomplish.

State and local health departments, although never flush with money, can be useful places to start seeking funding. Community partners such as corporate supporters, parks and recreation departments, and departments of transportation and education all can be investigated. Private foundations and local community foundations often have a keen interest in health and health promotion. Partnerships are particularly attractive to funders because of their naturally inclusive nature. Another resource would be a local college or university. Faculty members frequently have additional ideas about securing sources of funding. In short, there is no one way or source for funding partnerships in physical activity and public health. You are limited only by your energy and your creativity.

---

from advocating for more school physical education, to improving access to places to be active, to improving city planning to create environmental supports for physical activity.

Shilton (2008) made the case that physical activity public health intervention has convincing scientific evidence and a broad support base to justify its global advocacy. As he reported, the World Health Organization (1995) defined **advocacy** for health as "a combination of individual and social actions designed to gain political commitment, policy support, social acceptance and systems support for a particular health goal or program" (Shilton 2006, 119). In keeping with the public health theme (as opposed to the medical model, which focuses on the individual), Shilton also noted that "the key goal of physical activity advocacy is not individual behavioral change, but achieving advances in political commitment, policy support, infrastructure, funding, and systems changes" (Shilton 2006, 766).

One need not reinvent the wheel to be a successful advocate for physical activity; lessons learned from successful public health initiatives can enhance advocacy efforts for physical activity. For example, Shilton (2008) summarized the results of antismoking interventions that have positively affected public health policies globally (as first reported by Yach, McKee, Lopez, and Novotny 2005). The dramatic decrease in tobacco smoking

in the United States is an outcome of years of multifaceted work; advocacy assisted in this process tremendously.

Lessons Learned From Successful Antismoking Campaigns

1. A small group of dedicated, persistent, media-savvy, and politically astute leaders and agitators can have a significant effect on public policy.
2. Broad-based support and well-networked coalitions are needed.
3. Commitment to a comprehensive package is crucial—in this case, a 10-point plan.
4. Interventions known to be effective must be fully implemented.
5. The issue of individual versus environmental action must be addressed early, often, and well.
6. Acknowledging the evidence of harm is necessary but not sufficient for policy change.
7. Decades of effort may be required.

A persuasive case can be made that as long as a disconnect exists between the scientific evidence regarding how physical activity improves health and the lack of programs and policies that support physical activity for the health of populations, advocacy strategies for physical activity promotion should be a priority. As covered in previous chapters of this textbook, the evidence that physical activity improves health is substantial and growing. We also know the types of strategies that work to promote physical activity. Partnerships must now advocate for a long-term commitment to such strategies to improve public health.

How does one go about becoming an advocate? Advocacy in public health is more art than science, but several key steps have been identified. Figure 16.1 illustrates five critical steps to successful physical activity advocacy.

The first step to effective public health advocacy is to establish the urgency of the problem in a way that decision makers and other advocates can understand it. That urgency is most credibly established with a solid grounding in the research of a particular problem (i.e., the science base). The stronger the science, the easier the case for urgency can be made. In the case of physical activity and public health, the science base that has established physical inactivity as a leading cause of death, various noncommunicable diseases, and disability in the world is a very important starting point in expressing the urgency.

Unfortunately, amassing a science base does not, by itself, make the case for change. It is necessary, but not sufficient. Even the strongest public

WHO IS AN ADVOCATE?

The answer to this question is—anybody. Anybody can be an advocate for change for increasing physical activity opportunities, places, and access. Decision makers to influence aren't just elected officials. They can be school principals, workplace supervisors, or neighborhood associations. Decision makers can even be family members. Change doesn't happen by accident—advocates make changes faster than they might happen naturally. One program—the Active Life Movement in Austin, Texas, USA—teaches adolescents and teenagers how to be effective advocates for healthy change (www.activelifemovement.org).

health science is not usable unless it is understood by decision makers and policy makers as well as other advocates. Although a municipal legislator is unlikely to read the most current peer-reviewed paper on determinants of physical activity from a scientific journal, she is likely to read fact sheets, websites, and summaries that succinctly and credibly convey the urgency of the problem.

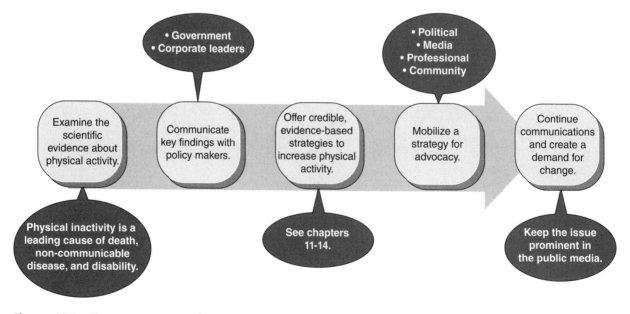

Figure 16.1 Five steps to successful physical activity advocacy.
Adapted from Shilton 2008.

Once the urgency of the problem has been established in an understandable fashion, it must be creatively communicated to people who can make a difference. This may be as simple as making a presentation to a corporate human resources director regarding how much money in health care costs might be saved if the corporation began a physical activity promotion program. Alternatively, it could be a multipronged, systematic approach to inform governmental agencies about the problem. For example, people involved with health and health promotion could be approached with a summary of the improvements in health that might be realized if the prevalence of people meeting physical activity guidelines increased by a certain percentage. People involved with transportation and city planning might be approached with estimates of how much traffic congestion may be reduced in a certain neighborhood if bicycle lanes and sidewalks were added to encourage physical activity.

Communicating the urgency of the problem to people who have the ability to do something about it is not sufficient to effect change. Advocates must be ready to offer credible solutions that preferably have been shown to work elsewhere. Once decision makers have been convinced of the need to do something, they need guidance on what that something is. Evidence-based strategies that have been shown

to increase physical activity are reviewed in chapters 11 through 14 of this textbook.

Evidence of successes in other physical activity initiatives can also be helpful to advocates. What are other people doing or saying about physical inactivity? Are there other examples of advocacy leadership that can assist? One very appropriate example is the Toronto Charter for Physical Activity (Bull et al. 2010). The Toronto Charter, published in 2010, is a call for more political and social commitment on the national level to support health-enhancing physical activity. It advocates the following four actions: (1) implementing a national plan for physical activity promotion; (2) introducing policies that support physical activity; (3) reorienting services and funding to make physical activity a priority in education, transportation and planning, the built environment, worksites, sports, parks and recreation, and health care; and (4) developing effective partnerships for action. The Toronto Charter asserts that physical inactivity is a global problem and must be addressed throughout society.

Because change does not usually happen by chance, the fourth step in physical activity advocacy is to develop and enact an advocacy strategy. Such a strategy can involve both the political and media arenas. Communicating with organizations in these arenas should be frequent to raise the profile of the issue and create a demand for change. Although less

What happens to teamwork and performance when you lose a rower or two?

KEY LEADER PROFILE

Jacqueline N. Epping, MS

Why and how did you get into this field?

Courtesy of Jacqueline N. Epping.

Trained as a physical education teacher, I planned to pursue teaching as a career. However, around the time that I completed graduate school, physical activity and its relationship to chronic disease prevention and health promotion began receiving increased attention in public health research. In particular, the National Institutes of Health funded several studies to examine the effects of increasing physical activity in school settings. The National Heart, Lung, and Blood Institute (NHLBI) funded the Child and Adolescent Trial for Cardiovascular Health (CATCH) study, which examined the effects of increasing physical activity and improving nutrition on risk factors for heart disease among children.

That experience prepared me to develop and direct another NHLBI-funded school-based physical activity intervention as part of the Lifestyle Intervention for Activity Project (LEAP) study, a multicomponent intervention to increase physical activity among high school girls. The intervention research experience I gained resulted in a position in the (then relatively new) Physical Activity and Health Branch at the CDC, where I have been since 1999.

Did any one person have an overriding influence on you?

I have been influenced by a number of outstanding colleagues and mentors, but probably most strongly by two individuals. Mr. Lloyd Huval, my gymnastics instructor in college, taught me that I could accomplish what I wanted even when I was afraid (think back handspring and pommel horse). This was a huge lesson that not only induced an ever-expanding spiral in the growth of my own self-confidence to achieve a variety of goals, but also demonstrated that this lesson, learned through physical activity, was one that was available to just about everyone. Anyone who has the ability to be physically active can learn confidence from movement that can then be applied in other areas of life.

The other person that most influenced me early in my training and career was Dr. Bobbie Eason. Dr. Eason convinced me to select the thesis option for my graduate degree and to conduct a clinical study for the thesis. This was invaluable in providing me with the experience necessary to understand firsthand every step in the conduct of clinical research.

What are your current research interests?

Using dog walking to promote physical activity and health. This is a relatively new field of inquiry, which has seen remarkable growth in the past several years. With nearly 40% of U.S. households including dogs, and as many as half of owners not regularly walking their dogs, there is potential for a significant impact on increasing physical activity and improving health at the population level.

What drives you as a researcher and activist?

I believe that the myriad of benefits from physical activity may make it the single most important health behavior in which we can engage.

What are one or two key issues to be addressed by 2022?

We need to recognize that no single societal sector—including government—can possibly affect all that is necessary to create environments in which physical activity is the default, or easy, behavior. These changes will only be accomplished through sustained and resourced efforts across all sectors of society at multiple levels—national, state, and local.

In terms of burden of disease, a key issue is that of getting the least active segment of our population moving.

frequently addressed, professional networks and community organizations with similar advocacy interests can amplify the message. Enlisting more advocates can be very beneficial to a cause.

Finally, translating and communicating the urgency to decision makers (step 2) may not be sufficient. Shilton (2008) calls for creative persuasive communication tactics that keep an issue prominent in the media. Creative use of the popular media such as television, the Internet, and print can go a long way in getting the attention of decision makers and policy makers. Creating a demand for change can often drive that change faster than the most passionate advocate.

THE TORONTO CHARTER FOR PHYSICAL ACTIVITY

A PDF of the Toronto Charter is available online at www.globalpa.org.uk/pdf/toron tocharter-eng-20may2010.pdf. Additional free resources related to the Toronto Charter are available from the *Journal of Physical Activity and Public Health* at http://journals. humankinetics.com/jpah-supplements-special-issues.

CHAPTER WRAP-UP

What You Need to Know

- Partnerships (relationships that focus on mutual cooperation and responsibility) are needed to promote physical activity and exercise.

- Partnerships can be categorized as cooperation, coordination, or collaboration.

- There are several steps to building partnerships in physical activity and public health.

- The goal of physical activity advocacy focuses less on individual behavioral change, and more on achieving advances in political commitment, policy support, infrastructure, funding, and systems changes.

- Active Texas 2020 is a state-level plan that provides measurable goals and a framework; it is an adaptable set of tools that local leaders can use to design their own community plans based on their unique priorities, assets, needs, risks, opportunities, and challenges.

- Community physical activity plans should include leadership and collaboration; goals and the tools to measure them; strategies and resources to reach goals and targets; and implementation and evaluation plans.

- Physical activity advocates should communicate the urgency of the problem, offer solutions, provide evidence of the successes of other initiatives, use an effective advocacy strategy, and communicate that strategy well.

- The Toronto Charter is a physical activity advocacy tool at the local, regional, and national levels and is a valuable resource for influencing political leaders, decisions makers, and colleagues.

Key Terms

partnerships

cooperation

coordination

collaboration

advocacy

Study Questions

1. What three types of partnerships are commonly seen in physical activity and public health? Define them.

2. What are five key questions that increase the success of a public health partnership?

3. What is Active Texas 2020?

4. What are the eight principles of Active Texas 2020?

5. What are four Active Texas 2020 leadership strategies that local leaders can use to increase physical activity in their communities?

6. What is the U.S. National Physical Activity Plan?

7. How can the lessons learned from the anti-smoking campaigns be applied to physical activity advocacy?

8. What is physical activity advocacy?

9. What are five imperatives for successful physical activity advocacy?

10. What is the Toronto Charter for Physical Activity, and how can physical activity practitioners use it?

E-Media

Explore issues related to physical activity, exercise, and public health at the following websites:

U.S. Department of Health and Human Services: Physical Activity Guidelines for Americans	www.health.gov/PAGuidelines
International Society for Physical Activity and Health	www.ispah.org
Active Texas 2020	www.ActiveTexas2020.org
The Toronto Charter for Physical Activity	www.globalpa.org.uk/pdf/torontocharter-eng-20may2010.pdf
U.S. National Physical Activity Plan	www.physicalactivityplan.org
Active Life – Organizing the Movement for Healthy Change	www.activelifehq.org

Bibliography

Active Texas 2020. 2011. Taking Action to Improve Health by Promoting Physical Activity. www.shapingpolicyforhealth.org/resources/pageContent.aspx?ID=7. Accessed 15 August 2011.

Bull FC, Gauvin L, Bauman A, Shilton T, Kohl HW III, Salmon A. 2010. The Toronto Charter for Physical Activity: A global call for action. *Journal of Physical Activity and Health* 7: 421-422.

Calise TV, Moeti R, Epping JE. 2010. Developing partnerships. In Brown DR, Heath GW, Martin SL, eds. *Promoting Physical Activity: A Guide for Community Action*, 2nd ed. U.S. Department of Health and Human Services,

Public Health Service, Centers for Disease Control and Prevention, National Center for Chronic Disease Prevention and Health Promotion, Division of Nutrition and Physical Activity. Champaign, IL: Human Kinetics.

Pate RR. 2009. A National Physical Activity Plan for the United States. *Journal of Physical Activity and Health* 6 (Suppl 2): S157-158.

Shilton, T. 2006. Advocacy for physical activity: From evidence to influence. *International Union for Health Promotion and Education* 13 (2): 118-126.

Shilton, T. 2008. Creating and making the case: Global advocacy for physical activity. *Journal of Physical Activity and Health* 5: 765-776.

World Health Organization. 1995. *Report of the inter-agency meeting on advocacy strategies for health and development: Development communication in action.* Geneva, Switzerland: World Health Organization.

Yach D, McKee M, Lopez AD, Novotny T. 2005. Improving diet and physical activity: 12 lessons from controlling tobacco smoking. *British Medical Journal* 330: 898-900.

PHYSICAL ACTIVITY IN PUBLIC HEALTH SPECIALIST

This chapter covers these competency areas as set forth by the National Society of Physical Activity Practitioners in Public Health:

1.1.1, 1.1.2, 1.1.5, 1.1.6, 1.2.1, 1.2.2, 1.4.3, 2.2.1, 2.2.2, 3.4.1, 3.4.2, 3.4.3, 3.6.2, 3.6.3, 3.6.4, 4.2.5, 4.3.3, 4.5.1, 4.5.4, 4.6.4

INDEX

Note: Page numbers followed by an italicized *f* or *t*, indicate that a figure or table will be found on the page, respectively.

from cancer 140-141, 140t
due to atherosclerotic heart
disease 174
leading causes of 5, 6t, 47, 47f
decisional balance 213t, 215
Declich, S. 64
deliver alternatives, assessment of
245
depression 155
detraining
description of 23
examples of 35
development and promotion of walking trails (case study) 230
diabetes
blood glucose levels 90
fasting blood glucose concentration 88
prevalence of 86
as a risk factor for CVD 77
risk factors for 87
tests used to diagnose 88-89
diabetes mellitus 75
Diabetes Prevention Program 89-90
*Diagnostic and Statistical Manual of
Mental Disorders* 154
diaries 61
Dietary Guidelines for Americans
(2010) 20
Dill, D.B. 52
DiPietro, Loretta 112
direct observation techniques
60-61, 61f
diseases. *See also* cancer; cardiovascular disease; diabetes; mental
health disorders
atherosclerotic heart disease 174
Black Death 4-5
chronic 5
germ theory of 4-5
infectious 5
metabolic, risk factors for 86-87
nutritional 5
overweight and obesity as 96
distance run tests 82
distress 156, 160, 160f
dose-response analysis 251
doubly labeled water technique 58
dual-energy X-ray absorptiometry
(DXA) 107-108, 127
Dunn, A.I. 158
dynamic physical activity 25
dysthymia 155

E
early diagnosis of cancer 147
Eason, Bobbie 267
ecological model 50
economic costs
of cancer 141
of mental health disorders 154-155
of metabolic disease 86
of osteoarthritis 119
of osteoporosis 118-119
of overweight and obesity in the
U.S. 103, 103f
economy of movement 106
effectiveness studies 48
efficacy trials 48
Ehsani, Ali 36
elevated inflammation biomarkers 77
E-Media. *See* websites
empowerment and inclusion 9
endurance, muscular 126
energy balance equation 100f
energy balance related to obesity
104-105
energy expenditure
decline of 52
electronic measurements of
58-60, 58f-59f
energy balance and 111
weight management and 100-
101, 101t
energy intake, excessive 105
environment, definition of 7
environmental conditions 171
environmental health 7-8
environmental health sciences 46
epidemics
Black Death 4-5
HIV/AIDS 9
epidemiology 6-7, 44-45, 46f
Epping, Jacqueline N. 258, 266
essential fat 107
evaluation
designs 251-252
logic model for 247
observational studies 251
questions 246
strategies 243
evidence-based public health 182
Evonuk, Eugene 191
excessive energy intake 105
exercise. *See also* physical activity or
exercise
dynamic 25

kinesiology-based training model
19, 19f
kinesiology disciplines and 20
static 25
taxonomy of 18
training theory and principles 23
exercise/heart hypothesis 45-46
exercise physiology 48, 122, 122f,
175
experimental designs 251
experimental evaluation studies 251
external factors, uncontrollable 249

F
falls
injury as a result of 172
prevention of 134
fasting blood glucose concentration
88
fasting plasma glucose (FPG) test 88
fat. *See* body fat
Fiatarone, Maria 121
Fick equation 78
fitness. *See* physical fitness
fitness assessments. *See* assessments
fitness evaluations 23, 33
Fitnessgram 200
FITT variables 23-24, 26, 33-35
Fleming, Alexander 44
food safety 44
formal written codes 236
formative evaluations 249
Franklin, Barry 254
Friedenreich, C.M. 144-145
functional assessment measure
(FAM) 134
functional health 22-23
frequency for achieving 31t
musculoskeletal 131-132
recommendations for 133
risk factors for 132
time for achieving 31t
functional independence measure
(FIM) 134
funding public health partnerships
263

G
gait assessments 127
Galen 20
general recommendations. *See* recommendations
genetic
and individual variation 33, 33f

V

vaccine development 44
Vainio, H. 141
Van Dusen, D.P. 198
variation, genetics and 33
VERB campaign 187
vigorous-intensity physical activity 66t
vital statistics system 4-5
volume and dose response 23-25, 24f
volume of exercise and incidences of CVD 84
$\dot{V}O_2$max. *See* maximal oxygen uptake
Vuori, Illkka 254

W

waist circumference measures 109
waist-to-hip ratio (WHR) 109
Walk a Hound, Lose a Pound project 249-250, 250f
walking, relative intensity of 27t
walking trails, development and promotion of (case study) 230
walk-to-school program 245
Wallace, Andy 176
Wang, G. 103
web-based media 188

websites
Agita Mundo 135
BMI calculators 97
Community Guide 182-183
Dietary guidelines for Americans (2010) 20
Journal of Physical Activity and Public Health, additional resources 267
physical activity for the Americas 135
SOPLAY/SOPARC 60
Task Force 183
teach adolescents and teenagers how to be advocates 264
Toronto Charter for Physical Activity 267
U.S. National Physical Activity Plan 52
weight-bearing muscular endurance tests 126
weight loss 105, 105f
weight loss and stability, psychological effects of 106
weight maintenance 105
weight management counseling 106
West Virginia Walks (case study) 252

WHO. *See* World Health Organization
Williams, R. Sanders 176
Wilmore, J.H. 20
Wingate anaerobic power test 127
Winkelstein, W. 4
Wolin, K.Y. 144
Wood, Peter 52
working conditions 5
workload anaerobic method 31
World Health Organization (WHO)
clinical criteria for metabolic syndrome 86, 87t, 88
definition of mental health 154
leading causes of death 5, 6t
principles of health promotion 8-9
statistics on cardiovascular disease 76
on worldwide obesity 96
worldwide obesity 96
written standards 236

Y

Yach, D. 263
Yancey, Antronette K. 222
YouTube 188

Z

zoning 230
zoning regulations, municipal 236

ABOUT THE AUTHORS

Harold W. (Bill) Kohl, III, PhD, FNAK, FACSM, is a professor of epidemiology and kinesiology at the University of Texas School of Public Health Austin Regional Campus and the University of Texas at Austin. Before this appointment, he served as lead epidemiologist and team leader in the Physical Activity and Health Branch of the Division of Nutrition and Physical Activity at the Centers for Disease Control and Prevention in Atlanta.

He has worked since 1984 in the area of physical activity and health, including conducting research, developing and evaluating intervention programs for adults and children, and developing and advising on policy issues. He earned his doctorate in epidemiology and community health studies at the University of Texas Houston Health Science Center School of Public Health and a master's of science degree in public health at the University of South Carolina. Kohl's other areas of specialization are biostatistics and health promotion.

His research interests include current focuses on physical activity, exercise, fitness and health, and sports medicine surveillance systems for musculoskeletal injuries. In his recent efforts, he has concentrated on national and international physical activity surveillance and epidemiology issues as well as program development and evaluation studies for the promotion of school-based physical activity for children and adolescents.

Kohl has served as an elected trustee and is a fellow of the American College of Sports Medicine and is a fellow in the National Academy of Kinesiology. He is the founding president of the International Society for Physical Activity and Health. He has served in an editorial capacity for several scientific journals and is currently coeditor of the *Journal of Physical Activity and Health*. He has published more than 150 articles, chapters, and monographs in the scientific literature.

Tinker D. Murray, PhD, FACSM, is a professor of health and human performance at Texas State University in San Marcos. He earned his PhD in physical education from Texas A&M University in 1984. His research interests include school-based and clinical-based youth physical activity interventions for the prevention of obesity and diabetes, continuing education opportunities for coaching education, and personal fitness and training applications related to exercise physiology.

From 1982 to 1984, Murray served as director of cardiac rehabilitation at Brooke Army Medical Center, where he was twice recognized for his exceptional performance. Since 1984, he has been at Southwest Texas and Texas State University, where he served as the director of employee wellness from 1984 to 1988 and director of the exercise performance laboratory from 1984 to 2000. He was a volunteer assistant cross country and track coach at Southwest Texas from 1985 to 1988 and helped win three Gulf Star Conference titles.

From 1985 to 1988, he was a subcommittee member for the Governor's Commission on Physical Fitness that developed the Fit Youth Today Program. He served as lecturer and examiner for the USA Track and Field Level II Coaching Certification Program from 1988 to 2008 and as the vice chair of the Governor's Commission for Physical Fitness in Texas from 1993 to 1994. He has worked with the Texas High School Coaches Association (THSCA) since 2003 as a facilitator with the Professional Development Cooperative, which promotes continuing education opportunities.

Murray is a fellow of the American College of Sports Medicine (ACSM) and certified as an ACSM program director. He was a two-time president of the Texas regional chapter of ACSM (1987 and 1994). He served on the national ACSM Board of Trustees from 1998 to 2001. In the fall of 2003, he was a guest researcher at the Centers for Disease Control and Prevention (CDC) Division of Nutrition and Physical Activity. He has been a member of the International Society for Physical Activity and Health (ISPAH) since 2009 and has attended all three biannual meetings of the International Congress on Physical Activity and Public Health.